D1626309

IEE TELECOMMUNICATIONS SERIES 48

Series Editors: Professor C. J. Hughes
Professor J. O'Reilly
Professor G. White

Telecommunications Quality of Service Management

from legacy to emerging services

Other volumes in this series:

Telecommunications Quality of Service Management

from legacy to emerging services

Antony Oodan

Keith Ward

Catherine Savolaine

Mahmoud Daneshmand

Peter Hoath

The Institution of Electrical Engineers

Published by: The Institution of Electrical Engineers, London, United Kingdom

British Library Cataloguing in Publication Data

Telecommunications QoS (quality of service) management : from legacy to emerging services. (IEE telecommunications series; no. 48)
1. Telecommunication – Quality control
I. Oodan, A. P. (Antony P.) II. Institution of Electrical Engineers
384′ .043

ISBN 0 85296 424 2

Typeset in India by Newgen Imaging Systems
Printed in the UK by MPG Books Limited, Bodmin, Cornwall

Contents

Foreword

In the burgeoning competitive marketplace of modern telecommunications, gaining and retaining customers is essential to the success of telecommunications operators. We must never forget that the customer is king and has choice, which will be exercised if the perceived service is not satisfactory. Customer satisfaction must therefore be the cornerstone of our strategy. We must put the customer at the heart of everything we do and ruthlessly focus on enhancing the customer experience by refining the underpinning processes and developing our people around profitably serving our customers. Profit, of course, and its effect on share price, is the criteria of success for any company but sustainable profits are impossible unless you can hold on to and grow your customer base. It is therefore critical to seek out projects that achieve the classic double whammy of improving customer satisfaction whilst at the same time reducing costs and improving revenue generation. Indeed getting these things in line frequently delivers a triple whammy as it also improves morale and in some cases can even deliver cash flow benefits as well.

It is not possible to guess what the customer wants. Detailed, sector-by-sector, surveys are required to provide information by transaction [provision, repair, usage and complaint etc] on how we are performing and what really matters to customers. This must be followed by rigorous root cause analysis and action to improve performance.

Improving our processes, network performance, systems, products and tools are all important elements in driving up customer satisfaction. But the vital ingredient that really makes the difference is how we behave when dealing with our customers. Exhortations towards staff are never enough. We must set in place the appropriate training and culture change, together with reward and recognition schemes, to breed a working environment of self-motivation so that customer care becomes second nature to all employees. It must start at the top with the visible behaviour of the CEO and MDs (Managing Directors), and flow through the company to where our people interact with customers. The basic rules in dealing with customers are simple – be friendly, really listen, show you understand, give the customer the benefit of the doubt, don't be afraid to say you are sorry, don't leave the customer in the dark, don't blame other parts of the company, take responsibility, agree what happens next and always keep your promise.

We are in an era of constant change. The portfolio of services is rapidly increasing and we face the challenge of moving to a world of convergence of telecoms, entertainment, information services and eCommerce. This will impose unique requirements on the quality of service delivered to customers who are becoming increasingly dependant on us for their quality of life and the success of business. This requires us to constantly evaluate and refine the quality of service that we deliver to meet the rising expectations of our customers.

I welcome the publication of this book. It is timely and relevant. The content provides a good coverage of the elements that constitute the theory and practice of this vital subject. We cannot predict what will happen in the future, but the importance of customer satisfaction will not diminish and I am sure that the principles outlined in the book will be enduring.

Patricia Vaz, OBE
Managing Director, BT Retail Customer Service
Autumn 2002

Preface

In the ever more competitive environment of telecommunications, the Quality of Service [QoS] perceived by customers is becoming increasingly important in the battle to win market share. This led to the publication of the book 'Quality of Service in Telecommunications' in 1997, the success of which highlighted the level of interest in this important topic. The rapid development of telecommunication services, however, coupled with its convergence with eCommerce, entertainment and information services has imposed new and unique QoS requirements. This has prompted the authors to produce a new edition of the book aimed at meeting the needs of the existing and future environment. Expert authors have been engaged to deal with the new topics, which include Internet supported services, mobile communications, fraud and security, ergonomics and requirements for the disabled. However, since in most of the world legacy services dominate, these have also been covered in reasonable depth.

The emphasis of the book is on the customer QoS requirements and how this should be delivered and managed by service providers. Network performance has a major influence on customer QoS but is only treated superficially in this book because a great deal has already been published on this topic. Clearly it is not possible to be prescriptive about future services but a framework is suggested for the definition and identification of customer's QoS requirements together with guidelines for its management. Above all, the intention is to stimulate the thinking about how to identify and meet customer QoS requirement of future services.

The objective of the book is to provide a comprehensive view of the subject. The scope of the book, therefore, is large and deals with many aspects that impinge on customer QoS. Hence it has not been possible to deal with any topic in depth. The intention is to provide a sufficient overview of each topic to provide the reader with a basic understanding of the topic and the key issues, so as to enable them to pursue it in sufficient depth to meet their needs. The target audience for this book includes telecommunications managers from network and service providers, users from corporate and medium size companies, standards bodies, manufacturers, consultants, regulators, researchers and students of telecommunications.

The book is logically structured into seven main sections each dealing with specific aspects of QoS, namely:

Section I introduces the topic of QoS by defining quality and its parameters together with a discussion of the relationship of QoS with telecommunications.

Section II introduces a framework with the necessary models to facilitate the objective study and management of QoS in telecommunications.

Section III addresses QoS issues for legacy and emerging networks and services including the relationship between network performance and customer QoS, real and non-real time Internet based services, mobile and satellite services.

Section IV deals with the impact on the customer and includes customer relationship management, numbering, billing, ergonomic considerations and the requirements of those with special needs.

Section V concerns those external drivers that have a major impact on QoS including the role of regulation and standards bodies together with various user groups.

Section VI focuses on the management of QoS, paying particular attention to comparative studies, economics of QoS, fraud and security.

Section VII concludes by proposing an enduring international framework for the study and management of QoS. A promising start has been made with ITU-T Recommendation G 1000 published in October 2001. This is dealt with in the book. Finally, there is speculation on future services and their QoS issues.

Where appropriate, individual chapters have been supplemented by exercises designed to promote the study of the application of the principles outlined in the book. There are no model answers to these exercises because these very much depend on the individual reader's application.

Subject matter experts have written individual chapters. The main authors are established leading experts in their fields and come from both sides of the Atlantic reflecting the global nature of telecommunications and the universal concepts of QoS. Additional experts who have contributed to important aspects in the book include: John Bateman on customer service, John Buckley and Millie Banerjee on numbering and billing, Prof. Ken Eason on ergonomic considerations, Peter Hoath on security and fraud, Paul Tomlinson the requirements of the disabled, Dave Wisely on mobile communications, Chris Seymour and Millie Bannerjee on satellite communications.

It is important to recognise that, from a customer viewpoint, the level of quality delivered by a service provider is dependent on the perception of individual customers. The management of QoS is therefore both an art and a science and this is reflected in the contents of the book. The intention has been to discuss all aspects of QoS so that readers can appreciate the totality of the picture and understand how the concepts presented can be applied to their field of study relating to QoS. The principles and framework outlined in this book are thought to be of an enduring nature and it is therefore hoped that they will stimulate further study as new services are developed.

The Authors
Autumn 2002

Acknowledgements

The authors wish to acknowledge grateful thanks to the following contributors additional to those mentioned in the Preface:

A W Mullee for his contribution on subjective elements of QoS in Chapters 3 and 7; Paul Ebling for his comments on auditing procedures (Chapter 6); Prof. Andy Valdar of University College, London for making use of his diagrams in Chapters 8 and 9; members of the Service Performance Assurance Division in AT&T for their help in preparing Chapters 11 and 12, and in particular Charles Appel, Jim James, Rajiv Keny, Hiren Masher, Dave Ramsden and Marianne Tavani; Vivian Witkind-Davis of National Regulatory Research Institute (NRRI) for the contribution on QoS in the USA in Chapter 20; Michael Clements for his comments on the QoS matters in USA in Chapter 20; Peter Scott of the European Commission for the contribution on EC policy on QoS matters in Chapter 22; Sudol Maxine from Australian Communication Authority for information leading to the QoS reported in Australia in Appendix 8; Lilia Perez Chavolla of NRRI for information on QoS statistics of States of the US in Appendix 9, and John Nolan for his edited version of the DTI document on sustaining service in Appendix 11.

Acknowledgement is made to the European Foundation for Quality Management, Brussels for use of the EFQM model in Chapter 1; PBI Media Ltd for allowing John Bateman full access to their CRM & Customer Care training material during the production of Chapter 15; Bannock Consulting of the UK for their work on Internet Access QoS criteria as detailed in Appendix 6; the OECD for use of a table from their publication "Communication Outlook" 2001 in Appendix 10 and the standards bodies ETSI and the ITU-T for quoting their documents.

We wish to thank Prof. Gerard White who reviewed the book. His encouragement made the completion of the book less of an uphill task.

No acknowledgement will be complete without bringing to the attention the friendly, helpful and business like help given by various members of the publishing department of IEE in the various aspects of the production of the book which made the interface with the authors much more palatable.

The Authors
Autumn 2002

Glossary

Terms and definitions

Characteristics: Distinguishing features that may be quantitative or qualitative. A characteristic may be inherent or assigned. Where pertinent characteristics could be classified under categories e.g. physical, sensory, behavioural, temporal, ergonomic, functional etc. (adapted from EN ISO 9000-2000).

Criterion/ria: A feature or set of features which describe uniquely identifiable need/s of a user. For example 'resolution of complaints' would be a criterion but not a parameter (see parameter for definition).

Entity: An entity is an item (part, device, functional unit, system or subsystem) that can be individually described and considered. It can also be a 'product', a 'process', or an 'organisation'.

Feature: A group or collection of distinguishing capabilities that have a logical meaning.

Measure: Unit in which a parameter is expressed in. For example the error rate may be expressed in number of errors per second.

Parameter: A criterion defined unambiguously and with defined boundaries and with measure/s in which the parameter may be specified. For example the parameter time 'time for resolution of complaints' may be expressed in terms of working time or calendar time, including or excluding holiday time, instant of commencement of effective time, instant of ending of effective time and other pertinent factors. The definition for a parameter would be definitive and not open to different interpretations by different people.

Service: A set of functions offered by an organisation to enable the user to make and transfer information over a telecommunication connection between two points A and B. A given set of such functions may be given a specific service name e.g. basic telephony over PSTN, VoIP, supplementary service etc.

Quality: Totality of characteristics of an entity that bear on its ability to satisfy stated or implied needs.

Quality of Service: Collective effect of the performance levels of all parameters considered pertinent to a service. The set of parameters for a given service may have different priorities and performance level requirements by different segments of users.

Service feature: A group or collection of distinguishing capabilities or functions that have a logical meaning and related to the operation of a telecommunication service.

Acronyms

AAA	Authentication, Authorization and Accounting
AAR	Automatic Alternative Routing
ABS	Australian Bureau of Standards
ACA	Australian Communication Authority
ACF (model)	Availability, Continuity, Fulfilment
ACTS	Advanced Communication Technologies and Services
ADSL	Asymmetric Digital Subscriber Line
AMPS	Advanced Mobile Phone Service
ANFP	Access Network Frequency Plan
ANSI	American National Standards Institute
API	Application Programming Interface
APQC	American Productivity and Quality Centre's Clearinghouse
ARMIS	Automated Reporting Management Information System
ASEAN	Association of Southeast Asian Nations
ASP	Application Service Providers
ATIS	Alliance for Telecommunications Industry Solutions
ATM	Asynchronous Transfer Mode
ATOL	Air Travel Organiser's Licensing
ATTF	Air Travel Trust Fund
B2B	Business to Business (eCommerce)
B2C	Business to Consumer (eCommerce)
BABT	British Approvals Board for Telecommunications
BICC	Bearer Independent Call Control
BRAIN	Broadband Radio Access over IP Networks
BSC	Base Station Controller
BSS	Business Support Subsystem
BTS	Base Transreceiver Stations
CAA	Civil Aviation Authority
CAST	Centre for Applied Special Technology
CATV	Community Area TV

CCITT	Former acronym for what is now International Telecommunications Union-T
CDMA	Code Division Multiple Access
CDR	Call Detail Records
CERT	Computer Emergency Response Team
CFCA	Communications Fraud Control Association
CLI	Calling Line Identity
CLR	Cell Loss Ratio
CMA	Communication Management Association
CMA	Computer Misuse Act
CN Iu	Core Network Interface with RAN
COBRA	An ACTS Research project
COIN	Community of Interest Networks
COU	Central Operations Unit
CPE	Customer Premises Equipment
CPI	Comparable Performance Indicator
CRM	Customer Relationship Management
CSOD	Central Switching Office Designation
CT	Cordless Technology
CT1	Early analogue cordless telephone technology
CWC	Cable and Wireless Communications
D-AMPS	Digital Advanced Mobile Phone System
DAR	Dynamic Alternative Routing
DAVIC	Digital Audio Visual Council
DC	Data Collector
DDA	Disability Discrimination Act
DDI	Direct Dialling In
DECT	Digital Enhanced Cordless Telecommunications
DIW	Defensive Information Warfare
DNS	Domain Name System
DP	Distribution Point
DPC	Destination Point Code
DPM	Defects Per Million
DQDB	Distributed Queue Dual Bus
DRU	Data Reduction Unit
DSL	Digital Subscriber Line
DTI	Department of Trade & Industry
DTMF	Dual Tone Multi-Frequency
EDI	Electronic Data Interchange
EFQM	European Foundation for Quality Management
EPROM	Erasable Programmable Read Only Memory
ERM	Enterprise Resource Management
ETNO	European Telecommunications Network Operators
ETR	ETSI Technical Report

ETSI	European Telecommunications Standards Institute
EU	European Union
EURESCOM	European Institute for Research and Strategic Studies in Telecommunications
FCC	Federal Communications Commission
FDDI	Fibre Distributed Data Interface
FDM	Frequency Division Multiplex
FEXT	Far End Crosstalk
FIINA	International Forum for Irregular Network Access
FIRST	Forum of Incident Response and Security Teams
FITCE	Federation of Telecommunication Engineers of the European community - in French)
GAO	General Accounting Office
GATT	General Agreements of Tariff and Trade
GEO	Geosynchronous Earth Orbit
GGSN	Gateway GPRS Support Node
GMSC	Gateway Mobile Switching Centre
GPRS	General Packet Radio Service
GSM	Global System for Mobile
HLR	Home Location Register
HTTP	Hyper Text Transfer Protocol
HUD	Head Up Display
IA	Information Assurance
IAAC	Information Assurance Advisory Council
IANA	Internet Assigned Numbers Authority
ICANN	Internet Corporation for Assigned Names and Numbers
ICDS	Internet Content Distribution Service
ICSTIS	Independent Committee for the Supervision of Standards for Telephone Information Services (in the UK)
IDC	Internet Data Centre
IDN	Integrated Digital Network
IEC	International Electrotechnical Commission
IEEE	Institute of Electrical and Electronic Engineers
IETF	Internet Engineering Task Force
ILEC	Incumbent Local Exchange Carrier
IMAP	Internet Mail Access Protocol
IMEI	International Mobile Equipment Identity
IN	Intelligent Network
INA	Individual Number Allocation
INTUG	International Telecommunications User Group
IOCA	Interception Of Communication Act
IP	Internet Protocol
IPDR	Internet Protocol Detail Record

ISDN	Integrated Services Digital Network
ISI	Inter-Symbol Interference
ISO	International Standards Organisation
ISP	Internet Service Provider
IT	Information Technology
ITU	International Telecommunications Union
ITU-T 7	Common Channel Signalling System, formerly known as CCITT Number 7 or SS7
ITU-T	International Telecommunications Union - Telecommunications sector
IVR	Interactive Voice Response
IW	Information Warfare
IWQoS	International Workshop on QoS
KDD	Kokusai Denshin Daiwa (Japanese Telco)
LAN	Local Area Network
LE	Local Exchange
LEO	Low Earth Orbit
LMDS	Local Multipoint Distribution System
MAC	Medium Access Control
MAP	Mobile Application Part (of the ISUP)
MDT	Mean Down Time
ME	Main Exchange
MEO	Mid Earth Orbit
MFN	Most Favoured Nations
MIPS	Millions of Instructions per Second
MNC	Multinational Companies
MOS	Mean Opinion Score
MPEG	Motion Picture Experts Group
MPLS	Multi Protocol Label Switching
MSC	Mobile Switching Centre
MSS	Mobile Switching System
MT	Mobile Terminal
MTBF	Mean Time Between Failures
MTTR	Mean Time to Repair/Replace/Restore
MUA	Mail User Agent
NANP	North American Numbering Plan
NARUC	National Association of Regulatory Utilities Council
NASA	National Aeronautics and Space Administration
NEXT	Near End Crosstalk
NFU	Network Field Units
NHSTUG	National Health Service Telecommunications User Group
NICC	Network Interoperability Consultative Committee
NISCC	National Infrastructure Security Co-ordination Centre

NMC	Network Management Centre
NOU	Network Operations Unit
NPA	Numbering Plan Area
NRA	National Regulatory Authority
NRRI	National Regulatory Research Institute
NSC	National Security Council
NSIE	Network Security Information Exchange
NSP	Network Service Provider
NSTAC	National Security Telecommunications Advisory Committee
NTP	Network Termination Point
NTSC	National Television Standards Committee
OECD	Organisation for Economic Co-operation and Development
OFCOM	Office of Communication (UK- to replace Oftel in the UK)
OFDM	Orthogonal Frequency Division Multiplex
OFGEM	Office of the Gas and Electricity Markets
OFTEL	Office of Telecommunications (UK)
OLO	Other Licensed Operator
OMC	Operations and Maintenance Centre
ONA	Open Network Architecture
ONP	Open Network Provision
OPC	Originating Point Code (ITU-T 7)
OSGi	Open Services Gateway initiative
OSI	Open Systems Interconnection
OSS	Operational Support System
PARLAY	An open multi-vendor consortium formed to develop open technology-independent Application Programming Interfaces (APIs)
PBX/PABX	Private (Automatic) Branch Exchange
PC	Private Circuit, Personal Computer
PCM	Pulse Code Modulation
PCP	Primary Cross Connect Points
PDC	Personal Digital Cellular
PDH	Plesiochronous Digital Hierarchy
PDN	Public Data Network
Phreak(er)	Telephony fraudster
PIN	Personal Identification Number
PLC	Packet Loss Concealment
POP	Post Office Protocol **or** Point of Presence
POTS	Plain Old Telephone Service
PROM	Programmable Read Only Memory
PSC	Public Service Commission
PSD	Power Spectral Density
PSTN	Public Switched Telephone Network
QDU	Quantisation Distortion Unit

QoS	Quality of Service
QSDG	Quality of Service Development Group of the ITU
QUASIMODO	Quality of Service Methodologies
R2	CCITT Signalling System
RAN	Radio Access Network
RED	Random Early Detection
RGA	Remote Gathering Agents
RIPA	Regulation of Investigatory Powers Act
RLR	Receive Loudness Rating
RME	Remote Method Execution (protocol)
RNC	Radio Network Controllers
RRM	Radio Resource Management
SCP	Service Control Point
SDH	Synchronous Digital Hierarchy
SET	Secure Electronic Transaction
SIP	Session Initiation Protocol for establishing multimedia sessions and Internet telephone calls
SLA	Service Level Agreement
SLR	Sending Loudness Rating
SMDS	Switched Multimegabit Digital Service
SMTP	Simple Mail Transfer Protocol
SPM	Subscriber Private Metering
SSL	Secure Sockets Layer
STMR	Side Tone Masking Rating
STP	Signal Transfer Point
STQ	Speech Transmission and Quality
TACS	Total Access Communications System
TAM	Technology Acceptance Model
TCP	Transmission Control Protocol
TCP-SAC	TCP Selective Acknowledgement
TDD	Time Division Duplex
TDM	Time Division Multiplex
TDMA	Time Division Multiple Access
TE/MT	Terminal Equipment/Mobile Terminal
Telco	Telecommunications Operator
TIPHON	Telecommunications and Internet Protocol Harmonisation Over Networks
TLD	Top Level Domain
TMN	Telecommunications Management Network
TQM	Total Quality Management
TRS	Text Relay Service
UDP	User Datagram Protocol
UDPM	Unified Defects per Million

ULE	User Lost Erlang
UMTS	Universal Mobile Telecommunication System
UNIRAS	Unified Incident Response and Alert Scheme
URL	Uniform Resource Locator
USO	Universal Service Obligation
UTRAN	UMTS Terrestrial Radio Access Network
VAD	Voice Activity Detection
VAN	Value Added network
VC	Virtual Channel
VLR	Visitor Location Register
VMB	Voice Mail Box
VoIP	Voice over Internet Protocol
VP	Virtual Path
VPN	Virtual Private Network
VSAT	Very Small Aperture Terminal
WAN	Wide Area Network
WAP	Wireless Application Protocol
WLAN	Wireless Local Area Network
WRC	World Radio Conference
WTO	World Trade Organisation
WTP	WAP Transport Protocol
WWW	World Wide Web
X.25	ITU-T data protocol based upon the first three layers of the ISO OSI model

Authors

Antony P. Oodan

After 12 years in thermionic tube industry Antony moved to telecommunications. In a career spanning 24 years with BT he worked on various projects on transmission, logistics, switching, service specification and QoS. He was the Secretary to a study commission under FITCE on 'Study of network performance considering customer requirements'. His consultancy work includes determination of user's QoS requirements for BT and Telfort of Netherlands. He has advised Uzbekistan Telecommunications Authority and nine East European countries on selection of parameters and setting up of monitoring and reporting systems for QoS, under TACIS and the PHARE programs respectively, for the European Commission. He is a member of the ETSI User Group and the Speech and Transmission Quality Group. He is an ETSI Expert currently studying the User's QoS criteria for Internet Access in Europe.

He is one of the co-authors of the predecessor of this book under the title, 'Quality of Service in Telecommunications'. He lectures on QoS to MSc courses of London University. Antony has published papers on matters related to thermionic tube development and on QoS.

Professor Keith E. Ward

Keith Ward joined British Telecom [BT] in 1948 as an engineering apprentice. In 1982 he became Chief Engineer responsible for planning BT's trunk telephone network together with the development, procurement and installation of switching and transmission equipment. He subsequently assumed a similar responsibility for the whole of BT's inland network. His final job with BT was to design, launch [in 1990] and run the BT Telecommunications Masters Programme, validated by the University College London [UCL] for the award, to successful delegates, of an MSc in Telecommunications Business. On retirement from BT in 1992 he became Visiting Professor of Telecommunications Business at UCL, where he continued to run the BT Masters Programme as well as lecturing on other graduate and post graduate courses, but his commitments are now only part-time. He continues to lecture and undertake consultancy work in many parts of the world. Keith is a Chartered Engineer and Fellow of the Institution of Electrical Engineers, he is also a Fellow of the City and Guilds of London Institute and Member of the Chartered Management Institute.

Catherine G. Savolaine

Catherine G. Savolaine, manager of the Service Performance Assurance Division of AT&T, is responsible for quality planning and evaluation on AT&T service offerings. This encompasses analysing proposed services, identifying key quality impacting parameters, and designing and implementing statistical experiments to evaluate service performance. Current responsibilities are focused on service offerings over IP networks and hybrid IP/circuit switched networks, including voice, data, web hosting, email services and streaming media. Her division develops methodologies and systems, and uses them to measure service quality from an end-to-end customer perspective.

Ms. Savolaine joined AT&T-Bell Labs in 1970. She is active in international standards bodies and serves on the organizing committees of several international technical conferences. She has published numerous articles on software development and testing, technology management and ensuring quality in telecommunications networks.

Ms. Savolaine received a BA in mathematics from Grove City College and an MS in mathematics from Drexel University.

Dr. Mahmoud Daneshmand

Dr. Mahmoud Daneshmand is Technology Leader, Chief Scientist Organization, AT&T Labs. He has more than 25 years of teaching, research, and management experience in academia and industry including Bell Laboratories and Stevens Institute of Technology. Mahmoud has a Ph.D. and MA in Statistics from the UC Berkeley and an MS and BS in Mathematics from the University of Tehran.

As a Technology Leader, Mahmoud supports the AT&T Labs Chief Scientist in the creation of technology strategy for AT&T, and in understanding and articulating the effect of disruptive technologies on AT&T's Networks, Services, and Operations. He is a recognized expert in the quality and reliability of IP-based services and applications. He has published more than 50 papers, co-authored two books, co-edited the publication of the Seventh IEEE International Conference Proceedings on Computer and Communications and contributed to the standards and regulatory work of the ITU, T1, and FCC.

He founded and served as the Head of the Department of Statistics and Dean of the School of Informatics and Management Sciences of the National University of Iran.

Peter Hoath

Peter Hoath is Head of e-Security in BT's Retail line of business. After periods in PSTN switch design, hardware and software, and data networking he switched to a security role in 1985. A spell leading the team providing technical support services to BT's in-house investigations team led to numerous witness appearances in court culminating in becoming BT's lead expert witness for telecommunications crime prosecutions. Peter went on to become Head of Network Integrity in BT's Wholesale organisation. He chaired an Oftel/Industry task group which produced a set of guidelines for network security and integrity which was published in 2002 and has received international acclaim. Peter has been an active member of the International Forum for International Network Access (FIINA) since 1990 and has been a member of its Executive Committee for the past 3 years. He is a member of the British Computer Society, where he has been vice-chair of its Security Committee and continues as a member of its Security Experts Panel. Peter is a Chartered Engineer, a Member of the Institution of Electrical Engineers and holds an MBA.

Section I

Introduction to Quality of Service

Quality can be defined. It is expressed in terms of parameters that indicate benefits to the user. These parameters may be quantitatively or qualitatively expressed. A set of quality criteria will encapsulate the benefits of a product or a service to the user. Quality of Service (QoS) in telecommunications is expressed by a unique set of parameters (qualitative and quantitative) on a service by service basis. A grasp of these concepts will facilitate a better understanding of the issues related to the study and management of QoS in telecommunications.

Chapter 1

Quality

1.1 Introduction

An understanding of the basic concepts on quality and its management is essential for the professional management of Quality of Service (QoS) in telecommunications. Quality can be defined for physical products and with less precision for services. QoS in telecommunications could be developed from the basic concepts on quality. The concepts on quality may also be applied to any product or service or organisation.

1.2 Definition of quality

1.2.1 Terminology

Various people have defined the term 'quality' over many years. Crosby [1, 2], Deming [3] and Juran [4] are well known in the industry for their contributions to the understanding of quality. Their approaches may, to some extent, be described as a bottom-up approach. The standardising body has taken a top-down approach and the current definition for quality by the International Standards Organisation (ISO) is [5]:

Degree to which a set of inherent characteristics fulfils requirements.

Characteristics may be defined as distinguishing features. These may be inherent or assigned. There are various classes of characteristics; e.g. physical (mechanical, electrical, chemical or biological), sensory (e.g. related to smell, touch, taste etc.), behavioural (courtesy, veracity and honesty), temporal (punctuality, reliability, availability), ergonomic (physiological characteristic, or related to human safety-functional e.g. maximum speed of aircraft).

For the purpose of understanding quality and from it the understanding of QoS in telecommunications, it is considered easier to use an earlier definition by the ISO [6] which defines quality as:

> Totality of characteristics of an entity that bear on its ability to satisfy stated and implied needs

An *entity* is an item that can be individually described and considered. An entity may be an activity or a process, a product or an organisation – indeed anything that has a quality attribute.

This definition of quality draws much from the work of Lancaster's theory of consumer demand [7–11] in relation to the characteristics or attributes of products. Under Lancaster's approach, a product is viewed as a bundle of characteristics. Consumers derive utility from the characteristics embodied in the product rather than from the product *per se*. Characteristics are defined as 'those objective properties of things that are relevant to choice by people'. Subsequent work by various authors indicates that, for analytical purposes, individual characteristics may be disaggregated into sub-characteristics and several characteristics may be aggregated into an aspect of a service quality. For example, several characteristics of airline services such as cabin temperature and humidity, aircraft stability, level of crowding and seat comfort may be aggregated into the aspect of on-board comfort. Similarly, the characteristic of seat comfort may be disaggregated into sub-characteristics such as cushion softness, seat width, seat pitch, reclining features and fabric smoothness.

Each product potentially possesses a large number of characteristics but the operational use of Lancaster's model depends on the ability to confine the analysis to a relatively small number of characteristics with measurable properties. Lancaster, therefore, proposes that practical studies should be limited to the relevant characteristics. Identification of the characteristics and assigning values would then lead to the definition of quality. The characteristics may be expressed quantitatively or qualitatively. This concept is applied in formulating a methodology for the determination of quality criteria for any product or a service (Section 1.3).

1.2.2   Expression of quality: quality parameters

The terms, quality criteria and parameters are used in the management of quality.

When quality characteristics are defined unambiguously with clear boundaries, these are termed *parameters*. Parameters are useful in expressing the quality by allocating a figure to represent the level of performance indicated by the parameter. Distance and temperature are examples of parameters.

Parameters may be expressed in one or more *units of measurement*. The parameter of temperature may be expressed in Celsius or Fahrenheit scales. Values of 25 and 77 represent the same temperature on the Celsius and Fahrenheit scale. Similarly the parameter of distance may be expressed in 'miles' and/or 'kilometres'.

Not all quality characteristics defined as parameters can be measured objectively. Quality criteria such as courtesy, flexibility, professionalism etc. are expressed by qualitative parameters such as subjective performance ratings. Flexibility offered by a service provider may be rated by the customer on a 1–5 rating scale, 1 being poor and 5 being excellent. In this case the question asked of the customer becomes the 'parameter' and the rating 'the measure'.

> The stated and implied benefit to the users (quality) of a product, service, process or an organisation may be expressed by a set of relevant parameters in defined units of measurement. Values measured for these parameters indicate the level of quality.

1.2.3 Examples of quality in everyday life

The following examples in daily life illustrate the application of the above concept; a product, a service and a process (manufactured product). The quality characteristics and, therefore, the quality parameters of a physical product can usually be specified in quantitative (objective) terms. The corresponding parameters of a service are usually specified in qualitative (subjective) terms. A manufactured product can have a mix of objective and subjective quality characteristics. An example of each is given below.

Quality of a product: diamond – quantitative (objective) quality parameters

The quality parameters for diamond are Cut, Colour, Clarity and Weight.

Cut may be categorised by shape, proportions and finish. Shape is the outline of the diamond combined with the facet distribution. Proportions represent the relationship between various parts of a diamond and the girdle diameter. Finish deals with the exactness of its shape and the arrangements of the facets and also the quality of the polish. *Colour* is expressed in twelve categories. *Clarity* has two standards and is divided into fifteen categories. These categories are given in Appendix 1. *Weight* of a diamond is expressed in *carats*. One carat is 0.2 g. A carat is divided into 100 points and the weight of a diamond piece is expressed in carat and fractions of it.

Quality parameters for diamond are objective. Even though cut, clarity and colour are measured by the human eye, for all practical purposes these are considered in the industry as objective measurements. Categorisation of diamond quality is considered uniform throughout the world by the industry. The price is fixed in relation to the four quality parameters.

Quality of a service: legal service – qualitative (subjective) quality parameters

The following quality characteristics may be considered pertinent to a lawyer: ability to apply legal knowledge to a particular case, ability to discern truth from falsehood with minimum of evidence, communication skills, interpersonal skills, ability to deal with people, understanding of human nature, balanced judgement etc.

The above quality characteristics are predominantly qualitative or subjective. These criteria cannot be quantified (objectively) as in the case of a physical product and one has to resort to a meaningful qualitative (subjective) rating. Qualitative parameters would have to be expressed in subjective terms.

*Product of a process: a motorcar – has quantitative (objective) and
qualitative (subjective) quality parameters*

A motorcar is an example of an item, which has both objective and subjective quality
parameters. Examples of objective parameters are reliability, engine performance,
fuel consumption etc. and those of subjective nature are aesthetic appeal of the car,
driving comfort, after sales service etc.

In summary, the quality of a product or service may be expressed by a set of
parameters that indicate the usefulness/meaning/benefit to the user. The determination
of such parameters should reflect all pertinent characteristics of the product or service,
both implied and stated.

1.3   Framework/methodology to determine quality parameters for products and services

A convenient method of identifying the quality parameters of an entity is to partition
its functional components and quality criteria pertaining to the characteristics, which
produce benefits to the user. Thus in Table 1.1 the functional components (on the
y-axis) aggregated cover the utility of the product, service or the process and any
associated activities relating to its delivery or use. For example, in the case of airline
flight from point A to point B the functional components (F_1 to F_x) could be broken
down to the following:

- purchase of ticket
- travel to the airport
- check-in
- waiting to board
- flight
- disembarkation
- baggage clearance
- post flight support
- etc.

The quality criteria (Q_1 to Q_n on the x-axis) are related to the benefits of each of
the functional components. Examples of such criteria are:

- flexibility of booking agent/s to tailor clients ticket needs
- seating preference allocation
- waiting room facility
- quality of food
- information availability
- passenger assistance (porter, guidance, interpreter etc.)
- in-board comfort
- flight comfort
- etc.

Table 1.1   General matrix to illustrate determination of quality parameters

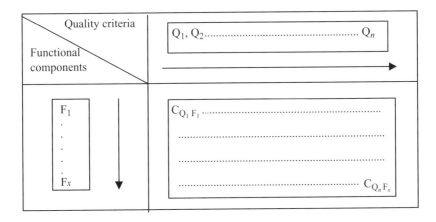

In the matrix against each functional component the relevance of quality criteria is checked and relevant airline flight quality characteristics identified are noted down. Thus functional component F_1 is checked against quality criterion Q_1 and any identifiable parameters are noted. Certain cells could have more than one quality criterion. This process is carried out for every cell. Not all cells (n times x in Table 1.1) of the matrix will be populated. However, based on an iterative process and brain storming sessions with knowledgeable people of the industry, it will be possible to develop a set of quality parameters which cover most, if not all quality considerations. It may be necessary to arrive at the set of parameters in a two-step process. In the first step 'the airline flight quality characteristics' are determined and in the next step these characteristics are specified unambiguously these become parameters. It is possible to have more than one parameter for one cell. Application of this principle is used for telecommunications services in Chapter 4.

The above principle, partitioning the functional components of an entity and the relevant quality criteria may be used to determine the quality parameters. The granularity of the quality parameters can thus be controlled. For basic telephony service one can identify as many as 43 QoS parameters, but in practice only around 12–15 need to be regularly monitored.

1.4   Importance of quality

Quality is important for the following reasons:

(a) Quality parameters and measures are necessary to provide an indication of how well a product, service or the process is.
(b) Quality measurements are necessary to study the steps to be taken to optimise resources with revenue.
(c) Quality could be the differentiator when other factors are equal.

1.4.1   *Quality as an indicator of excellence*

To evaluate a product, service or an organisation it is necessary for its quality parameters to be specified together with the corresponding performance values. In its
absence no value can be attached to it. The price of a diamond is dependent upon
its quality. The quality of a school, hospital, a car or any item needs to have its
quality parameters specified and values attached to its performance. Performance
assessment is an essential requirement for the functioning of all aspects of every day
life. Agreement on the choice and definition of parameters are necessary for such
evaluations. This will form the basis for specification when purchasing goods from a
supplier. It is necessary for both parties to agree on the parameter and the method of
measurement to avoid disagreement. The role of standards is to specify the important
parameters.

1.4.2   *Quality as a factor to optimise resources with revenue*

In a free market there will usually be differentiated levels of quality requirements.
This is usually due to the stratified purchasing power of individuals in the market
and differences in the level of application or use of the product or service. In these
circumstances it would be necessary for the supplier to determine the quality requirements of the market place and produce goods to match, as closely as possible, these
requirements. The car market illustrates this situation. Car manufacturers produce
vehicles to match different purchasing requirements. Some of the features may not
be strictly related to quality; however, the concept to be understood is that where
there is a quality–price relationship it is necessary to have an understanding of this
before the supplier diverts its resources to produce and supply goods. Ideal matching
of the quality requirements of the market and the supplier's production to match this
may not be accurately achieved; however, it is far better to aim for this ideal than
produce goods relying on luck to match the market requirements.

  The conventional argument for quality is that it saves unnecessary expenditure of
resources. Poorly engineered goods may have to be repaired more often resulting in
expenditure of financial, time and human resources. This could slow down the output
arising from the use of the produce, service or the process. A good quality product
will satisfy the user's needs with minimum (or optimised) expenditure of financial,
time and human resources. This is one of the fundamental tenets in the understanding
of quality.

1.4.3   *Quality the differentiator*

When price and availability of a product is similar among competitors, the determining
factor to the discerning customer would be quality. The supplier with an enhanced
quality item is in a better position to increase its market share. With the emergence
of competition among the telecommunication suppliers it is important for the role of
quality of service provided to be understood and appreciated.

1.5   The EFQM Model

Recognition of the importance of quality and its management is acknowledged by the formation of European Foundation for Quality Management (EFQM) [12] in 1989. It was founded by 14 large European companies to promote the practice of TQM and to cross-fertilise the quality management ideas applied across many industry sectors. Today there are more than 750 members from most European countries and most sectors of activity. Regular meetings are held to promote learning and practice of quality management. It has the backing of the European Commission. Some of the European telecommunications service providers who are members are, BT, France Telecom, Deutsche Bundespost Telekom, Telecom Italia, and Telefonica.

The EFQM Excellence Model, illustrated in Figure 1.1, is a non-prescriptive framework based on nine criteria. Five of these are 'Enablers' and four are 'Results'. The 'Enabler' criteria cover what an organisation does. The 'Results' cover what an organisation achieves. 'Results' are caused by 'enablers'. Definitions of the individual boxes are as follows:

Leadership: How leaders develop and facilitate the achievement of the mission and vision, develop values required for long-term success and implement these via appropriate actions and behaviours, and are personally involved in ensuring that the organisation's management system is developed and implemented.

People: How the organisation manages, develops and releases the knowledge and full potential of its people at an individual, team-based and organisation-wide level, and plans these activities in order to support its policy and strategy and the effective operation of its processes.

EFQM Model

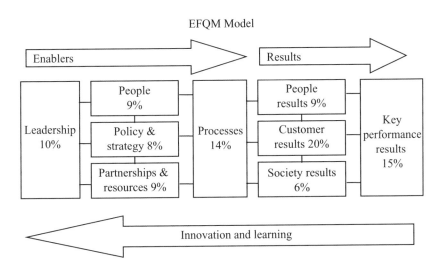

Figure 1.1   The EFQM model for rating performance of an organisation (Acknowledgement is made to EFQM, Avenue des Pleiades 15, Brussels, Belgium to reproduce the model)

Policy and strategy: How the organisation implements its mission and vision via a clear stakeholder focused strategy, supported by relevant policies, plans, objectives, targets and processes.

Partnerships and resources: How the organisation plans and manages its external partnerships and internal resources in order to support its policy and strategy and the effective operation of its processes.

Processes: How the organisation designs, manages and improves its processes in order to support its policy and strategy and fully satisfy, and generate increasing value for, its customers and other stakeholders.

People results: What the organisation is achieving in relation to its people.

Customer results: What the organisation is achieving in relation to its external customers.

Society results: What the organisation is achieving in relation to local, national and international society as appropriate.

Key performance results: What the organisation is achieving in relation to its planned performance.

A scoring process has been described by the EFQM. The closer an organisation is to the 100% mark the better run it is. Readers are recommended to familiarise themselves with the process.

The principal point to be noted is that the quality related criterion under the heading 'Customer results' has the single largest weighting with 20% of the overall score. Examination of the sub-criteria under this heading show that all criteria are related to quality of the product or service offered by the organisation. This reflects the importance of quality of the products and services offered by an organisation.

A contribution made by the EFQM, apart from the model and the award to successful companies practising TQM, is the encouragement given to the member companies to apply the successful practices across industry sectors. EFQM is a respected organisation within Europe and its contribution could be noted by all service providers for their own benefit. A detailed examination of its work is outside the scope of this book and for more detailed information on its activities and achievements the reader is advised to contact its office in Belgium.

1.6   Studies on quality

Large numbers of studies have been carried out on quality worldwide. It is outside the scope of this book to review these. The enquiring reader is invited to search the Internet websites through various search engines to review the work published in this field. Our interest is chiefly on the standards on quality and its related components. The latest document on quality, ISO 9000:2000 supersedes ISO 8402 along with other quality-related documents. In this chapter we have used the definition of quality given in ISO 8402 because it is considered easier to use for the derivation of basic quality parameters from fundamental principles. ISO 9000:2000 is more suitable for the management of quality. There is no basic conflict between the old ISO definition of quality and that in ISO 9000:2000.

1.7   Summary

Management of quality of any product, service or the output of any process (e.g. organisation, manufactured article, a system etc.) can be made easier and professional if both the bottom-up and top-down approach to the understanding of the subject can be made to result in a set of defining statements. This chapter has shown how the application of the basic concepts on quality can be applied to determine the quality parameters of an entity. The remainder of the book develops the quality parameters from these concepts for the understanding and management of QoS in telecommunications.

1.8   References

1   CROSBY, P.B.: 'Quality is free: the art of making quality certain' (McGraw Hill, 1979)
2   ——: 'Quality is still free: making quality certain in uncertain times' (McGraw Hill, 1996)
3   DEMING, W.E.: 'Out of the crisis: quality productivity and competitive position' (Cambridge University Press, 1986)
4   JURAN, J.M.: 'Juran on quality by design: the new steps for planning quality into goods and services' (Free Press, 1992)
5   ISO 9000-2000: Quality management systems – Fundamentals and vocabulary.
6   ISO 8402: Quality management and quality assurance – vocabulary, 1994.
7   LANCASTER, K.J.: 'A New Approach to consumer theory', *Journal of Political Economy*, 1966, April, pp. 132–57.
8   ——: 'Consumer Demand: A New Approach' (Columbia University Press, New York, 1971)
9   ——: 'The measurement of changes in quality', *The Review of Income and Wealth*, 1977, June, pp. 157–72.
10   ——: 'Variety, equity, and efficiency' (Columbia University Press, New York, 1979)
11   ——: 'The economics of product variety: A survey', *Marketing Science*, 1990, **9**(3), pp. 189–206.
12   European Foundation for Quality Management, Avenue des Pleiades 15, B-1200 Brussels, Belgium.

Exercises

1   Prepare a presentation to illustrate the criteria for the choice of diamond for (i) a betrothal and for (ii) industrial use. Explain how you will educate the audience on the relationship between quality and customer appeal/requirements (you determine the level of technical content).
2   Develop a quality matrix for the determination of quality of any utility (e.g. water, gas, electricity supply, health care).
3   Develop a quality matrix and determine a set of quality criteria for a family car.

Chapter 2

Quality of Service in telecommunications

2.1 Introduction

Quality of Service (QoS) management is an essential function in the provision and maintenance of a telecommunication service. It is necessary to understand the role of management of QoS and the main features. Effective management of QoS would reflect in optimum use of network, human and financial resources of any network or service provider. This chapter attempts to address these issues, review some of the studies carried out on QoS to date, evaluate the shortcomings of the present method of management of QoS and proposes a framework for the study of QoS management.

2.2 QoS in the management of a telecommunication service

The following factors are pertinent to the study and management of QoS for telecommunication services:

– meeting market's requirements on quality
– optimisation of resources of the network/service provider
– use of quality as the differentiator in the market.

2.2.1 Meeting market's needs on QoS

Availability of complex terminal equipment and software has resulted in the availability of sophisticated applications. In turn, this has resulted in equally sophisticated QoS requirements by different segments of the population. Meeting these stratified QoS requirements is a key factor which will have to be addressed by a service/network provider. Being able to provide the precise level of quality or better is essential to attract customers and maintain their loyalty. If a supplier does not provide the desired/preferred quality and another provides it the customer is more inclined to switch its custom to the other party.

2.2.2   *Optimisation of resources of the network/service provider*

Basic economic theory requires the optimisation of resources to the provision of goods for maximum profit. Analysis of the telecommunications market has shown that different customer segments have different quality requirements resulting in varying rankings and levels of performance of QoS parameters. The supplier of a telecommunications service or the network needs to be aware of this information to plan the optimisation of resources in order to reduce the risk of over or under engineering of the hardware, human and financial resources. Customers require a level of QoS at a competitive price. In an increasingly competitive industry optimisation of resources, to provide service at minimum cost is essential for effective management of a telecommunication service.

2.2.3   *Quality as the differentiator in the market*

There are two principal arguments for quality being the differentiator in the market. First, when service features and price are similar or comparable, quality will be the differentiator for the customer. Second, quality may be used by service providers to earn the image of a 'trusted and respected' provider. There is usually always a segment of the market that goes in for quality and reliability and is willing to pay more for this type of service. Service providers who offer quality and reliability (both in network and organisational support to customers) will enjoy a more favourable elasticity of customer loyalty in times of economic difficulty and downturn. Even though traditionally it is the more established and older service providers who offer quality service this need not be so. Newer entrants to the industry can equally well provide quality and reliable services and support to customers.

2.3   Features of QoS in telecommunications

Quality of service in telecommunications has the following features:

QoS is measured and expressed on an end-to-end basis.
They are service specific.
There could be separate industry and customer oriented performance measures.
QoS performance parameters would have different priorities for different user sectors.
Levels of performance may be different for different user sectors as well as among users within the same user sectors.
Quality requirements are not necessarily static. They are dynamic, that is, priorities of parameters and their levels could vary with time.

These features are elaborated in the following paragraphs.

2.3.1   *End to end*

Telecommunication services usually comprise different segments of the network interconnected to carry a particular service for the intended duration of use. These elements are 'disengaged' at the end of the connection use. As far as the user is concerned the quality of the service is experienced when all the necessary network elements are connected together to make the service possible. While the individual element performance may be of interest to the service and/or network provider the user experiences quality of a service on an end-to-end basis. Hence, there ought to be a mechanism to either measure the end-to-end performance or to estimate it and to specify it.

2.3.2   *QoS is service specific*

The technical performance of a service may be expressed by a set of parameters, which are unique to that service. Cell loss ratio (CLR) as applied to the ATM platform would be meaningless in a circuit switched analogue transmission network. Non-technical parameters such as speed of provision and time to repair may be common to many services. For a given service, a set of performance parameters that are uniquely applicable to that service needs to be identified and specified.

2.3.3   *Industry and customer oriented parameters and measures*

Within the industry outage ought to be specified in terms of

– number of outages;
– distribution pattern of outages;
– magnitude of these outages; and
– outages of the individual links.

However, presenting this information to the non-technical customer needs translation into terms such as the maximum number of outages in one year and the maximum outage period or any one outage on an end-to-end basis.

2.3.4   *Prioritised set of QoS parameters*

A set of parameters ranked for importance for a given service may be different for different customer populations. International news agencies usually require highly reliable and error free connections on a 24/7 (24 hours a day, 7 days a week) basis. High street retail businesses perhaps have little requirement for such connections save when data is being transmitted at off-peak periods. For travel agents, fast response time would be very high on the order of ranking, perhaps unlike news agencies that transmit information mostly in one direction. It is essential for effective management of QoS to identify the sectors of industry that have unique performance parameter rankings.

2.3.5   *Levels of preferred performances for the parameters*

In addition to the ranking of parameters in the order of importance (as described in 2.3.4) different segments of the customer population may require different levels of performance for the same parameter. For example, a residential customer may tolerate larger response times in an interactive service compared to a travel agent but with different attitudes to tariff. The preferred levels of performance for each parameter need to be identified for each uniquely identifiable segment of the population.

2.3.6   Dynamic nature of QoS requirements

Customer's ranked parameters and the levels of preferred performance could change with time. With the advances in technology and changes in circumstances of the company or the individual, the preferred quality requirements may change with time. For effective management of QoS it is necessary to identify when these changes take place and to be able to record these.

2.4   Parties in the management of QoS

Identification of most, if not all, of the key issues may be facilitated by a matrix made up of the parties involved in the supply, use and monitoring of these services, separated into 'national' and 'international'. The quality issues may be different in character under the demarcation 'national' and 'international'; hence telecommunication services are conveniently divided into these categories. The matrix is shown in Table 2.1.

The principal parties in the supply, use and monitoring of telecommunication services are the network provider, service provider, equipment manufacturer, user, standards bodies and the regulator. Other bodies such as universities and research establishments may be considered as playing supporting roles to any one (or more) of the above players.

2.4.1   Network provider

The issues related to QoS that should be addressed by the network provider on a *national* basis are as follows:

(a) Aspects related to Service Level Agreements (SLAs) between customers and service providers. The issues related to SLAs are discussed in Section 5.4.1.3.
(b) Where interconnection exists, aspects related to SLAs with other network providers.

Table 2.1   Relevant parties in the management of QoS

Parties	Services – national	Services – international
Network provider	Sec. 2.4.1	Sec. 2.4.1
Service provider	Sec. 2.4.2	Sec. 2.4.3
Equipment manufacturer	Sec. 2.4.3	Sec. 2.4.3
Customer/user	Sec. 2.4.4	Sec. 2.4.4
Standards bodies	Sec. 2.4.5	Sec. 2.4.5
Regulator	Sec. 2.4.6	Sec. 2.4.6

(c) The specification of quality related technical performance of network systems required of equipment suppliers.

In the *international* area, the network provider has to address issues on quality in the following areas:

(d) Relevant standards on performance in interconnection with foreign networks.
(e) SLAs with foreign network providers, which may or may not be additional to the requirements from the standards.

2.4.2 Service provider

In the provision of *national* telecommunication services, the service provider has to address issues related to QoS in the following areas, which are discussed in more detail below.

(a) Management of QoS indicators for:
  • internal business use,
  • for the benefit of customers,
  • meeting regulator's requirements, and
  • as a differentiator in a competitive market.
(b) General management of quality, including Total Quality Management (TQM) and how management of QoS fits in to the overall TQM (see Section 1.5).
(c) Influence on the standardising bodies for optimum benefit to service provision.
(d) Support of research into quality issues.
(e) Determination of customers' requirements on QoS and its management.
(f) Determination of customers' perception of QoS and its management.
(g) Specific needs on quality of special interest groups, for example, the disabled, low-income groups, etc.
(h) Economics of QoS.
(i) Societal aspects.

For *international* services the following specific issues have to be addressed:

(j) QoS matters in SLAs with foreign administrations on the correspondent provision of service between host and foreign country.
(k) QoS aspects in relation to competitors in the global market, where a global service provider is bypassing the traditional bilateral correspondent relationships for the provision of international services.
(l) SLAs with multinational companies.
(m) Influence on standards bodies on quality-related matters for international services.

(a) Management of QoS indicators

In the case of network providers who are also service providers, the network-related performance parameters are usually measured by the network providers within the network systems. Network providers who are not service providers have to supply the service provider relevant network-related performance for the service provided. For non-network-related performance parameters the service providers have had to institute their own performance systems. The principal shortcoming in this area has been the choice of performance parameters, which are mainly of interest to providers rather than customers. Choice of QoS indicators to meet providers' needs and additional indicators, where necessary, to meet customers' requirements should satisfy all parties.

(b) General management of quality

Management of QoS would be most effective if carried out within the context of company-wide TQM process and not in isolation. There is adequate information available on the topic of TQM to enable any company to apply the fundamental principles and to benefit from its implementation. TQM is discussed briefly in Appendix 2.

(c) Service provider and standards bodies

The service provider has to address the issue of when to involve standardising bodies to develop performance standards for new services. It also has to address the issue of what types of performance parameters are to be standardised. A current shortcoming is a tendency to focus on the technical issues without taking full account of their implications on customers' requirements. Standards are covered in Chapter 21.

(d) Research on quality issues

With the development of new applications leading to increasingly sophisticated services, it became necessary for more performance standards to be specified. With increased complexity in technology, the parameters became more complicated. Many of the service providers in the Organisation for Economic Co-operation and Development (OECD) countries had their own research departments who carried out fundamental research into many aspects of telecommunications, including quality. Nowadays, due to privatisation, research establishments are increasingly being run on a commercial basis. These are funded to address specific issues to improve the commercial viability of the service providers by better exploitation of technology. Nevertheless, studies on quality are being carried out, sometimes for the mutual benefit of many providers. An example is the European Institute for Research and Strategic Studies in Telecommunications (EURESCOM). This organisation has the common interests of participating European network and service operators. However, the research establishments have also concentrated on the service providers' interests, despite some excellent work for the benefit of customers. There has been criticism from users that not sufficient has been done to meet their interests.

(e) Determination of customers' QoS requirements

Determination of customers' QoS requirements is a necessary activity on the part of the service provider, especially if it wants to provide the quality that will satisfy customer needs. After the emergence of competition outside USA there is an increased awareness of the customers' QoS requirements. Despite this awareness, there is no formal mechanism to capture these nor is there an agreed methodology to manage this exercise. User's involvement is further covered in Chapter 19.

(f) Customer perceptions of quality

Customer perception surveys have been carried out both by the service providers as well as by user groups representing telecommunications companies. Surveys carried out by service providers are of two kinds, one for public consumption and the other strictly for internal use. It is perhaps no secret that some of the more sensitive issues and in-depth findings are not made available for public consumption. The surveys carried out by user groups ought to be taken seriously by service providers and their findings acted upon, particularly when these findings are credible and useful. In addition to obtaining 'free consultancy' service from these groups it will enhance the possibility of user groups carrying out more of these surveys in areas requiring action.

(g) Quality issues of special interest groups

In many of the more developed countries, the need of special interest groups, for example, the disabled, low-income groups etc. have to be identified and addressed, especially by the dominant service provider, often at the request of the regulator or to enhance the image of the service provider. This topic is covered in Chapter 18.

(h) Economics of quality

The cost of providing a level of quality, the revenue generated and the effects of competition are some of the considerations that should be addressed by service providers. No qualitative measures have been developed to determine the balance of QoS benefit and cost. This topic is covered in Chapter 23.

(i) Societal aspects

The dominant service provider in any country is most likely to face societal issues. These include provision of unprofitable services to remote areas, attitudes towards the countryside, arts and heritage of the country, sponsorship of sporting events, commitment towards the community, education of the younger generation, etc.

The issues under (j), (k), (l) and (m) for international services are similar in nature to the corresponding ones in the national scene, but with different implications.

2.4.3   Equipment manufacturers

The following are the principal areas where equipment manufacturers should be involved with QoS issues:

(a) Design and manufacture of individual elements of telecommunications networks in order to meet specified end-to-end performance.
(b) Design of test equipment and support systems to monitor performance of individual elements of a telecommunication system and, in certain cases, of delivered end-to-end performance.
(c) Influencing standardising bodies on realistic performance standards.

The quality issues to be addressed by the manufacturers are unlikely to be significantly different for national and international provision of services. Their principal roles are to meet the performance statements specified by the purchasers, the service providers and network providers. Additionally, they could enlighten the standards bodies on what is realistically achievable and influence the timing of standards for the common benefit of all parties in the telecommunications industry.

2.4.4   Customers/users

In relation to *national* telecommunication services, users are likely to be concerned with the following issues on quality:

(a) Publication, by service providers, of delivered QoS on parameters which are meaningful to customers.
(b) Quality needs of specific groups being met by the service provider. Examples of specific groups are:
  • Banks and certain newspaper publishers who require zero outage during certain periods of the day.
  • Broadcasters who require large call handling capacity from service providers during periods of televoting.
  • Quality requirements of the disabled, low-income groups etc.
(c) Businesses who require customised performance in contractually binding SLAs with service providers and network providers.
(d) Consultation by network and service providers on their requirements of performance.

In the provision of services in the *international* scene the following quality issues will be of concern to users:

(e) SLAs between multinationals and global alliances of service providers on quality aspects. These will be more complex than SLAs for services provided within a country.
(f) Consultations with users and user groups on their quality requirements for sophisticated services.

The principal shortcoming in the above areas is lack of adequate representation of the customers' interests by the service providers.

2.4.5   Standards bodies

There are national, regional and international standardising bodies. Although there is only one international standardising body for telecommunications, the Telecommunications Standardising Sector of the International Telecommunications Union (ITU-T), the convergence of telecommunications, Information Technology (IT) and information industries is resulting in other international standards organisations influencing the quality of service delivered over telecommunications networks, for example ISO.

The issues to be addressed by the standards bodies are:

(a) Development of end-to-end performance standards for an international connection and determining how the national part can be deduced from this.
(b) Development of standards for specific issues, for example, relationships between degrading factors on transmission quality and their effects on the perception of customers.

Role of standards bodies are covered in Chapter 21.

2.4.6   Regulators

Where regulators exist, their influence is confined to within the boundary of a country. The only current exception is the European Commission but the World Trade Organisation (WTO) is beginning to take an interest in global telecommunications regulation. The key issues on QoS to be addressed by the national regulators are:

(a) The influences on quality under monopoly and competitive environments.
(b) Determination of QoS parameters on which delivered quality data are to be published by the service and network providers.
(c) Audit of published QoS data.
(d) Whether there ought to be a 'reward and punishment' policy towards providers in relation to delivered quality.
(e) The timing of introduction of quality parameters for new services.
(f) Steps to maintain delivered quality by service providers.

The regulators appear not to concern themselves with international services. However, the European Commission does have an intra-Community interest in the harmonisation within the European Union member countries. This concerns international services between service providers in different member countries of the European Union.

2.5   Studies in QoS

A number of organisations carry out studies on QoS in telecommunications, perhaps due to the ubiquitousness of the service. Yet another reason for worldwide coverage of studies is the fact that this industry is not technologically mature. Telecommunications offers exciting opportunities for the researcher. A few of the entities that carry out studies on QoS are listed below. Further information on their work may be obtained by visiting their web-sites (URLs given in the References).

2.5.1   ITU-T

The Telecommunications sector of the ITU carries out studies on a wide range of topics related to Telecommunications with a view to producing a recommendation (a de facto 'standard') for use by the industry. Recommendations on QoS and related aspects of it reflect the studies carried out by ITU-T in the past. Many of the QoS studies are carried out in Study Group 12 though some QoS studies take place in other study groups (Study Groups 4, 7, 9, 11, 13, 15 and 16). Of particular interest is the most recent study on 'Communications Quality of Service: A Framework and Definitions' being issued as ITU-T Recommendation G 1000. (See also Section 27.5.) The theme of this recommendation is reflected in this book.

2.5.2   ETSI

The European Telecommunications Standards Institute (ETSI) was founded in the 1980s (from the CEPT) to focus on the telecommunications standards needs unique to Europe. This regional body carries out work on the lines of ITU-T and issues documents under the categories of Standards, Guides, Technical Specifications, Technical Reports, Special Reports, Technical Bases for Regulation and other miscellaneous documents. The QoS studies are carried out in many technical sub-groups, but mainly in STQ (Speech Transmission and Quality). Among the original contribution by ETSI sub-groups are TIPHON (Telecommunications and Internet Protocol Harmonization Over Networks) document TR 101 329 and Networks Aspects (NA): General aspects of QoS and network performance (NP) document No ETR 003. ETSI subgroups are continually studying problems that could be of interest to the European telecommunications industry.

2.5.3   Regional standards bodies

Regional standards bodies exist to study and develop standards which meet the regional requirements. The ANSI (American National Standards Institute) caters to the North American market, particularly to the USA. A comprehensive list of standards bodies may be found in Macpherson [1].

2.5.4   IETF

The Internet Engineering Task Force (IETF) members carry out studies on a 'discussion' basis and submit the 'final' document as a prelude for standardisation. Though the IETF issue their own documents in the form of RFCs many of the documents are taken by the ITU-T for issue as Recommendation. Their web site is a source of information on QoS topics among other topics.

2.5.5   EURESCOM

The best known research organisation in Europe for telecommunications is perhaps the EURESCOM a body founded by 20 service providers from Europe in 1991. Now there are a few more members. They carry out collaborative studies on all topics that are of mutual interest to these service providers and include QoS related topics. Recent studies on QoS are:

- P-806: A Common Framework for QoS/Network Performance in a multi-Provider Environment.
- P-906: QUASIMODO – QUAlity of ServIce MethODOlogies and solutions within the service framework: Measuring, managing and charging QoS.

2.5.6   NRRI

The NRRI was established by the National Association of Regulatory Utility Commissioners in 1976 at The Ohio State University and is the official research arm of NARUC (National Association of Regulatory Utilities Council). They provide research and assistance to state utilities including telecommunications. Some of the studies are on QoS related topics. A visit to their web sites is useful both for the manager and the regulator. See also Section 20.4.

2.5.7   European Commission funded programmes

Perhaps the best-known government funded programme on QoS in Europe is the one funded by the European Commission. A recent study was commissioned on the determination of customer's Internet Access QoS parameters. The consulting company Bannock of the UK produced their findings in August 2000. They have identified 21 access parameters which are said to be of prime interest to the user. These are listed in Appendix 6.

2.5.8   Conferences, seminars and symposia

Conferences, seminars and symposia held in different parts of the world at various times during the year often attract papers describing work carried out on various aspects of QoS. Due to the nature of these events, the work reported is of recent nature and is meant to be work not made public before. It is rare for a regular event to be dedicated to QoS with the exception of the International Workshop on Quality of Service – IWQoS; most of the telecommunications related events usually provide some papers on matters related to QoS. Three events worthy of note are given below.

IWQoS held annually since 1993. The papers presented are predominantly on computer communications and are based on work mostly carried out in the academia.

ISOC/IEEE hold an annual conference and here too there is a selection of papers describing work related to QoS. There is a strong emphasis on IP related topics.

FITCE (Federation of Telecommunication Engineers of the European Community) hold an annual conference and there is a selection of papers on topics related to QoS. These papers tend to cover both IP related and the traditional circuit switched communications, both land and mobile based.

In addition or in lieu of presentation of work at conference type events findings may also be presented in journals. Readers are recommended to keep track of the current conferences, seminars, symposia and the journals to keep up to date with the state of the art in the studies on QoS related topics.

2.5.9   *Service provider's laboratories (e.g. BT's Adastral park, AT&T's research division, CNET etc.)*

Most large service/network providers have research facilities to address problems associated with the provisioning and maintenance of services. It is known that studies on QoS are part of these. However, these reports are usually treated in commercial confidence, unless they are of public interest and commercial interests have been addressed. Detailed investigation of these studies is outside the scope of the book.

2.6   Issues on QoS facing the industry

The main issues related to QoS facing the telecommunications industry are as follows:

1   Lack of an internationally agreed set of terms and definitions to embrace the different aspects of quality and QoS from the customer's and service provider's viewpoints have resulted in various sets of assumptions and ad hoc definitions. These assumptions and definitions varied with authors and each had a valuable contribution to make. However, due to the multiplicity of these assumptions and definitions there has been a noticeable lack of synergy and collective benefit than would have been the case had these studies been undertaken within an agreed framework using an agreed set of terms and definitions. It would not be stretching the point too far to say that a certain amount of inertia is creeping into the discussions and papers on QoS. An internationally agreed on set of terms, definitions and an architectural framework would benefit the researchers, standardising bodies, users, regulators and the manufacturers of telecommunications equipment. This has been resolved in the form of ITU-T Recommendation G 1000. However, this recommendation need to be mapped to existing recommendations and areas requiring further work need to be identified and pursued.

2   A set of QoS parameters needs to be developed for each of the principal services on an end-to-end basis. This would enable the end users to compare performances offered by different suppliers.

3   For the evolving services, for example, 3G mobile services, Internet supported services etc. standards need to be developed for optimum application by various players in the industry. The problem is compounded by the speed of the evolution of technology and applications. Perhaps there is a catalytic role to be played by the industry to optimise standards and provision of new services.

4   Identification of the QoS requirements of emerging services needs to be carried out and the changes in requirements that take place have to be identified. Elsewhere in this book we have shown that planning of service quality based on user's quality requirements will prove to be a principal factor in the optimisation of resources of the service provider. This is an area where, traditionally, the industry has had less than perfect involvement in the past.

5   Privatisation of telephone service is a continuing trend around the world. Once most telephone companies were government agencies, government monopolies

or regulated monopolies. Private companies were prohibited from offering services to customers. This began to change in the late 1970s. The largest breakup came in 1984 when AT&T in U.S.A was divested of the local operating companies. Private companies other than AT&T began to provide long distance service. The customer had more choices but also more difficulty. No longer was there one company responsible for the end-to-end service and its quality. Trouble isolation and resolution has become increasingly difficult for customers as services have become more fractionalised. When customers buy a service from one company they prefer to view that company as the sole provider of the service. Many companies, however, may be involved in delivering the service, all of whom must be managed from a QoS perspective to meet the customers' quality expectations.

A new consideration arising from the difficulties experienced by customers from service providers due either to unconventional accounting practices or collapse of trading is the service availability to existing suppliers. This new topic is addressed briefly in Appendix 11.

This book attempts to address the key issues in the management of QoS by offering a framework for its management and guidelines for its implementation.

2.7   A framework for the study and management of QoS

Any framework, to be useful, must meet the following requirements:

(a) it must identify and describe all principal components (functions or activities);
(b) it must explain the principal characteristics of each component from which one could identify the key issues;
(c) it must provide guidelines for the management of the key issues; and
(d) it must be practical to avail the framework to a wide spectrum of users.

A framework designed to fulfil the above requirements is shown in Figure 2.1.

The four viewpoints of QoS, described in Chapter 3 provide a means of establishing criteria relating the customer needs and perceptions to what a Service provider offers and delivers. Detailed criteria and parameters can be established using the QoS matrix, described in Chapter 4, which was developed in the era of telephony services provided from a circuit switched platform. Whilst still appropriate for Internet derived services, the technical service functions have been expanded by the use of the ACF Model, used in the IP chapters (10, 11 and 12) and developed by the IETF (Internet Engineering Task Force) to present the computing industry view of QoS. Finally, when discussing the future aspects of QoS in Chapter 28, it has been necessary to use the Four Markets Model to take account of the interaction between the players in the value chain and their contribution to QoS which must take account of the total customer experience of, for example, e-commerce services.

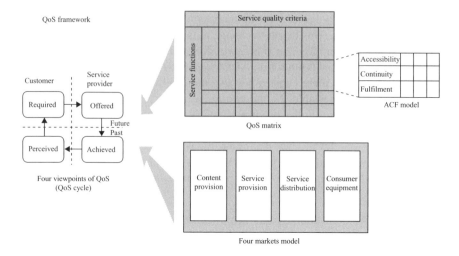

Figure 2.1   QoS framework

2.8   Summary

Using the concepts on quality from Chapter 1 the features of QoS in telecommunications have been established. After reviewing the state of the art the shortcomings of the present method of managing QoS have been identified and a framework for its study is introduced. The rest of the book deals with different aspects of the framework for the identification, study and management of the key issues and guidelines on how to address these.

2.9   References

1   MACPHERSON A.: International Telecommunication Standards Organisations, Artech, 1990.

Web sites

1 ITU-T	**http://www.itu.ch**
2 ETSI:	**http://www.etsi.org**
3 ANSI	**http://www.ansi.org**
4 IETF	**http://www.ietf.org**
5 EURESCOM	**http://www.eurescom.de**
6 NRRI	**http://www.nrri.ohio-state.edu**

Exercises

1   Identify the key parties/players in the provision and maintenance of telecommunications services in your country. Establish the relationships between these parties

and identify problem areas that require investigation and recommendations for improvement.

2   Identify the relationships between the key parties/players in telecommunications in your country with similar parties outside the national boundary. Evaluate the importance of these relationships in the welfare of the telecommunications industry and people of your country.

Section II

Framework

Frameworks and models are helpful to address logically and systematically a particular issue. Development of frameworks and models and their application bring a significant element of objectivity and professionalism in addressing an issue. In this section, a framework (with models) is offered for the study and management of QoS in telecommunications. Due to the dynamic nature of QoS, arising out of frequent development of sophisticated applications, amendments and refinements to the framework could be envisaged in the future. However, the basic concept of framework offered ought to withstand the test of time.

Chapter 3

Four viewpoints of Quality of Service

3.1 Introduction

Discussions in the previous chapter led to a framework for the study and management of Quality of Service (QoS) in telecommunications. This chapter describes the four viewpoints (the quality cycle), its purpose, and the characteristics of each viewpoint and the relationships of the four viewpoints within the model. Developed particularly for the legacy services, the principles can be equally applied to IP supported services with the support of the ACF and Four Market Models described elsewhere in the book. Chapters 4–7 cover the management issues arising from the application of each of the four viewpoints.

3.2 The four viewpoints of QoS

The two principal parties concerned in the management are the service providers and the customers. The customers' viewpoints are their QoS 'requirements' and their 'perception' of received performance. The service provider's viewpoints are the QoS that are planned (and therefore the 'offered') and the QoS actually 'achieved' or 'delivered'. All QoS described are that experienced by the customer on an end-to-end basis for any service. Network element performance would be of primary interest to the network or the service provider but not necessarily to the user. The above concept is illustrated in Figure 3.1.

Classification of the four viewpoints of QoS has the following advantages:

(a) Decoupling and identification of the management issues: this facilitates easier focusing on the issues and due consideration is given to the process of resolving these issues.
(b) Sufficient clarity is given to the four QoS data as and when these become available. The data have different meanings and are to be interpreted in the context of these meanings.

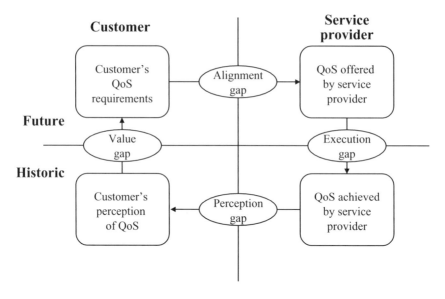

Figure 3.1   The four QoS viewpoints or the quality cycle

Management of QoS becomes clearer, without confused intermingling of QoS figures and the relevance and importance of the interrelationships between the various data can be interpreted. It becomes much easier then to focus on improving QoS by focusing on the remedial steps to be taken on the pertinent resource. The characteristics of each viewpoint are described in the following subsections.

3.2.1   Characteristics of customers' QoS requirements

QoS requirements by the customer are the statement of the level of quality of a particular service. The level of quality may be expressed by the customer in technical or non-technical language. A typical customer is not concerned with how a particular service is provided, or with any of the aspects of the network's internal design, but only with the resulting end-to-end service quality. The management of customers' QoS requirements are covered in Chapter 4.

Customers' QoS requirements have the following principal characteristics:

 (i) The requirements are expressed in customer understandable language and not that of the supplier.
 (ii) Quality is expressed on an end-to-end basis.
(iii) Quality is expressed separately on a service-by-service basis for the principal services offered by the service provider.
(iv) Factors influencing requirements are:
   • awareness by customers of service features and their knowledge of telecommunication technology and practice;
   • relationship between price of the service and quality;

- variation of quality requirements with time;
- variation of quality levels with the level of quality experienced;
- the value of telecommunications to their life style or business performance.

3.2.1.1 Expression of QoS requirements in customer understandable language

For sophisticated services, for example, services using frame relay, customers are likely to be more knowledgeable on telecommunications. Requirements are, therefore, likely to be expressed in sophisticated terminologies, by the telecommunications managers of businesses, in terms similar to those used by the supplier. However, the requirements of ordinary customers are more likely to be expressed in terms more meaningful to customers rather than the supplier's trade language. A customer may indicate that the acceptable number of occasions when moderate difficulty in call clarity during a telephone conversation is tolerated would be a maximum of one in a hundred calls. The service provider translates this requirement to network performance parameters and assigns target values to network elements to result in an end-to-end performance of not more than one call in 100 experiencing moderate difficulty. Such translation ought to be considered for every relevant parameter.

3.2.1.2 QoS requirements on an end-to-end basis

The customers' quality requirement is meaningfully expressed on an end-to-end basis. Customers cannot be expected to appreciate the specific characteristics of component parts of an end-to-end-service but can only evaluate the service based on actual use. See also Section 2.3.1.

3.2.1.3 QoS criteria on a service-by-service basis

Quality criteria could be different for different services. Even in cases where quality criteria or parameters are the same for different services, targets for these could be different for each service. For meaningful management of QoS it is essential that each principal service have its own set of QoS parameters specified and targets.

3.2.1.4 Factors influencing customers' QoS requirements

Factors influencing customers' QoS requirements may be grouped under the following titles:

 (i) type of application;
(ii) competition;
(iii) technology;
(iv) economics.

(i) Type of application

Figure 3.2 shows the QoS chain linking the provider and the customer. The network provider provides the network for the service provider who requires a level of QoS from the network provider. The customer served by the service provider, who could

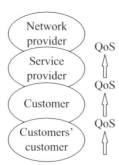

Figure 3.2   Chain of QoS requirements

be a reseller, requires a level of QoS from the service provider. Lastly, the reseller has customers of its own and they require a level of service from the reseller. This chain of QoS requirements could result in a number of different levels of quality.

At the customer end there would be different types of customers with specific requirements. Banking and publishing industries would require zero outage and error-free transmission during specific times of the working day. Those in continuous process industries may be able to tolerate short outages of 1 or 2 minutes a few times a day but not one long outage in a week. Some resellers may tolerate one long outage in a year but not many short outages in a day. The particular quality requirements for QoS parameters could similarly vary from customer to customer depending upon the particular application of the service and its importance. Recognition of this factor would assist the service providers in identifying the particular quality requirements of their customer base for various market sectors. Knowledge of the market and the particular customer requirements will assist in identifying the specific customer requirements on quality.

Medium and large organisations often have more sophisticated telecommunications requirements with specific performance requirements. These companies often require a Service Level Agreement (SLA) with the provider. The SLA would mention the performance requirements of the services provided by the service provider with penalty clauses for under-performing. SLA is discussed in more detail in Chapter 5 and in Appendix 3.

(ii) Competition

The bargaining power of the customers in a competitive environment will influence the level of quality they would aim for. In particular, major business customers require service providers to tender against specifications that include QoS requirements. Such influences work in the telecommunications industry due to the continually improving nature of technology and the increasing communications knowledge of business customers. In cases where customers are satisfied that one performance parameter has reached an acceptable level, other parameters could take priority. Thus, quality requirements could be influenced by the improvements offered by competitors and what is realistically achievable from technological improvements.

(iii) Technological improvements

Customers' expectations on quality are influenced by their perception of the potential benefits of technological advances. The two principal sources of this perception are improvements in customer premises equipment and development and availability of new services based on enhanced communications platforms such as frame relay and ATM. Business customers are constantly on the look out for improved means of communication to improve the efficiency of their telecommunication applications and make them more competitive. Voice over IP is now possible. Customers will expect this to be commercially available in due course. Quality of speech will then be expected to be as good as conventional digital communications.

(iv) Economics

Considering the relationship between price and quality, the residential customer is not too concerned with high levels of quality but cost could be important. At the other extreme, large businesses whose primary activity is dependent upon transactions of accurate information will require high-quality connections whenever they wish to use the service and cost will be less important. The relationship between cost (to customer) and quality is a complex one. It is an area where further research is required. Unlike retail industries, where competition has existed for a very long time and the climate for such researches has existed, there has been very little investigation to understand the relationship between quality and the price, a customer is willing to pay for different usage of telecommunications services.

Different market sectors require different levels of quality. Banking and publishing industries that require zero outage and error-free transmission during certain periods may pay a high price for such quality. These types of customers are, within reason, willing to pay 'any price' for quality. Other businesses, which require high quality, but are cost conscious, may only be willing to pay a lower price. For such customers, while quality is important, it is not vital to their survival. The residential market, which traditionally makes fewer calls, may be content with poorer quality, but is also cost conscious. Customer differentiation is shown in Figure 3.3.

The service provider needs to be aware of the different dimensions of customers' QoS requirements and the factors that influence these. Such awareness will differentiate the service providers in their approach to quality.

3.2.2 Characteristics of QoS offered by the service provider

QoS offered by the service provider is a statement of the level of quality expected to be offered to the customer by the service provider. The level of quality is expressed by values assigned to QoS parameters. The service provider may express the offered QoS in non-technical terms, where necessary, for the benefit of customers and in technical terms for use within the business. The characteristics of QoS offered by the

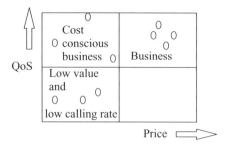

Figure 3.3   Customer population differentiated by 'value'

service provider is covered in this section and the management of this aspect covered in more detail in Chapter 5.

The principal characteristics of offered quality are as follows:

(i) It is usually in the de facto language of the telecommunications service and network provider and not the customer's.
(ii) It is separately specified on a service-by-service basis.
(iii) There could be different levels of performance for each service.
(iv) It has a relationship with customers' quality requirements.

3.2.2.1   Supplier language

Terminologies exist within the organisation of the service and network provider to express the performance and features of a service. These terms have evolved over the years and, even though there is a commonality among many service providers, there is no agreed set of definitions for the expression of QoS or service performance. Today, global and multi-national customers are faced with the task of interpreting various forms of service offerings from different service providers. It is therefore impossible for customers to compare service offerings or calculate the true end-to-end service performance and compare it with their requirements. Therefore, it is vital that there is an internationally agreed set of service quality definitions (but not including targets of performance).

3.2.2.2   Offered quality is expressed on a service-by-service basis

For the quality to be meaningful the performance levels of each service should be specified individually, because customers usually buy a service and not a portfolio of services. In order to sell a service, the quality level must be quoted so that customers know what is being bought and they can compare the service offering from a number of service providers. Different services could also have their own unique performance parameters. Additionally, the various parameters of quality can have a different impact on different services. For example, error rates will have a major effect on data services but voice service can stand relatively high levels of error with no apparent effect on listeners. The reverse of this example, would apply to the parameter 'delay'.

3.2.2.3   Different levels of performance for the same service

Due to varying requirements for the levels of performance by different customer groups, it is necessary for service providers to aim for correspondingly different service levels for the same service, for different customers. Therefore, service levels need not necessarily be the same for all customers; they are more likely to be tailored to meet individual requirements. Variations in performance can include not only network performance (e.g. high-quality circuits versus normal-quality circuits) but also different levels of service support and other non-network-related parameters. However, it is difficult for a service provider to differentiate levels of service for network-related parameters, especially if the services are provided on a ubiquitous network. This may ease for future variable-bit-rate broadband networks based on ATM technology where, for example, priority on delay and bandwidth can be given to individual customers. However, it is relatively easy for a service provider to offer different levels of quality for non-network-related performance parameters to different customers for the same service.

3.2.2.4   Relationship with customers' requirements

The offered quality must have a relationship with the customers' requirements in the form of an audit trail back to the original customer requirement. Where there is absence of such a relationship, it is evident that the service provider has decided the quality level for the customers, probably based on its capability, or what it can afford, and not customer requirements. The relationship with customer requirements has two features:

 (i) The offered quality must be *derived* from the customers' requirements in so far as every effort is made to meet as much of the customers' requirements as possible. The relationship would be the degree or percentage to which customer requirements were met.

(ii) The parameters with which the offered quality is *described* would have a similarity with the parameters expressed by the customers. The correspondence may be direct or indirect. For example, the customers may express the call quality qualitatively and the service provider may express it in technical parameters to achieve the customers' requirements.

3.2.3   *Characteristics of QoS achieved by the service provider*

Quality of Service achieved by the service provider is a statement of the level of quality provided to the customer. This is expressed by values assigned to parameters, which should be the same as those specified for the offered QoS so that the two can be compared. These performance figures are summarised for specified periods of time, for example, for the previous three months. For example, the service provider may state that the achieved availability for the previous quarter was 99.95% with five breaks of service of which one lasted 65 minutes. The characteristics of QoS delivered are covered in this section and the management of this viewpoint of QoS is covered in more detail in Chapter 6.

The principal characteristics of the delivered quality are as follows:

- It is objective (i.e. it is a measure of the actual delivered performance from measurements, and therefore non-contentious).
- It may be expressed in either of two forms, as 'element performance' or as 'end-to-end performance'.

Delivered performance is usually estimated from measurements carried out on the network supporting the service and its relevant organisational support functions. Certain network performance parameters may be monitored on a sampled basis, for example, transmission performance. Others, such as exchange outage, are usually continually monitored to cover all nodes. Performance data on parameters, which are non-network related, for example, time to repair, must be computed from 100% of incidents. If sampling is carried out on these parameters there is a risk of not providing customers with a true and fair view of achieved performance. The network or the service provider carries out the measurements. The parameters chosen for measurements must have a direct relationship to the parameters describing offered quality in order that they can be compared.

The measured performance data may be presented in two ways:

 (i) The delivered performance may be expressed on an element-by-element basis. This will be of particular benefit to the network provider and the service providers who can monitor the delivered performance against planned or specified element performance and take any necessary remedial action. Element performance is of little interest to customers.
(ii) The delivered performance may also be expressed on an end-to-end basis. This will be of particular use to customers and regulators, who may not be concerned with the element performance within networks. To the service provider it is an indication of how well or otherwise the service performance is delivered to the customer.

3.2.4   *Characteristics of QoS perceived by the customer*

QoS perceived by the users or customers is a statement expressing the level of quality they 'believe' they have experienced. The perceived QoS is expressed, usually in terms of degrees of satisfaction and not in technical terms. Perceived QoS is assessed by customer surveys and from customer's own comments on levels of service. For example, a customer may state that on unacceptable number of occasions there was difficulty in getting through the network to make a call and may give it a rating of 2 on a 5-point scale, 5 indicating excellent service. The QoS should include all aspects of quality, not merely the network-related aspects. Customers cannot be asked to state their network-related and non-network-related QoS criteria separately. QoS performance figures would be more meaningful if specified on a service-by-service basis. Some parameters would have the same definition when applied to other services, for example, provision of service. Other parameters, especially technical ones, could vary in definitions and targets and may have to be specified for each service. The characteristics of customer perception of QoS are covered in this section and the management of this viewpoint is covered in more detail in Chapter 7.

The principal characteristics of customer's perceived quality are as follows:

- It has varying degrees of subjectivity.
- It is expressed as opinion ratings, quantitatively.

3.2.4.1 Subjective elements

Customers' perceptions could be influenced by a variety of factors, which will affect their judgement of quality. Examples of these influencing factors are:

 (i) customers' 'awareness' of telecommunications services;
 (ii) customers' expectations;
(iii) experienced quality;
(iv) customers' recent experiences;
 (v) advertising;
(vi) nature of the opinion survey;

(i) Customers' awareness

There are many elements in the complex array of influences which enable increased customer awareness. Business customers have well developed, clearly defined needs and their choice of service provider will be principally governed by their level of confidence in an operator to satisfy and/or delight them and hence support their business activities. Conversely, personal customers may be less discerning and accept more basic, reliable communications services. They are more influenced by softer 'service surround' issues such as courteous customer service and an efficient, effective repair service and ergonomically designed terminal units. Price is an issue for them but few personal customers would go to the inconvenience of switching operators on the basis of a nominal price differential alone. Any such move would be either opportunistic, such as during a house move, or be prompted by an emotional reaction to a negative, unsatisfactory experience with the service provider.

Figure 3.4 shows some of the influences, which combine to shape customer awareness.

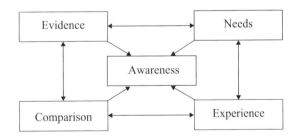

Figure 3.4   Influences on customer awareness

Source: Adapted from Mullee and Faulkner [1].

Let us briefly consider the major areas of influence.

Evidence: The customer is confronted with many forms of evidence, from external publications and media, to publicity, advertising and management reports produced by the service providers themselves.

Badly presented bills or management reports awash with meaningless information will evoke a negative reaction from the customer. By paying care to those aspects of the customer experience concerned with the generation and consumption of 'evidence' the service provider has a powerful tool through which to influence, win and retain customers in a competitive environment.

Comparison: The criteria by which one operator is compared with another will vary over time and with the needs of the customer in question. As competition increases, so customer's comparisons will have a greater effect on their perception. The two most likely areas customers will attempt to compare are price and quality.

Needs: The specific needs of a customer will have an impact on perception. A personal customer may need to receive separate bills for a number of individuals (i.e. teenage children) who all make calls from a single telephone line. During certain periods in their life, such as family illness, it may be important for them to have 100% service availability. Hence their needs will vary with their circumstances.

A business customer may need a structured, customised communications plan, requiring services and service combinations outside of the service provider portfolio or technical capability. They may need a particular style, frequency or delivery of billing information to assist in dynamic business accounting.

Experience: The most direct, and therefore the most potent form of influence on customer perception is that of personal experience. Research has shown that customer's mood, involvement level and the quality of the experience with a product, service or service provider have a significant effect on their future buying intentions. Customers in good moods tend to evaluate good experiences still better, and customers in a bad mood tend to exaggerate bad experiences and make them even worse.

Each customer will have certain expectations against which they will be most sensitive to the quality of their experience with a product or service. If the service provider is not adequately attuned to those expectations then the quality of the experience for the customer will be low.

(ii) Customer expectations

In a study by Zeithaml *et al.* [2], they show that perceived quality is related to expected quality as shown in Figure 3.5. The model defines the key determinants of expected quality as 'word-of-mouth communications', 'personal needs', 'past experience' and 'external communications' from the service provider.

Word-of-mouth: This covers what customers hear from other customers, that is, recommendations, which may be good or bad and are directly outside the control of the service provider.

Personal needs: This embraces a multitude of criteria based upon a customer's personal characteristics and circumstances, which are not totally under the control

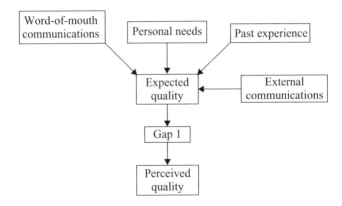

Figure 3.5   Customer's expected and perceived quality gap

Source: Adapted from Zeithaml et al. [2]

of the service provider. A service provider will not be able to meet the totality of a customer's needs (e.g. total customisation) economically.

Past experience: This will encapsulate an individual customer's entire past experience of a particular service provider. Whether this is a good or bad experience will be a matter of fact and the starting point for all future dealings. Therefore, this determinant will be outside the direct control of the service provider.

External communications: This covers all the contact between the service provider and the customer including advertising, media coverage, promotional material, sales and enquiry contact. All of these, except media coverage are under the direct control of the service provider.

This model shows the gap between expected and perceived quality. It follows that good management of the key determinants by the service provider can reduce this gap. However, the level to which the gap can be reduced will depend upon the service provider's control over the determinants. For example, external communications can be made up of advertising – directly under the service provider's control and media coverage – not directly controlled by the service provider.

According to Gronroos [3], good perceived quality is obtained when the experienced quality is the same as the expected quality, as illustrated in Figure 3.6. From this model it can be seen that the expected quality is linked to the traditional marketing activities undertaken by a service provider. Therefore, to some extent the service provider can set the expectations of their customers and this will then have an effect on the perceived quality. However, image, word-of-mouth and customer needs are not entirely under the control of the service provider as explained earlier.

The experienced quality is affected by a number of determinants as shown in Figure 3.6.

Achieved Quality: Achieved quality is made up of the following areas – technical performance and service quality. Technical performance is a technical measure of how a product or service performs during its life cycle (e.g. does it meet all its technical

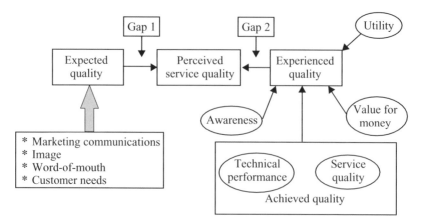

Figure 3.6   Perceived quality model

Source: Adapted from Gronroos' perceived quality model [3].

performance parameters and targets?). The magnitude of this determinant will depend upon an individual customer's bias towards the technical aspects of the product or service. Therefore, the influence of this determinant upon experienced quality may vary from customer to customer. Service quality is the aspect of experienced quality, which reflects how well the product, or service is supported. The aspects of service quality, which are important to customers and affect this determinant, will again vary from customer to customer.

Awareness: Experienced quality will also be influenced by a customer's awareness of similar products and services on the market. However, the effect of this on experienced quality will be influenced by the factors forming 'expected quality' explained in relation to Figure 3.5.

Value for money: This is a subjective judgement made by the customer when considering all the aspects of quality including price. If a customer feels that they have, or have not, experienced value for money, then it also follows that subjectively their experienced quality may be affected.

Utility: This refers to the 'usefulness' of the product or service (e.g. does it do all the things it is supposed to do?). The limits of this determinant can be set by the service provider with careful consideration of their external communications.

It can be seen from Gronroos' model [3] and the adaptation in Figure 3.6 that the perceived quality is a function of expected quality and experienced quality. If customers' expectations are unrealistic the total perceived quality would be low, even if the experienced quality were good. This shows that it is important to understand the linkages and the determinants, which make up both 'expected' and 'experienced' quality. The model serves as a useful tool to focus a service provider's attention on the key critical determinants which need to be carefully managed in order to improve customers' perceptions of their products and services.

(iii) Experienced quality

In an ideal world, the perceived quality should correspond directly to the quality experienced (which includes achieved quality) by the service provider. Even though customers may require improvement in quality, their perception is more likely to correlate with the quality actually experienced especially if they have an SLA with the service provider. Customers are more likely to monitor the experienced quality when there is an SLA with the service provider and their expectations of quality will reflect the contents of that SLA.

(iv) Recent experiences

A customer's perception will be influenced by their most recent experience with the service provider, the timing of that experience and its perceived value to the customer (i.e. was it a good or bad experience).

A recent bad experience is very likely to degrade a customer's opinion, sometimes unduly. Similarly, a good experience may negate recent bad experiences depending upon the quality of the good experience. For a bad experience to be forgotten a number of counteractive good experiences may be required. In telephone interviews and personal interviews it is easier to assess the depth of these adverse experiences and assess how much these have influenced the opinion of a customer. Such an assessment would enable a more realistic correlation to be made between the perceived and experienced quality.

(v) Advertising

It has been found that advertising has an impact on customer perception. Advertising is an excellent tool with which to influence perception because the messages are directly under the control of the service provider. Image and 'word-of-mouth' are only indirectly controlled by the service provider, because they are based upon the past performance of a product, service or of the service provider, by either the customer or someone close to the customer. Research in the retail industry has shown that comparative advertising (advertising that compares product performance) and price has an effect on perceived quality. In the work carried out by Gotlieb and Sarel [4], three hypotheses were put forward with supporting evidence. These were as follows:

(a) *A direct-comparative advertisement will have a more positive impact on perceived quality of a new brand than will a non-comparative advertisement.* This impact is particularly true for a new brand and when the comparison is made against a market leader.

(b) *There will be an interaction effect of source credibility and price on the perceived quality of a new brand.* The source here is the person who delivers the message in the advertisement, the person's credibility being the influential factor. When the source is highly credible, the consumers are less likely to discount the message of the advertisement. Consequently the influence of price is moderated and they are more likely to use multiple cues, for example, product attributes, price, source credibility, to judge quality of a new brand. Conversely, when the source has low credibility, the influence of price on perceived quality is less likely to be moderated by other cues in the advertisement and consumers are more likely to discount the message.

(c) *There will be a three-way interaction affecting the type of advertisement (i.e. direct-comparative versus non-comparative), source credibility and price on purchase intention toward a new brand.*

The applicability of the above hypotheses to telecommunications services is unknown. However, there is reason to believe that the first two hypotheses would be relevant in a customer's perception of quality of a telecommunications service.

(iv) Nature of survey

Customer surveys are subjective by nature and, therefore, the results should be interpreted with caution. The three most popular ways of conducting customer perception surveys are, postal questionnaire, telephone interview and person-to-person interview. The wording of a questionnaire, the way questions are asked and the mood of the interviewer and interviewee will have a significant impact upon the answers given and the ratings associated with those answers. A customer will also be likely to give slightly different opinion scores depending upon the viewpoint of the interviewer. To reduce 'voice fatigue' of the interviewer it may be useful if the order of questions is changed to keep the interest and attention of the interviewer. Depending on whether the interviewer is the supplier, user group, regulator or an independent consultant, the interviewee's opinions and corresponding opinion ratings could be significantly different.

3.2.4.2   Customer opinion ratings

Customer's perception of quality is expressed qualitatively (e.g. I think the service is good), but in order to turn this type of statement into meaningful data covering a large number of customers, this perception must be expressed quantitatively. For example, Figure 3.7 shows the concept of illustrating qualitative statements with quantitative values suitable for further analysis.

The numbering attached to statements such as 'good' enable the service provider to perform mathematical computations on the qualitative data for further analysis.

3.2.5   *Relationship between the four viewpoints of QoS and the 'gaps'*

The customer's QoS requirements may be considered as the logical starting point. A set of customer's QoS requirements may be treated in isolation as far as its capture is concerned. This requirement is an input to the service provider for the determination of the QoS to be offered or planned. The service provider may not always be in a position to offer customers the level of QoS they require. Considerations such as cost of quality, strategic aspects of the service providers' business, benchmarking (world's best) and other factors will influence the level of quality offered. The customer's requirements may also influence what monitoring systems are to be instituted for the determination of achieved QoS for the purpose of regular reports on achieved quality.

Voice Quality

The service provider will attempt to provide 100% of sufficient quality to enable you
and the person at the other end of the connection to understand each other without
difficulty. However, this may not always be possible. How would you rate the quality
of your voice service?

Excellent	Good	Fair	Poor	Bad
5	4	3	2	1

On a rating of 1–5 please indicate how important this parameter is to you (ignore the
present performance you receive)

5	4	3	2	1
Very important	Important	Neutral	Somewhat important	Not important

Figure 3.7   Example of quantitative values for qualitative data

It may not always be possible for the service provider to offer what the customer
requires. The reasons are discussed in Chapter 5. The gap between the customer's
requirements and QoS offered by the service provider is called 'alignment gap'.

It may not always be realisable for the service provider to achieve the level of
QoS planned for a given service. The reasons for this are discussed in Chapter 6. The
gap between QoS offered or planned and what is delivered is called the 'execution
gap' and is of concern to the service provider.

In an ideal world the customer perception and the delivered quality ought to have a
1 : 1 correspondence. In practice this is not always so. The gap between perceived and
achieved quality is called 'perception gap' and is of interest to the service provider.
This is discussed in more detail in Chapter 7.

The gap between customer's perception rating and their previous stated requirements is called the 'value gap'. Customer's stated rating implies their perceived value
of the quality they believe they have experienced. Customer's perception ratings are
subjective and, therefore, influenced by various factors. This is dealt with in Chapter 4.

The combination of relationships between the viewpoints and the gaps, termed
the Quality Cycle, forms the basis of a practical and effective management of quality
of telecommunications services. This is particularly the case for legacy services. For
Internet services the ACF model, as described in Chapter 2, is added and discussed
in Chapter 10.

3.3   A methodology for the management of quality

3.3.1   The four viewpoints and basic telephony

Identification of management issues of basic telephony service can benefit from the
four viewpoint model. The ACF model and the Four Market Model are not appropriate

to this and most of the legacy services. The ACF model is covered in Chapter 10 and the Four Market Model in Chapter 28.

3.3.2   Methodology for the management of QoS

The diagram in Figure 3.8 illustrates, schematically the principal activities required for the management of QoS.

Step 1: Customer's QoS requirements are ascertained. This exercise may have to be carried out on a service-by-service basis and, depending upon the granularity of information required, customer segmentation may have to be carried out since different customers may have different QoS requirements. The topic of determination of customer requirements is dealt with in Chapter 5.

Step 2: Customer's requirements are translated into the language usually used by the service and/or network providers as determination of customer's requirements is usually carried out using terms more meaningful to the customers.

Step 3: The service provider determines what level of quality is to be offered to the customers. Considerations such as cost of implementation, competitive aspects, future of the services etc. are taken into account before a level of quality is decided upon. This topic is dealt with in more detail in Chapter 6.

Steps 4, 5, 6 and 7: The offered QoS is translated into internal specifications. The network-related and non-network-related aspects of quality are separated and respective planning and implementation documentation produced. Where the service

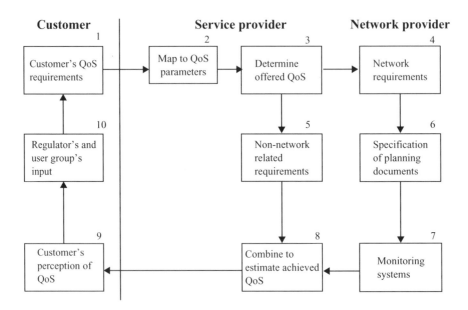

Figure 3.8   Principal activities in the study and management of QoS (based on the four viewpoints model)

provider and the network provider are the same, the documentation may be produced by the same source. Where these two entities are different, the network-related aspects are passed on to the network provider by the service provider for appropriate action. The monitoring systems for the network aspects may involve specification of test equipment, details on the frequency of measurement, sample size etc. These are dealt with in Chapter 7.

Step 8: The raw data from monitoring systems are gathered at regular intervals and the end-to-end performance produced on a regular basis. In addition, performance of individual elements in the network is passed on for study within the network provider. This is dealt within Chapter 7.

Steps 9 and 10: The service provider and/or the regulator or an equivalent body could carry out customer surveys to determine how satisfied the customers are with the service they have been receiving. This is dealt with in more detail in Chapter 7.

The above ten steps, when carried out, would ensure that a close watch and control is kept on the service management. Quality is unlikely to be a major issue of contention with the service providers if customers' input is regularly carried out and the quality cycle is reviewed on a regular basis. The subsequent chapters in this book deal with various aspects of this management cycle and associated activities.

3.4   Summary

Effective management of QoS can be made easier with the aid of the framework described in this chapter. Identifying the key issues for each of the four viewpoints enable a clear focus in their management. A clear focus on issues should enable a better attention to be given for the resolution of problems. This in turn could result in optimised quality, commensurate with the resources expended. The next four chapters deal in more detail with the four viewpoints of QoS.

3.5   References

1   MULLEE, A. W. and FAULKNER, R.: 'Planning for a customer responsive network', Sixth international network planning symposium, September 1994, pp. 337–51

2   ZEITHAML, V. A., PARASURAMAN, A. and BERRY, L. L.: 'Delivering quality service' (Free Press, New York, 1990)

3   GROONROOS, C.: 'Service management and marketing – managing the moment of truth in service competition' (Lexington Books, Mass, 1990), pp. 41–7

4   GOTLIEB, J. B. and SAREL, D.: 'The influence of type of advertisement, price and source credibility on perceived quality', *Journal of the Academy of Marketing Science*, **20**(3), pp. 253–60

Exercise

1   Based on the four viewpoints of QoS for telecommunications, develop a corresponding framework for the management of QoS for any one public utility.

Chapter 4
Customers' Quality of Service requirements

4.1   Introduction

The logical starting point for the study and management of QoS is the determination of customers' requirements. Customers of service providers are in both the business and residential market sectors. Not all service providers are network providers. Customers of network providers will include service providers. In this chapter, the applications, the principal players and the key issues relating to the management of this viewpoint of QoS (Figure 4.1) are covered. The characteristics of the requirements of QoS were covered in Chapter 3.

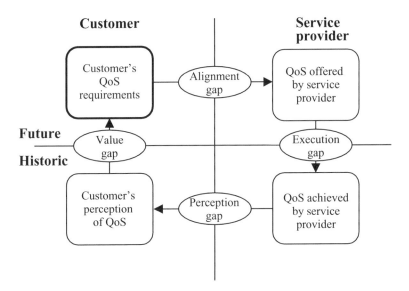

Figure 4.1   The four viewpoints (quality cycle): customers' QoS requirements

4.2   Applications of customers' QoS requirements

In an effective management of QoS, customers' requirements become the formal input to the quality cycle described in Section 3.2. Such an input will enable the supplier to attempt to ensure the network and operational aspects are neither over-engineered or under-engineered, but optimised for maximum benefit to both parties. In the monopoly era the supplier decided what was good enough for the customers. Service providers now have to change their planning procedures in response to the pressure from users and the competition. For an effective and meaningful dialogue on quality between customers and service providers, the requirements on quality become the starting point. In a competitive environment the customer has the room to negotiate with the service providers to achieve equilibrium between quality and price. While this dialogue may be restricted to larger business users, the principle may be equally well applied by the service provider to estimate the likely revenues against varying levels of quality for various segments of the customer population.

Customers' performance requirements could form the basis for the regulator for the specification of performance parameters on which to publish achieved results. The regulator may choose to specify these parameters on a service-by-service basis, as such publications are demanded by a significant proportion of the customers.

Some test equipment manufacturers take into account customers' requirements in their design of measuring equipments. Consideration of customers' quality requirements in the design of telecommunication switching and transmission equipment, by equipment manufacturers is marginal, as they are usually guided by the requirement set by service providers and network providers. In these cases the manufacturers do benefit, directly or indirectly, from customers' requirements on quality.

4.3   Relevant parties in the management of customers' QoS requirements

The following parties have varying degrees of involvement in the management of quality of a telecommunication service:

(i)   Service providers;
(ii)   Network providers;
(iii)   International standards bodies;
(iv)   User groups;
(v)   Regulators.

4.3.1   Service providers

The service provider is the principal party responsible for the management of the QoS. The large service providers own and manage their networks, for example, AT&T, BT, NTT and Telstra. The smaller service providers may lease network facilities, either in whole or in part, from a main network provider. The customers always deal with the service providers whose responsibility is to ensure that quality, whether network related or not, is of the desired level. The responsibility to determine levels of QoS the users want, order of priority of QoS parameters, requirements differentiated by relevant customer population sectors and/or segments, identifying areas of concern by users and forecasting quality-related issues for the future are examples of issues to be addressed by the service provider.

4.3.2   Network providers

The network provider, irrespective of whether or not a service provider, is responsible for the network-related quality criteria delivered to the service provider. Where the network provider is not the service provider, formal agreements (SLAs) with the service provider may exist for the levels of quality. Network providers could have agreements with several service providers. They have to address the issue of managing network quality, for possibly different levels of quality, demanded by the service providers. But this is not easy for an ubiquitous network.

4.3.3   International standards bodies

The standards bodies specify the technical performance of telecommunication networks and services. Most of the work within the standardising bodies has been carried out by the representatives from large service providers who also own their networks. These standards, when incorporated into network and service management should result in the prescribed level of end-to-end quality. To date, these specifications have been mainly geared to meet the needs of the network provider and not that of the customer. However, this situation is expected to change. More standards to meet customers' requirements are expected to be issued. Recently the first architectural framework for the study of QoS, based on the user requirements was approved for issue as an ITU-T Recommendation (G 1000). The role of standard bodies is further discussed in Chapter 21.

4.3.4   User groups

User groups exercise influence by stating the performance parameters that are preferred for the publication of achieved quality. Service providers have been accused of selecting QoS parameters for publication that give the impression of good overall performance. But, as the power of user groups increases in a competitive environment, it is quite likely that their requirements for published QoS parameters will be respected by the service providers in the future. The service provider could use user groups as a source of free or low cost consultancy by requiring them to provide QoS requirements of their member population. If this represents the industry then this information is invaluable for planning purposes for the service provider. The role of user groups is covered in Chapter 19.

4.3.5   Regulators

Since the principal role of the regulator is to ensure that customers receive value for money, they have become concerned with the quality provided by the suppliers and require QoS parameters to be published on a regular (usually quarterly) basis. For example, in the UK, Oftel requires delivered performance to be published every quarter. Even though their involvement in QoS has been to a great extent restricted to the publication of achieved quality by the suppliers, their influence has only marginally impinged on customers' requirements of quality levels. The regulator could act as a catalyst in persuading the service providers and network providers in optimum provision of QoS as required by the user population. The role of regulator is covered in more detail in Chapter 20.

4.4   Management of customers' QoS requirements

4.4.1   *The principal issues*

The principal issues to be addressed in the management of customers' QoS requirements are illustrated in Figure 4.2.

These are discussed in more detail in the following sub-sections.

4.4.2   *Derivation of service specific QoS parameters*

4.4.2.1   The matrix

The QoS parameters may be identified with the assistance of a matrix designed for this purpose. It appeared as an ETSI Technical Report, ETR 003 in 1994 [1]. A more up to date form has been approved for issue as an ITU-T Recommendation in year 2001 [2], namely G 1000.

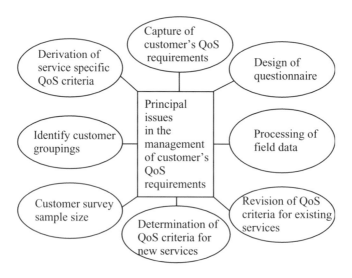

Figure 4.2   Principal issues in the management of customers' QoS requirements

Table 4.1   Service functions making up a telecommunication service

Service management	1 Sales, pre-contract activities
	2 Provision
	3 Alteration
	4 Service support
	5 Repair
	6 Cessation
Connection quality	7 Connection establishment
	8 Information transfer
	9 Connection release
	10 Charging and billing
	11 Network/service management by customer

By applying the principles in the determination of quality of a product or service as described in Chapter 1 (Section 1.3) a telecommunication service may be subdivided into uniquely identifiable 'service functions' whose sum constitutes the service. A set of quality criteria are also developed based on knowledge of the industry and iterative process of developing these criteria. The choice and grouping of the service functions are influenced by the sequence in the subscription to a service, that is, starting from sales and ending with the cessation of the service. The technical, charging/billing and network/service management functions are grouped separately. The service functions are shown in Table 4.1.

When a matrix is formed with the above sets of criteria, the resulting cells indicate QoS parameters for a telecommunication service. It may be necessary for QoS derived from cells to be tightly defined to enable these to be called parameters. Otherwise they will be QoS criteria for that service. This matrix is illustrated in Figure 4.3. The basic principles of the matrix were first published by Dvorak and Richters [3].

Quality criteria of concern to customers for telecommunication services may similarly be found under the following categories:

- Speed;
- Accuracy;
- Availability;
- Reliability;
- Security;
- Simplicity;
- Flexibility.

The service functions have been derived to represent the various components of a telecommunications service. Since the quality criteria are believed to be those that are most likely to concern customers and the service is subdivided into uniquely identifiable service functions, it should be possible, in theory, to identify all possible QoS parameters most likely to concern a customer. This is the basis for identification

Service quality criteria / Service function			Speed 1	Accuracy 2	Availability 3	Reliability 4	Security 5	Simplicity 6	Flexibility 7
Service management	Sales and precontract activities	1							
	Provision	2							
	Alteration	3							
	Service support	4							
	Repair	5							
	Cessation	6							
Connection quality	Connection establishment	7							
	Information transfer	8							
	Connection release	9							
Charging & Billing		10							
Network/service management by customer		11							

Figure 4.3   Matrix to facilitate determination of QoS criteria

of all QoS parameters for a telecommunication service. The tendency in the past has been for telecommunication network engineers to treat network quality parameters in isolation and segregate them from other parameters. However, from the customer's point of view, this separation is not desirable.

It is unreasonable to expect customers to separate the network and non-network-related quality parameters and to state the respective quality requirements separately. Confusion can also arise if separate customer surveys require customers to state the relative importance of technical and non-technical QoS parameters.

4.4.2.2   Service functions in the matrix

Service management

The first six service functions introduced in Table 4.1 are grouped under the title 'Service management' in the matrix and may be described as follows:

(1) *Sales and pre-contract activities.* In the business relationship between a service provider and the customer for supply, provision and maintenance of a service the first stage is the 'sales'. At this stage, the customer is provided with information on the service features, its limitations, and any planned changes in the service features or capabilities for the future, the initial cost to supply, rental (if applicable), maintenance arrangements and the level of quality that can be expected.

All information transactions regarding a service, before the customer agrees to hire a service from a service provider, should be included under the service function of 'sales and pre-contract activities'.

(2) *Provision.* Provision is the setting up, by the service provider, of all the components required for the operation and maintenance of a service, as defined in the service specification and for use by the customers.

(3) *Alteration.* During the life cycle of a business relationship between the customer and the service provider there could be occasions when the customer requires amendments to the service or service features. Such changes or alterations are more frequently demanded by business customers than residential customers. Such changes are best incorporated into the original contract for the service. All activities, such as contract amendment and logistics for the changes to service, are grouped together under this service function.

(4) *Service support.* The following are included in service support:

- enquiries on the operation of a service (e.g. queries on configuring a personal computer for access to the Internet service);
- documentation on service operation and any other relevant matter;
- procedures for making and following up on complaints;
- preventive maintenance.

(5) *Repair.* This service function includes all activities associated with repair from the instant a service does not offer one or more of the specified features to the instant these features are restored for use by the customer.

(6) *Cessation.* In the 'product life cycle' of a business relationship between the customer and a service provider the last logical service function is the cessation of the service. All contract clauses come to an end with the cessation of a service. Activities associated with this are: removal of the equipment from customer's premises, settling of accounts and updating of appropriate records.

Connection quality functions

The next three service functions are grouped under the title 'connection quality' and cover the technical quality of the connection. They are:

(7) *Connection establishment.* All activities associated with the establishment of a telecommunication service are included in this service function. It is the time elapsed from the instant the customer requests service (e.g. lifting of the handset of telephone) to the instant any of the following indications are received:

- ring tone or equivalent;
- destination engaged tone;
- call answered indication;
- any other signal, indicating the status of the called party or the network.

(8) *Information transfer.* This service function, probably the most sensitive to customers, groups together all the quality measures that influenced the transfer of information, covering the period from the instant the connection is established to the instant release is requested by the customer.

(9) *Connection release.* All activities associated with the release of the connection from the instant release is requested by the customer to the instant the connection is restored to its dormant state are covered in this service function. Performance measures include conformance to period specified (within limits) for release and the order in which the release is carried out by the different elements in the network.

(10) *Charging/Billing.* This service function deals with all aspects associated with charging and billing. Examples of its constituent parts are: accuracy of charging, retrieval of charging information, format of bills and specific requirements of customers on bill structure, format and payments requirements (e.g. by direct debit etc.).

(11) *Network/service management by the customer.* With increasing sophistication in telecommunication services some control of the service is given to the customer, for example, the control of routing arrangement for an advanced freephone service where a single, non-geographic (0800) number is used for multiple destinations with caller choice controlled by additional digits keyed under voice prompting. All performance measures associated with the service management by the customer are included in this service function. It also includes all activities associated with the customer's control of predefined changes to telecommunication services or network configurations.

4.4.2.3   Quality criteria in the matrix

The quality criteria in the matrix may be described as follows:

(1) *Speed.* This quality criterion is expressed by the time taken to carry out any of the service functions.

(2) *Accuracy.* This quality criterion deals with the fidelity and completeness in carrying out the communication function with respect to a reference level.

(3) *Availability.* Availability is the likelihood with which the relevant components of the service function can be accessed for the benefit of the customer at the instant of request (not be confused with the ITU-T definitions on 'Availability').

(4) *Reliability.* Reliability is the probability that the service function will perform within the specified limits for speed, accuracy or availability for a given period, for example, one year.

(5) *Security.* This quality criterion deals with the confidentiality with which the service function is carried out. No information is to be supplied to an unintended party, nor can information be changed by an unintended party.

(6) *Simplicity.* The ease in the application of the service function. This quality criterion concerns the ergonomic aspects with which the service feature is dealt with by the service provider. It also includes the customer's preferred requirements for a particular service. For example, customers may prefer a standard, simple and logical 'log on' procedure for access to an electronic mail service.

(7) *Flexibility.* This quality criterion groups together the customer's optional requirements associated with the service. Examples of this are choice of monthly or quarterly telephone bills, the time of the month when monthly bills are sent to the customer, the amount and format of billing information to the business customer etc.

4.4.2.4   Application of the matrix

Generic cell descriptions for the 77 cells are given in Appendix 4.2. The principal application of the matrix and the resulting cell descriptions is the derivation of relevant service-specific QoS parameters for any telecommunications service. An example of the derivation of QoS parameters for telephony is given in Table 4.2. Even though the

Table 4.2   QoS parameters for basic telephony derived from the matrix

Parameter number	QoS parameter (cell references from matrix in Figure 4.3 are given in brackets)
1	Speed of obtaining pre-contract information on service from provider e.g. tariff, service availability, service features, choice of telephone features etc. (Sales/Speed)
2	Professionalism with which a customer is handled by the supplier (Sales/Reliability)
3	Speed of complaints handling (Service support/Speed)
4	Speed of repair (Repair/Speed)
5	Repairs carried out right first time (Repair/Accuracy)
6	Accuracy of reaching destination first time, i.e. absence of misrouting (Connection set-up/Accuracy)
7	Availability of network resources when requiring to make a call (Connection set-up/Availability)
8	Respect of wishes made with regard to display of Calling Line Identity (Connection set-up/Security)
9	Delay in making long-distance calls, i.e. call-set-up time (Information-transfer/Speed)
10	Transmission delay especially on international calls (Information-transfer/Speed)
11	Call quality (combined effect of noise, clicks, sidetone, other degrading factors (Information-transfer/Accuracy)
12	Availability of network resources to keep call for the intended duration of the conversation (Information-transfer/Availability)
13	Connection release as specified by the supplier (Connection-release/Accuracy)
14	Charging and billing accuracy (Charging/Billing/Accuracy)

matrix was developed originally for the determination of customer's QoS parameters, it can be used for the determination of the offered QoS parameters to identify shortfalls and in the determination of parameters for the achieved performance in the context of specification of monitoring systems.

Derivation of the service-specific QoS parameters may be carried out by an interactive process. The generic matrix is explored, cell by cell, for possible performance parameters, comparing, where applicable, the existing or known performance parameters for the service in question. By this process, QoS parameters, not hitherto considered, are likely to be captured. With the help of the matrix, this may be used to capture a comprehensive range of QoS requirements. The granularity of the QoS parameters may be selected to suit the individual requirement of a service provider. For basic telephony finer granularity may be achieved by specifying quality parameters based on derivations from further cells. Up to 43 have been identified for basic telephony.

4.4.3   *Methods of capture of customers' QoS requirements for established services*

Capture of customers' QoS requirements may be attempted using a combination of the following methods:

   (i) Questionnaires;
  (ii) Face-to-face interviews;
 (iii) Telephone interviews;
 (iv) Analysis of complaints profile;
  (v) Case studies.

These are covered in more detail below.

4.4.3.1   Questionnaires

Questionnaires are probably the most obvious choice for the person seeking information. The questionnaire asks the respondents to fill in the required level of quality by quoting figures for quantitative parameters and state in their own words the qualitative levels for other parameters. The questionnaire must state the time frame to which the quality levels will apply in the future. It must not ask how satisfied the customers are for the service currently received. This will not capture their requirements, but will capture their perception of past service. The latter is of limited value for future planning purposes.

The percentage of respondents replying to questionnaires usually varies from 1% to 50% and sometimes higher. The determining factors for the level of replies include

- the level of loyalty felt towards the questioner;
- the level of benefit perceived by the respondent in returning a questionnaire;
- the level of interest in the content of the questionnaire;
- the amount of time available at the disposal of the respondent.

Seeking information by questionnaires has limitations. It is reasonable to assume that more questionnaires will be filled by people who have a 'complaint' to make. The results of a survey may, therefore, automatically introduce a bias into the estimation of the requirement of the whole population. This argument, while technically plausible, is not of great concern to this exercise as the people who make complaints are the opinion formers in the quality of a service. The people who have complaints to make are also, generally, those who have a critical requirement of the service. These users also happen to be, mostly, large business users and therefore their opinions cannot be directly equated to those of residential customers.

In returning questionnaires, the respondent does not usually have the means to clarify questions that they may not understand. This could result in vague answers. Some parts of the questionnaires may be left blank. There may also be some questions the respondents may have misunderstood. Notwithstanding these shortcomings, questionnaires are a powerful tool to capture customers' requirements. Many of the above shortcomings can be overcome by increasing the sample size and by intelligent analysis of the respondents' replies.

4.4.3.2 Face-to-face interviews

Face-to-face interviews, with a questionnaire, should provide the ideal alternative to the postal questionnaire. The shortcomings mentioned in the previous method can be eliminated. However, face-to-face interviews are time consuming and make a high demand on resources. Nevertheless, it is always useful to have some face-to-face interviews, if only for the analyst to acquire an appreciation of the customers' problems and their concerns not covered in the questionnaire. This should enable an intelligent interpretation to be made of the replies on questionnaires and may also help to interpret postal questionnaires. Face-to-face interviews are particularly useful for the capture of qualitative QoS parameters.

4.4.3.3 Telephone interviews

Telephone questionnaires are a compromise between postal questionnaire returns and face-to-face interviews. The principal shortcoming of this type of interview is the absence of body language of the respondent. This limits the understanding of the questioner of the respondent's concerns. This method is however better than postal questionnaire returns without any form of person-to-person contact.

Where possible, a combination of postal questionnaires, telephone interviews and face-to-face discussions should be carried out to achieve a good representation of customers' QoS requirements.

4.4.3.4 Analysis of complaints profile

Analysis of complaints, particularly related to the service in question, is a source of information on what customers expect in terms of quality. A customer usually complains when the level of dissatisfaction has reached an unacceptable level. By studying the pattern, the analyst can ascertain the QoS that are of most concern to the customers. This source should not be considered as the main one for the capture of customers' QoS requirements, but only to supplement the information gathered by the means described earlier.

4.4.3.5 Case studies

Ascertaining customers' QoS requirements from case studies may be resorted to whenever the opportunity arises. In many first-hand experiences of customers based on multiple observations of various tasks carried out by the suppliers, customers are able to provide valuable insights into the weaknesses of the service providers and what level of quality they normally expect.

4.4.4 Determination of customer groupings

For a given telecommunication service different customers can have different requirements. Groups of customers with similar requirements, that is, market segments, can be identified and their common requirements, where such groups exist, may then be obtained. For example, hospitals, retail trade, airline booking offices, etc. have their own unique QoS requirements. For hospitals it is availability throughout the clock throughout the year. For retail trade, it is availability during trading hours. For airline booking offices availability of network resources and fast response times are critical.

The conventional way of grouping according to market sector and segment is the preferred method. The population is first divided according to sectors, e.g. banking, electrical, transport, etc. The Standard Industrial Classification [4] may be used as a starting point and modified in the light of known data for the population under study. Segmentation may then be carried out. For example, segmentation into multinational, national and local segments of each sector may be necessary. These segments could have differing telecommunication requirements. In the light of experience these groupings may be revised to suit the particular characteristics of a service for a particular country.

These procedures are rigorous, but short cuts should only be taken when warranted by evidence that certain procedures may be eliminated. If short cuts are pursued on a normal basis, this practice must be weighted against missing unique customer requirements that may prove costly to rectify at a later stage.

4.4.5 Sample size

The sample size for administering the questionnaire to ascertain QoS requirements for a particular group of customers is dependent upon the level of accuracy required. The accuracy may depend upon the nature of the individual samples and the distribution of the samples in the population. If the individual samples are considered to be members of a normal distribution, the selection task is straightforward. It will be necessary for the service provider to carry out some basic research and analysis of the sample to establish if it is representative of the group. The actual sample size will depend upon the degree of confidence required. In the absence of any further information, the population's requirements may be assumed to follow a normal distribution. The formula for the range of mean of the total population estimated from the sample mean for given levels of confidence are found in standard textbooks on statistics. A mathematical treatment of sampling theory is not within the scope of this book. Standard textbooks on statistics should provide guidance in this area [5].

In practice, the judgement is also likely to be based on a qualitative basis in addition to a mathematical basis. It will be up to the service provider to assess, from the sample, whether it represents a consensus view of the group. This is probably a better way of estimating the requirements of the group than a mathematical method that excludes the subjective element. The subjective element, when it is an informed one, can be very meaningful.

4.4.6 Design of questionnaire

The following guidelines should be borne in mind in the design of a questionnaire to capture customers' QoS requirements:

(a) The questionnaire should be designed around the QoS criteria derived for the service from the framework, but there must be sufficient flexibility to accommodate

customers' answers expressed in their own words. These could later be translated to terms used within the service providers' organisation.
(b) Questions must be worded in such a manner that they reflect the customers' cultural thinking on the topic of the quality of a particular service. In other words, the questioner should relate to the respondent. In the absence of a person readily available to clarify enquiries the questions must be unambiguous and clearly impress upon the respondent's mind what exactly is being asked for.
(c) The questionnaire should also be geared to obtain the qualitative issues from customers.

The questionnaire should be aimed at finding out what the customers' QoS requirements are and not what the service provider thinks the customers' requirements should be. In the past, some service providers behaved as if they were graciously doing a favour to the customers by offering them a service, and this was sometimes reflected in the high-handed attitude of such companies in assuming they knew what was good for the customers. The introduction of competition is causing many such service providers to change their attitudes.

The design of questionnaires is a specialised subject and no attempt is given in this book to dwell on this in any detail. Telecommunications engineers must resist the temptation to dismiss the topic of questionnaire design as a task requiring low level of intellectual skills. Design of a good questionnaire requires an understanding of psychology, cultural behaviour of the respondents, customers' knowledge of the telecommunications services, quality implications, dynamics of market forces and the customers' expectations, and an awareness of competitors' products and services. Service providers, network providers, regulators and user groups should note that, with experience and professional advice, meaningful information could be obtained from adequately designed questionnaires. A poorly designed questionnaire could produce misleading results and waste customers' and the questioner's resources.

Four categories of responses may be captured from the customers relating to their QoS requirements. These are

(a) levels of quality for the quantitative performance parameters;
(b) narrative requirements for the qualitative performance parameters;
(c) priority of these parameters;
(d) any other quality issue relevant to the study not asked for by the questionnaire.

Levels of quality for each parameter may also be expressed by the customer qualitatively. Customers conversant with telecommunication terms (telecommunication managers of large companies) may be able to supply quantitative levels. It is possible that some concerns may not be picked up by the questionnaire. To capture these, customers may be asked to state any quality issues they wish the service provider to address, which is not covered in the questionnaire.

The most effective method of administering a questionnaire for capturing customers' QoS requirements is to interview the customer personally. However this may be more economical to carry out on large business customers and, to a lesser degree, with residential customers and those with special requirements, for example, the blind and slightly hard of hearing.

Although the practice of obtaining customers' requirements by means of questionnaires is popular, this method of obtaining results should be restricted to the minimum as maximum understanding of customers' requirements is obtained by personal interactive association with customers.

The task of ascertaining customers' QoS requirements is best given to a market research agency that is competent in telecommunications. A pattern will soon emerge after interviewing a small number of customers within the same group; thus after a certain number of interviews, further interviews may not be considered necessary. With interaction between the service provider and the agency, it should be possible to obtain a meaningful representation of the customers' requirements. In the design and administration of the questionnaire, it is essential for the service provider to be familiar with the cultural aspects of the customers. This will provide a more meaningful understanding of the answers given by the customer that could result in a more credible representation of the customers' requirements.

A typical question from a questionnaire on a quantitative measure is illustrated below.

Question on the maximum repair time

What is the maximum period within which you require any repair needed for the basic telephony service to be carried out for your particular applications?

_____ (days) (hours) (minutes) delete inapplicable. (Please use working time)

Please tick the relative importance of this parameter to you on the following scale:

1	2	3	4	5
(low)				(high)

A template questionnaire is given in Appendix 5.

4.4.7   Processing of customer responses

4.4.7.1   Estimation of preferred values

The quantitative answers given for each parameter are processed to estimate the customers' requirements. The following illustration shows how this may be carried out.

Distribution of the respondent's answers for the maximum time for repair:

1 h   1 h   1 h   1 h 1 h 1 h 1 h 1 h 1 h   1 h   1.5 h 1.5 h 1.5 h 1.5 h
1.5 h 1.5 h 1.5 h 2 h 2 h 2 h 2 h 2 h 2 h   2 h   2.5 h 2.5 h 2.5 h 2.5 h
2.5 h 2.5 h 2.5 h 3 h 3 h 3 h 3 h 3 h 3.5 h 3.5 h 3.5 h 4 h

The above figures may be grouped as in Table 4.3.

The cumulative percentage of the population of the sample may be plotted against the maximum repair time and is shown in Figure 4.4. It is seen that to satisfy 90% of the customers the service provider has to arrange for a maximum repair time of

Table 4.3   Customers' preferred maximum time for repair

Maximum time for repair (hours)	1 h	1.5 h	2 h	2.5 h	3 h	3.5 h	4 h
No. of customers	10	7	7	7	5	3	1
Percentage of population	25.0	17.5	17.5	17.5	12.5	7.5	2.5
Cumulative percentage	100	75.0	57.5	40.0	22.5	10.0	2.5

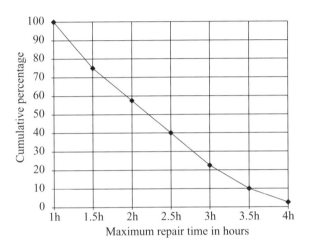

Figure 4.4   Cumulative percentage of customers and maximum repair time

around an hour and a quarter. This type of analysis may be carried out, on a parameter-by-parameter basis, from the replies received from customers for submission into the next stage of the quality cycle – the determination of the offered QoS.

In the treatment of customers' requirements, it must be appreciated that conclusions obtained from statistical information should not be treated in a clinical manner. Management of QoS is both a science and an art. Familiarity with customers' requirements, their past responses, if any, and industry knowledge will enable the analyst to conclude that any spurious, inconsistent or unreliable answers should be ignored. Meaningful customers' requirements can only be obtained by the experienced analyst and this is as much an art as well as a science. The contributory factors for this acquired interpretative knowledge are: familiarity with quality parameters, the currently achieved level of performance for these parameters, how these will affect the customers' attitude towards the supplier, what the competition offers, what level of performance is actually achievable, etc. The 'reading between the lines' to assess what customers really expect is of better value than mathematical analysis of the figures.

From the analysis of the responses on a parameter-by-parameter basis, a set of figures may be estimated to represent customers' level of performance required. For criteria where qualitative narrations are given by the customer, these are analysed to estimate the required performance. For example, with the following set of narrations from different customers, it is clear what the general message is.

Examples of narrations:

1   billing details should be simple and clear;
2   more consolidation – less bills;
3   more information on non-standard billing entries (one-off charges);
4   details of each line and one bill;
5   details could be more thorough;
6   presentation could be more thorough;
7   one bill format OK;
8   should include our cost centres;
9   bill every month as opposed to currently every two months;
10   default payments by service provider for not meeting SLA to be indicated.

Conclusion: customers require customisation of bills on an individually tailored basis.

4.4.7.2   Integration of customer responses

It is quite possible that different customer groupings have different needs for the same parameter. For example, availability would be crucially important over a 24-hour period of the day to the hospital sector, but only important to the retail industry over part of the day. Industry-specific requirements should be identified from the requirements and highlighted in the input to the quality cycle. Unique industry requirements should be treated separately to give these the attention they require in the quality cycle. A combination of parameters may be attempted should this produce no conflict of benefit to the customers.

4.4.7.3   Prioritisation of QoS requirements

In the example of a questionnaire in Section 4.4.6 the customer was asked to rate the importance of a parameter. With an adequate number of replies the mean of such responses would give a fairly accurate indication of how important this parameter is. With the computation of mean rating of the 14 parameters for basic telephony (as in Table 4.2) the parameters may thus be ranked (in the order of mean rating scores) resulting in a prioritised list. A precise mathematical interpretation ought not to be taken too seriously; however, this method of prioritising produces a fairly reliable guide as to how customers rate a particular parameter.

4.4.7.4   Editing of customer requirements to terms used by service providers

Certain QoS criteria will need to be translated into the service provider's language. For example, if customers are asked to state for how many calls in a 100 they are prepared to tolerate moderate difficulty in understanding during the conversation phase this has to be translated into meaningful technical parameters for the service provider to act upon.

Algorithms exist for the percentage difficulty customers are likely to experience with degrading parameters such as sidetone, echo and noise. Limits of performance of the network may be derived from these, should the service provider choose to meet the customers' requirements. Similarly, availability may sometimes be expressed both in qualitative and quantitative terms by the customer. This ought to be expressed mathematically for the service provider to act upon.

Conversion of customers' requirements into service provider's language entails familiarity with the service provider's terminologies and is to be attempted by the service provider rather than a non-telecommunication professional. The information is now ready for input to the next stage of the quality cycle.

4.4.7.5   Conclusions from the analysis of 'Value Gap' between QoS requirements and customer perception

Together with the survey researched information processed as described in the previous pages it is useful to analyse the gap between the customer perception and the QoS requirements, termed Value Gap. During the continued analyses of the various elements of the quality cycle (see Section 3.2.5) the findings forming the analysis of this gap may be relevant for inclusion in the summary of the QoS requirements. Customer's perception of quality may not have a 1 : 1 correspondence with what they say they require or expect. Ideally they would have a realistic expectation of what quality they would require based on their experienced or perceived needs. However, in practice, they may say what they think they want and this may be subjectively influenced by various facts. The service provider has to understand the local culture and acquire an intimate knowledge of the market in order to arrive at a credible set of QoS requirements based on all aspects of their clients' stated needs and perceived needs.

4.4.8   Revision of QoS requirements

The quality requirements of customers inevitably become more demanding with time; when their expectations are satisfied they look for improvements – this is a general feature for all customers. It, therefore, becomes necessary for the service provider to

  (i)   identify when revisions are required;
 (ii)   identify the nature of the changes; and
(iii)   ascertain the changes in priorities.

Intimate knowledge of the customers' needs, industry movements, competitive threats and technological developments will assist the service provider in the identification of these aspects.

4.5   Customers' QoS requirements and the quality cycle

The processed customers' requirements are now ready for input into the next stage of the quality cycle, which is the offered QoS. When varying levels of performance are required for a certain parameter, a dilemma is posed to the service provider. The service provider has to plan the provision of different levels of quality to optimise its resources. The different levels of quality requirement, the consolidated quality requirements, are submitted to the decision making process for the determination of the level of quality the service provider plans to offer for the next phase of the business cycle. This is discussed in Chapter 5.

4.6   Summary

The management of customers' QoS requirements has its own unique set of issues to be addressed. These are best identified and focused by decoupling management issues from other viewpoints of QoS. The chapter offers guidelines to address these issues. The approach may appear rigorous but there is no substitute for the 'fire-brigade' approach of remedying faults as and when they occur. Such an approach should be resorted to only for dealing with unanticipated problems and emergencies and should not be used as a substitute for professional management of determining and studying customer's QoS requirements.

4.7   References

1   ETSI Technical Report, ETR 003, October 1994, Second Edition, 'Network aspects (NA); General aspects of quality of service (QoS) and network performance'
2   ITU-T Recommendation G 1000: Communications Quality of Service: A Framework and Definitions, 2001
3   RICHTERS, J. S. and DVORAK, C. A.: 'A framework for defining the quality of communications services', IEEE Communications Magazine, October 1988
4   Central Statistical Office: 'Standard industrial classification' (HMSO, London)
5   CURWIN, J. and SLATER, A.: 'Quantitative methods for business decisions' (Chapman and Hall, 1994)

Exercises

1   Using the matrix determine as many QoS criteria as possible for the following services;

   • ATM platform;
   • mobile telephony;
   • mobile text messaging;
   • e-commerce (for any retail trade);
   • distance learning over the Internet;
   • medical diagnosis over multimedia network.

2   Develop an action plan for the determination of mobile telephony QoS criteria for the residential and business population. Include steps for prioritising QoS criteria for order of importance and the level of performance required, for 80% of the population, for the principal industry sectors of the population.

Chapter 5

Quality of Service offered/planned by the service provider

5.1 Introduction

If a formal specification of offered or planned quality to the customer existed, it would form part of the performance specification of a telecommunications service and also be the basis of the service provider's quality measures. A quality specification would be a subset of the normal service specification. With the onset of competition, privatisation and regulation in different parts of the world, telecommunications is

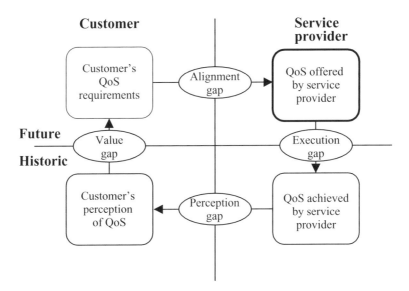

Figure 5.1   Quality cycle: offered quality

becoming a battlefield in which service providers attempt to survive. Survival will be based on differentiation and quality is fast becoming one of the principal vehicles for differentiation, particularly for the historically incumbent operator which cannot easily compete on price because of its overheads and universal service obligations.

This has resulted in the need for a formal specification of quality for each and every telecommunications service. Therefore, quality will in future form part of every service specification, along with other components such as the service description, service features, tariff, availability, service support and Service Level Agreements (SLAs). This quality specification will become important to customers, regulators, competitors and any other national or international service providers. This chapter looks at the applications of offered Quality of Service (QoS), identifies the main issues in its management and offers the briefest of guidelines to address these.

5.2   Applications of offered quality

The principal applications of the offered quality are illustrated conceptually in Figure 5.2.

5.2.1   Basis for service specifications

The offered QoS is the targeted level of quality the service provider is aiming to achieve. This information appears on the service specification. Customers go by this

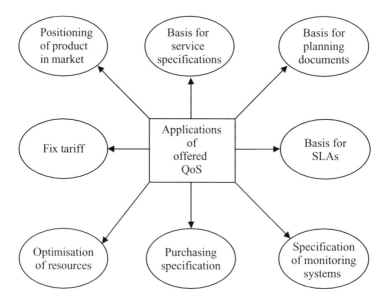

Figure 5.2   Applications of offered quality

information and the delivered quality (Chapter 6) to assess the performance of the supplier in the process of choosing a supplier for their communication needs.

5.2.2 Planning within the telecommunications company

The 'offered quality' will be used by the network provider and service provider in the planning and implementation of a service. The offered quality forms the basis for

- operations management, including items such as speed of provision, speed of repair, complaints resolution time, targets and costs, etc.
- design and dimensioning of the network,
- apportioning the performance of individual network elements to provide the desired end-to-end performance.

5.2.3 Service Level Agreements

Service Level Agreements between the supplier and a customer (usually a large business user) are usually contractual documents with agreed penalty clauses. The method of verifying the delivered quality in cases of dispute would be specified. The offered QoS is the basis of the quality part of the SLA. This subject is covered in more detail in Section 5.4.1.3 and Appendix 3.

5.2.4 Specification of monitoring systems

The monitoring systems to measure or estimate the delivered end-to-end performance are based on the offered QoS. These are covered in more detail in Chapter 6.

5.2.5 Basis for the purchasing specification

Telecommunications hardware and test equipment manufacturers use the information supplied by the service provider for design of their goods. The offered QoS is the benchmark to which the hardware and test equipment is designed.

5.2.6 Positioning of the product in the market

The service provider chooses the market segment in which it wants to offer its products and the level of quality it decides to offer determines the position in the market. Some organisations, by their philosophy, decide to offer the best quality while others offer what is most profitable. The positioning of quality is an important attribute of the service provider which the customer will judge the company by.

5.2.7 Fixing of tariff

Certain services are offered at different levels of quality and the tariff is fixed to match these. Different levels of access to the Internet are one example. While fixing of tariff may be done by this rationale it is also to be noted that there could be penalty clauses if the promised level of quality is not delivered.

5.2.8   *Optimising resources of the organisation*

The ingenuity of the service provider is displayed by how well it utilises the resources available to it, that is, plant, human and financial resources in offering quality service to the customers. Optimising quality and revenue for maximum profit is both a 'science' and an 'art'. In the battleground for the telecommunications market it is becoming an increasingly skilled activity to provide this match.

5.3   Parameters to express offered quality

In order to make effective use of its resources, it is operationally necessary for the service provider to specify as many as possible, if not all, of the parameters, to satisfy the majority of its customers and specify variants for specific customers. There is no internationally agreed set of parameters for the expression of 'offered quality'. For this reason, each service provider will propose its own set of quality parameters on which to specify network-related and non-network-related performance criteria. The only exceptions to this will be the technical, quality related performance parameters specified by recommendations from standards bodies such as the ITU-T and ETSI.

An illustration showing customers' quality requirements and the corresponding service providers' offered quality for basic telephony is shown in Table 5.1.

Appendix 6 shows typical QoS parameters for some telecommunications services offered by service providers.

5.4   Management of offered quality

The principal areas to be addressed in the management of offered quality are

(a) determination of the 'offered quality' parameter values;
(b) specification in planning documents for use by the service and network provider; and
(c) specification of monitoring and measuring systems.

The first two are dealt with in this chapter and monitoring management systems are explained in Chapter 6.

5.4.1   *Determination of offered quality*

The service and network providers consider customer requirements along with other internal and external business considerations in order to determine the optimum practical quality levels. The framework shown in Figure 5.3 illustrates how the offered quality may be determined.

Figure 5.3 shows the process to determine the offered quality from customers' requirements. There should be a dialogue between the customers and the service provider during the service provider's internal capability assessment, in order that a number of options with various quality levels and costs can be considered. In practice, this will normally take place between the service provider and a few large customers

Table 5.1   Examples of customers' quality requirements and the corresponding offered quality parameters: basic telephony service

Customers' quality requirements	Service provider's offered quality
1 Specified time to obtain pre-contract information on a service (i.e. tariffs, availability, features, options, offers etc.)	Target time to provide the pre-contract information.
2 Service provider to act in a professional manner at all times	Target number of customer satisfaction criteria which cover professionalism and are measured on a regular basis
3 Specified time for the resolution of complaints	Target time for resolution of customer complaints (e.g. 90% of all complaints to be resolved in 4 hours)
4 Specified time to repair the service	Target time for service repair (e.g. 90% of customers' reported faults to be repaired in 4 hours)
5 Faults to be repaired first time (i.e. repairs carried out right first time)	Target 90% of repairs to be completed correctly first time
6 Accuracy of reaching destination first time i.e. absence of misrouting	Probability of misrouting expressed as a percentage
7 Availability of the service for use by the customer	Target number of outages and the maximum duration of any outage
8 Calling Line Identity requirements (e.g. non-display of CLI)	A number of CLI display options
9 Delay limits on making long distance calls (i.e. call set-up time)	Maximum target call set-up time for international calls
10 Transmission delay on international connections	Target delays for different services expressed in units of time
11 Call quality requirements	Targets on performance parameters which contribute to call quality, such as loss, noise (different types), sidetone, echo, error rates, etc.
12 Service availability for the intended duration of use	Availability of end-to-end service for the intended duration of use expressed as a percentage, total number of outages and duration of longest outage
13 Flexibility in the provision of a service	Options to customise outside of the normal options
14 Charging and billing accuracy	Charging and billing accuracy specified in terms of: (1) maximum number of errors, and (2) maximum magnitude of any error

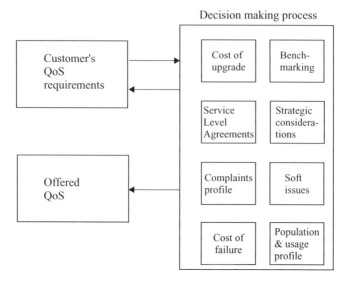

Figure 5.3   Process for the determination of offered quality

representing the market sector at which the service is directed. On completion of the internal capability process, the service provider will be in a position to decide on the level of offered quality. This may not meet all the original customer requirements, but it will be the most appropriate fit between customer requirements, cost and service providers' capabilities.

The principal considerations in arriving at the offered quality may be classified under the following headings:

   (i) cost of providing the required quality (e.g. by upgrading);
  (ii) benchmarking;
 (iii) SLAs;
 (iv) strategic considerations (time scales, competition, regulatory requirements, product life cycle, etc.);
  (v) cost of failure to meet offered quality (e.g. revenue rebates);
 (vi) population and usage criteria;
 (vii) complaints;
(viii) qualitative criteria (i.e. soft issues).

These are discussed in more detail below.

5.4.1.1   Cost of quality improvements

To satisfy many customer requirements, enhancements to quality may be required. In the unlikely event of offered quality being reduced, there must be an understanding of the potential penalties that may accrue, for example, loss of market share. With the complex service related cost structures available to most service and network providers, it will be difficult to produce an estimate of the relationship between quality, cost and financial benefits. However, it is sometimes possible to arrive at a relationship between increments in quality and cost of improvement. In the process of determining offered quality the incremental cost will be one of the inputs. The opportunity cost of not improving quality and the possible, resulting loss of revenue have also to be addressed. The network and service providers have a responsibility to develop a quality-versus-cost relationship in the future to obtain fair pricing of quality to their customers. (This is further discussed in Chapter 23.)

5.4.1.2 Benchmarking

Benchmarking is the comparison of a company's performance with that of competitors and like companies [1]. Often it results in a league table of parameters where the performance of a company is compared to others where individual parameters are considered best, for example, fastest in provision of repair of service, best customer satisfaction, lowest cost, highest connection per employee etc. Care must be taken to ensure that there is a valid comparison between service providers for each parameter, for example, a service provider with a lowest connection per employee may provide both local and trunk service, whereas one with highest connections per employee may only provide a local service. Once the benchmarks have been established, the service provider can set in motion activities to bridge the gaps between its performance and the 'best of breed'. The model for carrying out benchmarking is shown in Figure 5.4.

The use of 'Snake Charts', illustrated in Figure 5.5, are a useful way of graphically comparing performance against that of a competitor and highlighting areas for improvement. An alternative pictorial way to compare the performance of a service against a competitor is shown in Figure 5.6.

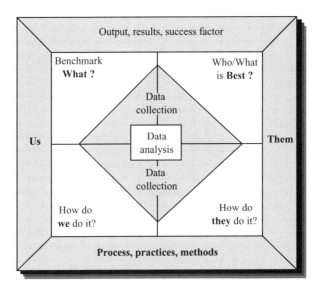

Figure 5.4   Rank Xerox Benchmarking Model [2]

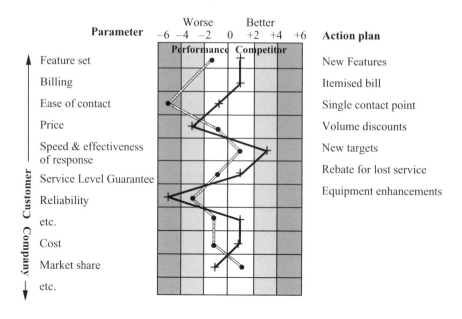

Figure 5.5   'Snake Chart' illustrating benchmarking

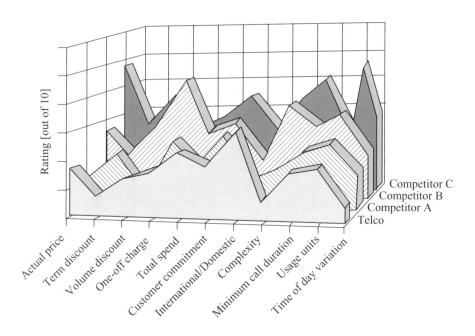

Figure 5.6   Competitor positioning

Benchmarking is also about improvement through change [2]. A research report quoted that, of the 1000 companies listed in the Times, 67% claim to have used benchmarking and that 82% of these used it successfully. The usefulness of benchmarking has been well recognised by the more successful companies. Telecommunications service providers are no exception; with increasing pressure to perform better with reduced resources, benchmarking is more widely becoming an input to the business planning and performance-assessment processes. Another dimension of benchmarking is adoption of the best practices of another organisation. This best practice need not be from the same industry; it may be translated to apply to the needs of another industry.

To benchmark, similar service performance statistics must be examined on a comparable basis. In practice, three principal difficulties exist. The service features are not always similar, performance definitions may differ and performance statistics do not conform to any commonly agreed methods of capture and presentation. True comparisons of performance become a combination of art and science. A comparative framework has to be developed by the individual service and network provider to give relevant weighting for various issues, such as market share, purchasing power in the local currency, service usage habits and so on, to enable a valid comparison to be made.

Another issue related to benchmarking is that it is not a constant performance figure, because most service and network providers aim to improve the quality of their services over time. If a service provider wishes to become the top service and network provider for quality then it must outdo other competitors in the long term.

Yet another issue to be addressed is the question of ethics. Benchmarking needs to be practised without recourse to industrial espionage. Information on performance may be closely guarded by service and network providers in the interest of competition. It will be a test of the integrity of a service or network provider to provide the level of service the customer wants and to differentiate itself from its competitors without breaking the normally accepted ethical practices in the industry. A useful reference on benchmarking ethics is 'The Bench marking Code of Conduct' developed jointly by the American Productivity and Quality Centre's Clearinghouse (APQC IBC) and the Strategic Planning Institute (SPI) Council on Benchmarking [2].

5.4.1.3   Service Level Agreements

SLAs are useful for the suppliers to enter into a formal (often contractual) relationship with:

 (i) other network providers abroad;
 (ii) other network and service providers in the same country with whom interconnection exists;
(iii) individual users;
(iv) business customers.

Increasingly, with more complex network interactions and sophisticated internetworked services, the agreement must also cover the possibility of one network compromising the integrity of another. (See also Chapter 9.) The principal management issues for each of the above categories are examined below.

(i) SLAs with other service and network providers abroad

As telecommunications services are provided on a global basis, it is necessary for a service or network provider to interface with service or network providers within other countries, either to provide a service to that country or to transit a service to a third country. Therefore, services can span more than one country and more than one foreign service or network provider.

There are two principal categories of issues to be addressed in dealing with quality on a global and end-to-end basis. The first is the inter-working (compatibility) issue and the second is the level of service quality that the two parties to the SLA plan to offer to their customers.

Inter-working between two networks

A network in one country connected to a network in another country has to conform to certain standards in order that they inter-work. Recommendations have been formulated in ITU-T and other standards bodies for all the essential interface requirements. What is lacking is a comprehensive list of all the end-to-end quality of service parameters and associated technical performance figures together with apportionment values between networks, in order that the individual service and network providers may design for their respective performance figures. Certain performance figures exist, for example, transmission delay in ITU-T G.114 [3]. Even though some of the major parameters are covered, there is a need to extend these to specify end-to-end performance and associated quality for a range of parameters for all the principal services. Chapter 27 makes out a case for this and proposes a framework for its study.

Mutually preferred end-to-end performance between two administrations

In some cases, a service/network provider may wish to liaise with a service/network provider from another country to improve the end-to-end performance of services between these two providers. This presents problems for both service or network providers. The first is to find another service provider with the best service fit, in terms of price and functionality. The second, is to establish the desired quality levels which will form the contents of an agreed SLA between the two suppliers.

This is a difficult task mainly due to the lack of internationally agreed definitions on QoS parameters. For an improvement in quality, the service providers will first need to agree upon a common set of quality definitions for the required quality criteria needing improvement. For example, one service provider may quote 'time to repair' as four hours, with the definition that this covers 'the instant the problem is reported by the customer, to the instant the problem has been resolved to the customer's satisfaction'. The other service provider may quote 'time to repair' as three hours, with the definition that this covers 'the instant that the problem has been reported to its operational staff to the instant the problem has been rectified to the customer's satisfaction'. However, if the elapsed time between the customer reporting the problem and the operating staff receiving notification of the problem is three hours, then real 'time to repair' from the customers perspective is six hours and not three hours as quoted by this operator. Therefore, achievement of improved quality levels dictated by market forces prompts the participating service providers to work out formal SLAs between themselves with agreed definitions to meet the desired level of quality.

In arriving at a SLA between two service providers the following considerations must be addressed. Cultural differences may influence the level of service required in each country. For example, the customer quality requirements for 'time to repair' a telephone service in the UK could be four hours and in France this could be six hours for the same service. However, this service may span both countries for the same (i.e. the multinational) customer. Therefore, a company having offices in the UK and France will be quoted two different repair times from two suppliers for the same service.

The vision and objectives of each service provider may differ. A service provider may be quite content to provide the basic level of service to its customers. This may be due to lack of competition, the service providers' resource capability and the maturity of the market (i.e. market penetration). For example, in a developing country where the market penetration is very low, the service provider's priority could be to provide service and increase penetration at the expense of quality. In this case, it will be very difficult for another service provider to agree a SLA covering tighter quality criteria for a service interconnecting both service providers' customers. This may result in both service providers agreeing a SLA with the minimum quality level acceptable within the ITU recommendations.

Therefore, for successful management of SLAs between two service providers from different countries, it is necessary to specify quality levels using the same performance parameters. Agreements must be reached between the two service providers on a common understanding of performance parameters. Where different methods of measurements and specification exist, for example, in speech loudness levels, agreement ought to be pursued on what constitutes mutually acceptable performance definitions and how these may be translated into understandable and meaningful levels of quality by both service providers. The service providers should then agree on a SLA, in writing, covering the levels of performance together with their definitions and should also specify their respective roles and responsibilities in the operation of the SLA.

(ii) SLAs with other national network or service providers

Service providers within a country may need to use each other's capability in order to reach the total customer base within that country. This is commonly known as 'interconnect'. The end-to-end performance will be dependent upon service providers capabilities and again, the quality offered to the customer will be based on an agreed SLA between the two service providers commensurate with their combined quality capability. In managing this type of SLA, the following points should be taken into account:

- the service providers concerned should agree, preferably contractually, on the parameters and their performance at their interface;
- they should specify these on a service-by-service basis;
- they should establish and agree on the procedures in the event of non-compliance with the SLAs.

In some countries, the regulator provides the guidelines within which the service providers should interwork and arbitrates in the case of disputes.

(iii) Individual customers

Residential or personal customers are not very discerning and generally accept a fairly basic level of quality in relation to price. This fact is exploited by service providers, since price is often the major issue and few customers will incur the inconvenience of changing providers on the basis of a nominal price differential. Any such move would be opportunistic, such as during a house move, or be prompted by emotional reaction to an unsatisfactory experience.

(iv) Business customers

At the other end of the spectrum (i.e. medium and large business customers) quality is a much larger issue, for example, near 100% reliability is required by customers such as banks, airline companies and those transferring financial data. For these types of customers, reliability is of utmost importance; consequently they will pay a premium price for high levels of quality. SLAs will usually be individually tailored to meet their requirements.

In a liberalised environment and an environment of ever-increasing litigation, significant financial damages may be levied against a service provider if a major customer, who is reliant on a service, sues the service provider for loss of earnings due to poor quality (e.g. a major outage). In a competitive environment, service providers often have SLAs with their major customers that are contractually binding and with large financial penalties for non-compliance. However, customers do not usually see these as ways to obtain rebates, but more as a measure of the service providers' confidence in the quality of their services. Customers do not normally wish to invoke financial compensation for breach of an SLA, but use breaches of an SLA to assess their choice of supplier (i.e. if an SLA is constantly breached are they with the correct service provider?). However, service providers must carefully consider all penalties (including financial ones) when agreeing on a SLA. For a service provider, poor performance can represent a significant loss of revenue either by direct payment or loss of a customer.

Appendix 3 shows a template for a typical SLA.

5.4.1.4   Strategic considerations

In the determination of quality to be offered the service provider should consider strategic business issues, such as:

- product life cycle of the service;
- effect of competition;
- effect on market share.

If a service were reaching the end of its useful life, as in the case of Telex, it would not be worthwhile to improve its quality. However, for new services, quality is likely to be a major contributor to market share in a competitive environment. For new entrants to the market, quality will be the prime differentiator, because the only way to increase market share is to offer equal functionality, be competitive on price and differentiate with quality. A new entrant may not wish to start a price war with the entrenched service providers, so the only viable differentiator is quality. The relationship between quality and market share will be an important input to the determination of the level of the offered quality.

5.4.1.5 Cost of failure to provide offered quality

In the consideration of SLAs earlier, an example was given illustrating how much revenue rebate may be returned to the customer should one of the quality parameters be found not to meet the agreed level. Revenue considerations aside, the competitive threat arising from not meeting the offered level of quality will have a detrimental effect on market share. The service provider's credibility will be lost if continued non-conformance to offered quality occurs, with the resulting loss of market share.

5.4.1.6 Population and usage criteria

The performance of a network service is often affected by the manner in which it is used. It is therefore necessary to understand the forecast population of users and how they use the services. For example, a usage profile that peaks during the day causes an overload that will severely degrade performance and hence the QoS. Also an unusual usage pattern that combines different services may cause undesirable feature interaction which degrades the performance of a number of services.

5.4.1.7 Complaints profile

Analysis of complaints can generate meaningful insights. Analysis of complaints is both an art and a science. Inferring meaningful insights from complaints requires knowledge of the market, cultural aspects of the customer population and their psychological behavioural patters. The areas of concern ought to be given priority while formulating the planned or offered level of QoS by the service provider. Resources spent on this area are particularly well spent when one considers the cost of regaining lost customers due to dissatisfaction. This information is an additional input to the decision making process in determining level of quality to be offered.

5.4.1.8 Soft issues

Any service provider has to take into consideration its impact on the society. This impact takes many forms, depending upon the size of the company and its philosophy. Sponsorship, its attitude towards quality, responsibility to the community, attitude towards the disabled and those with special needs, and its image all play an important part in how the company is perceived by the customers. These criteria have a quality element and are determinants in the arrival at the level of quality to be offered to the customers and public at large.

5.4.2 Alignment gap

More often than not, the offered QoS level is likely to be lower than what the customer or user had asked for. If the quality level offered has been improved to meet the levels of customer's and user's requirements the price charged to these would normally be higher than before unless the economics of the provider is able to absorb the incremental cost incurred in improving the service. The gap between what the customer and user wants and what is offered is the 'alignment gap'. The service provider would normally keep a track on this gap in relation to competition and price to enable them to assess the influence of these relationships in their market share for the service.

5.4.3   Specification of planning documents to meet offered quality

It was stated earlier in this chapter that the most intense use of offered quality is in the derivation of planning documents and operational processes to meet the targeted level of service. The key activities in the translation of the offered quality into planning documents is illustrated in Figure 5.7.

The offered performance criteria are separated into network-related and non-network related criteria. Where some criteria are a mixture of both, the two elements have to be separated and included in the respective category. The network-related criteria are broken down into element performance. The subject of specifying and managing network performance for legacy services is treated in Chapter 9.

Non-network-related performance criteria are broken down to unique and manageable functional elements. The non-network-related parameters (from Table 5.1) for basic telephony are developed in Table 5.2.

The principal management issues related to the above parameters to be addressed are listed below:

(i) Time for pre-contract enquiries on the service

Since basic telephony is a simple and ubiquitous service, customers will expect the service provider's point of contact to supply all the answers to their queries at once without any further internal consultation. The most common questions asked refer to: availability of the service, connection charges, tariff charges, terms and conditions relating to the service and quality. However, for new more complex services, customers may be prepared for pre-contract enquiries to take several days. Large

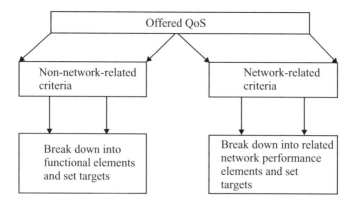

Figure 5.7   Offered quality planning process

Table 5.2   Non-network-related parameters for basic telephony

No	Parameter No in Table 5.1	Parameter for offered quality
1	1	Target time to provide pre-contract information on services to customer e.g. tariff, service availability, service features, choice of telephone features etc.
2	2	Target of a certain customer satisfaction score for the service provider's professionalism.
3	3	Target time for resolution of complaints e.g. 90% of all complaints to be resolved in 4 hours.
4	4	Target of e.g. 90% of customers' faults reported to be put right in y units of time (e.g. 4 hours)
5	5	Target of e.g. 90% repairs right first time.
6*	8	Call Line Identity (CLI) display options.

*This parameter has both network and non-network-related implications.
Source: Table 5.1

customers may even require service providers to tender competitively for provision of the service.

(ii) Professionalism

The service provider has to ensure that its customer-facing staff are trained to be courteous and business-like, sensitive to customer needs (e.g. special requirements such as loudspeaking telephones for the hard of hearing). Additionally, customer-facing staff should have a basic understanding of the service provider's entire portfolio of products and services. The service provider should attempt to measure the customers' opinion of their professionalism on a regular basis. Any dissatisfaction on the part of customers could be put right by the appropriate training.

(iii) Complaints resolution time

Customers are usually sensitive to the time taken for the resolution of complaints. The time will usually vary from country to country due to environmental conditions and the maturity of the market. However, the optimum time for resolution should be obtained by customer survey, then benchmarked with the competition and finally compared with the current time taken. Remedial action is then taken if necessary.

(iv) Time to repair

Repeated surveys indicate that 'time to repair' is among customers' top ten quality requirements. The issues to be addressed by a service provider are: current repair times within the business, world and principal competitor's benchmark, customers' requirements, resources to offer the required time to repair and the penalties likely to be incurred should the repair times not be met.

(v) Repairs not carried out right first time

A recent survey carried out in the UK and Europe indicated that about 50% of business customers have experience of repairs not being carried out right first time. This is wasteful of both the customers' and the service provider's resources. The service provider is paying for the same problem to be rectified more than once and the customer suffers both financial loss and the inconvenience of repeat visits from the service provider. The only beneficiary could be the service providers' statistics if they are measuring 'total number of faults cleared'. The service provider will then be shielded from this quality requirement until loss of the customer. The issues to be addressed by the service provider are similar to those of 'time to repair' with the addition of ensuring that the repair staff are competently trained and managed effectively to achieve the desired results.

(vi) Calling line identity requirements

This parameter has both network and service implications. The service-related issue is providing operational resources within the service provider to ensure that records are correct and up to date.

5.5 Summary

The management of 'offered quality' by the service provider is one of the key elements for the successful overall management of quality. It is perhaps in the provision of offered QoS where maximum decision making effort is expended. It is, therefore, necessary to ensure that all relevant information is given due consideration before the planned/offered quality of service targets are finalised. The quality of these decisions will affect not only the quality delivered but the service provider will be judged by the practical way in which QoS is seen to be managed. The next chapter covers the delivered QoS – how well the offered QoS has been achieved.

5.6 References

1   ZAIRI, M.: *Competitive benchmarking* (Technical Communications Publishing Ltd., Letchworth, United Kingdom, 1992).
2   CHERRETT, P.: 'Practical benchmarking', *British Telecommunications Engineering*, 1994, **12**, pp. 290–93.
3   ITU-T Recommendation G.114: 'One way transmission time'.

Exercises

1   Specify the quality of service parameters, the levels to be offered with tolerances for the travel industry for any data communication service of your choice.
2   You are given a set of QoS requirements for the customer population for one service. Produce an action plan identifying all the principal activities to achieve the required performance for all non-network-related performance criteria.

Chapter 6

Quality of Service delivered by the service provider

6.1 Introduction

Historically the *delivered* or *achieved* quality of service (Figure 6.1) of basic telephony has the oldest pedigree amongst the four viewpoints of quality. Of the four viewpoints, this is probably the best managed by the providers, albeit not as professionally as the customers would require. However, these were primarily related to technical performance, and were not quality parameters understandable by the majority of customers. Measurements on delivered performance of the more obvious technical parameters, such as transmission levels, noise levels in circuits, crosstalk etc. have been carried out on telecommunications networks for a long time. As progress was made in transmission capabilities, other parameters were added, such as delay and echo. As the quality of telephone instruments improved, sidetone was also added to the ever-growing list of quality parameters. Over the years, more and more parameter measures were added and a comprehensive set of performance parameters has evolved. However, these were based on objective measurements and were mainly associated with the service infrastructure (i.e. network technical performance). In due course other quality parameters were added, such as speed of provision, time to repair and similar operational parameters. Today, there is a need to understand clearly the meaning and significance of delivered quality and its relationship with the other quality viewpoints (i.e. the planned quality and the perception of customers).

In this chapter the applications of delivered QoS information are identified and the principal management issues are examined. As in Chapters 4 and 5 the backdrop of this chapter is basic telephony over circuit switched network. Chapters 10–14 cover, at a high level, the issues on the management of IP, mobile and satellite based services.

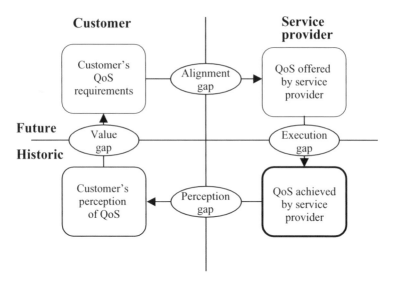

Figure 6.1   Quality cycle: Delivered or achieved quality

6.2   Applications of delivered quality data

6.2.1   Overview

The principal applications of delivered quality data are as follows:

(a) To monitor performance of network elements and operational capability of the service provider.
(b) To check the delivery of quality with agreed SLAs.
(c) To check if regulatory requirements have been satisfied.
(d) To use as publicity material.
(e) To use as a reference point against customer perception of quality.
(f) Benchmarking.
(g) Evaluation of planning and monitoring processes for optimum operational efficiency.

Each of these is covered in more detail below.

6.2.2   Monitoring performance of network elements and operational capability of the service provider

The data on delivered quality may be used by the network and service provider for an analysis of the variation of delivered from offered/planned level of performance. In the absence of reasonable explanations for unaccounted variations in delivered performance from planned quality, it will be necessary to carry out investigative work. It is the service provider's responsibility to study the delivered performance of operational functions associated with a service. If the end-to-end transmission quality of a national connection was designed to give a certain level of error-free performance and this level was not achieved, investigations may be initiated by the service provider through the network provider to identify the causes of this variation. If the target 'time to repair' was four hours for 99% of faults on leased lines and only 78% of such faults were repaired in time, there is reason for investigation, this time by the service provider, not necessarily involving the network provider. Analysis of variation of performance is based on the delivered performance and must be a standard activity for network and service providers. (See also Section 6.3.6 on execution gap.)

6.2.3   Check of compliance with SLAs

Delivered performance forms the basis for enforcing an SLA between the customer and the service provider. Audited performance data (see Section 6.3.5) may be used by the customers to pursue compensation claims in the cases of non-compliance to agreed performance levels within an SLA. The data may also be used by the provider to prove to customers that it is delivering to the targets set out in the SLA.

6.2.4   Check against regulatory requirement

In certain countries the regulator requires the service providers to publish data on delivered quality. The regulator may be more concerned with the publication of the data than the level of achievement for a particular service.

6.2.5   Use as publicity material

Delivered performance data may be used as publicity material in the following areas:

- for customers nationally;
- for international comparisons of performance;
- against competition.

Customers, naturally, want to know how well a service is performing. Such information may be published in the media. It may also be communicated individually by mail shots to organisations that have a special need for such information. Bodies such as the Organisation for Economic Co-operation and Development (OECD) [1] and the European Telecommunications Network Operators (ETNO) regularly publish performance statistics (see Chapter 22). These comparisons could be of use to customers (mainly large companies) when choosing a provider for their communication services.

Perhaps, the single most useful benefit to customers is to identify providers who offer the best quality. Professionally presented quality data will be welcomed by the customers. Those service providers who are reluctant to publish data on quality may be seen by the market as having to hide poor quality.

6.2.6   Use as reference point against customer perception of quality

The delivered quality forms the basis for the service provider to evaluate the customers' perception of quality. Since delivered quality is non-contentious, if monitored professionally and with an audit trail, unexpected customer perceived quality ratings could be investigated. The evaluation of customer perception and the relationship to delivered quality is dealt with in Chapter 7.

6.2.7   Benchmarking

When a service provider wishes to ascertain the world's best (benchmark) for performance, delivered quality data is the most often sought. This data is compared with their own performance to determine the gap that may need to be closed. Benchmarking is dealt with in more detail in Section 5.4.1.2.

6.2.8   Evaluation of planning and monitoring processes for optimum operational efficiency

Comparison of achieved performance against offered (and therefore planned) performance will enable the service provider to assess whether any improvement in the planning or monitoring systems is feasible and desirable. More effective means of monitoring could result in equally meaningful performance data perhaps at less cost and inconvenience. With increasing multiplicity of services being supported in the network the most efficient means of monitoring performance criteria needs to be under constant review.

6.3   Management of delivered quality

6.3.1   General

The responsibility for the management of delivered quality lies with the service provider. It is the service providers' responsibility to liaise with the network providers for network-related performance.

The principal activities in the management of delivered quality are

- Specification of monitoring systems;
- Determination of end-to-end performance;
- Publication of delivered quality statistics;
- Audit of delivered quality data;
- Review of delivered performance;
- Roles and relationships between service providers and network providers.

Each of these is set out in more detail below.

6.3.2   Specification of monitoring systems

The activities in the specification of monitoring systems are illustrated in Figure 6.2. Specification of monitoring methods involves the following principal steps:

(a) Determination of parameters to be monitored.
(b) Specification of monitoring systems for each parameter.

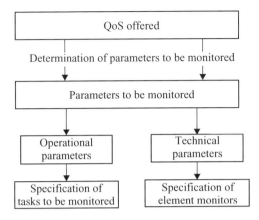

Figure 6.2   Process for the specification of monitoring systems

(a) Determination of parameters to be monitored

Determination of parameters to be monitored depends on whether the parameters are primarily for use outside or inside the service provider's organisation. Choice of performance parameters for external use would further be influenced by

- Customer preferred parameters on which delivered quality data is to be published on a regular basis, including those required for monitoring SLAs.
- Requirements of regulators.
- Service providers' obligations towards organisations for international comparisons of performance.

Customers' preferred QoS parameters for regular publication are best captured during the determination of their requirements on quality (see Section 4.4.3). A typical set of such parameters is illustrated in Table 6.1

A number of parameters on which customers require delivered performance to be published may be obtained on a service-by-service basis. Some of these parameters may be combined, for example, 'repair time' and 'resolution of complaints', even though the publication of statistical information may require separation of data on a service-by-service basis. In addition, individual customer requirements may have to be accommodated. This is particularly true for SLAs. Business customers who insist on SLAs will state parameters on which delivered performance are to be reported. The service provider will be faced with an array of quality parameters to cover the complete customer base. However, through careful combination of activities it should be possible to achieve economies of scale in allocating resources to establish the relevant monitoring systems.

The requirements of regulators would in many cases reflect the customers' requirements. However, the regulator's requirements often lag many of the leading businesses' requirements and the service provider could be faced with a regulatory demand for publication of performance parameters that could be different from the requirements of its customers.

Table 6.1   A typical set of performance parameters on which customers require delivered results to be reported on a regular basis

No	Parameter
1	Time to repair
2	Billing accuracy
3	Speed of provision
4	Fault rates
5	Service availability
6	Complaints resolution time
7	Adherence to CLI requirements
8	Number of complaints
9	Connection quality
10	Professionalism

A third consideration is the service providers' commitment or obligation to provide data for organisations such as OECD [1] and ETNO who attempt to publish comparative performance data from their member countries. Neither of these bodies has yet succeeded in obtaining performance statistics from all their member countries on agreed definitions of performance parameters. So far most service providers have supplied information matching closely the suggested definitions.

(b) Specification of monitoring systems

After the determination of performance parameters on which delivered performance should be published, they must be segregated under non-network-related and network-related categories. Certain parameters may contain both operational and network components.

Monitoring systems for non-network-related parameters
From Table 6.1 the non-network-related parameters are:

> Time to repair
> Billing accuracy*
> Speed of provision
> Fault rates*
> Complaints resolution time
> Adherence to CLI requirements*
> Number of complaints
> Professionalism

* These have network implications.

The guiding principles to establish monitoring systems for operational parameters are relatively straightforward. For example, to produce statistics related to 'time to repair' the following raw data are needed:

- Record of time at the instant the fault was reported.
- Record of time at the instant the fault was cleared, that is, when the customer was notified of the capability to use the service again.
- Geographic identification of the fault incidence.

From these records a number of statistics can be determined. Examples are:

- mean time to repair;
- percentage of repairs carried out within various times (e.g. 5, 4, 3, 2, 1 h);
- standard deviation of the distribution of repair times;
- performance on a national basis and for selected geographical areas;
- hot spots (areas where long times for repair exist).

Production of this statistical information need not be expensive. The basic data may be submitted to a computer program to produce the above statistical information.

The guiding principles for monitoring systems for non-quantitative parameters (e.g. professionalism) are more tenuous. The obvious method of expressing the professionalism of a service provider is to quote a customer satisfaction rating for this particular parameter (customer perception is dealt with in Chapter 7). Customer ratings may also be used for other non-quantitative parameters such as flexibility, courteousness and politeness of staff, sensitivity to customers' needs etc. on the part of service providers. However, wherever possible, quantitative measures should be used and backed up by customer ratings.

Specification of monitoring systems for network-related parameters

Network-related parameters are described in Chapter 9. When specifying monitoring systems, the following criteria need to be addressed:

- Identification of the unique elements whose performances are to be monitored.
- Estimation of the sample sizes and frequency of measurement.
- Method of measurement.

Consider the performance parameter, availability. Figure 6.3 illustrates conceptually an end-to-end connection. Links E_3, E_5 and E_7 may be traffic routes, which are usually monitored against call congestion and outage, or transmission routes which are monitored for selected transmission technical performance, for example, error and outage, or signalling routes which are monitored for signalling overload and outage. E_1 and E_9 are local access lines (or local loops) and it is impractical for dedicated monitoring systems to be employed due to the potentially high cost and the relatively low incidence of faults per line. However, in special circumstances monitoring equipment could be installed in the local access network and outages monitored. In most cases, only reported outages are used in the calculation of end-to-end performance. Links E_3, E_5 and E_7 carry a high volume of traffic and are not confined to providing service to one customer. Hence, these must be monitored for

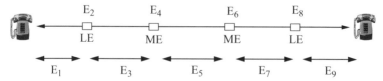

E$_1$ = Local line from customer to Local Telephone Exchange
E$_2$ = Local Telephone Exchange
E$_3$ = Link between Local Telephone Exchange and Main Telephone Exchange
E$_4$ = Main Telephone Exchange
E$_5$ = Link between Main Telephone Exchanges
E$_6$ = Main Telephone Exchange
E$_7$ = Link between Main Telephone Exchange and Local Telephone Exchange
E$_8$ = Local Telephone Exchange
E$_9$ = Local Line between the Local Telephone Exchange and the Customer

Figure 6.3   Elements of a hypothetical connection

outage. Nodes E$_2$, E$_4$, E$_6$ and E$_8$ need to be monitored for outages on a regular basis. Some exchanges have their own internal monitors to indicate the outage durations. Where this is not built in, it is necessary to provide an external device, to be triggered by the outage and reset by service resumption to note the duration and date of the outage. All nodes are monitored for traffic congestion and when thresholds are exceeded they trigger management action.

6.3.3   Determination of end-to-end performance from observations

Determination of end-to-end performance delivered for non-technical parameters is straightforward. Repair, complaints, provision time, billing enquiries need not involve any technical measurements of elements of the network. However, to compute end-to-end technical performance based on individual element performance may not be straightforward or economical. Due to the different traffic levels carried by different portions of the network, complicated modelling methods may have to be carried out to estimate the end-to-end performance based on measurements on elements, which is based on sampling. Making a one hundred per cent observation of an element may not make the end-to-end computation much easier either. This leaves the service provider with the only viable alternative of making end-to-end measurements at specific points as close to the customer as possible. Element performance may be carried out for analysing the performance of individual elements and to monitor their predicted performance.

6.3.4   Publication of delivered quality statistics

The style and publication format would depend upon the audience the information is aimed at. The publication of statistical information on delivered quality would benefit the following bodies:

(a) service provider;
(b) users at large;
(c) regulators;
(d) bodies responsible for comparisons of performance.

The service provider would require the statistics on delivered performance perhaps broken down on an element basis as well as on an end-to-end basis. This will enable the manager to carry out a variance analysis with the planned objectives, if necessary.

The statistical information intended for the user may have to be translated into meaningful terms easily understandable to the user. The publication may be carried out to suit local cultural preferences. Choices are booklets, condensed information on flyers and on the web. The frequency of publication would reflect local needs. Such supporting information could include any abnormal performance which may have occurred during the period of the report and their causes. For example, major exchange outages caused by localised fire, flood, earthquake etc. and serious weather related operational difficulties which contribute to poor service availability may be stated. If the availability performance were averaged out for the whole country the localised effect would be masked, to no-one's benefit. If a detailed explanation is given by the service provider and the national figures with and without these outages are stated, these could be more meaningful to the customers. The requirements of business and residential segments may differ. Business needs may reflect the SLAs and may not be public knowledge.

The information required by the regulators is usually based on the comparability between various operators to enable the consumers to make comparisons. In certain cases, for example, in Europe, the European Commission may require a set of parameters to be reported on, to defined parameter and measure formats, at specified frequency.

For international and national comparisons it is necessary for agreed definitions to be used. This topic is covered in more detail in Chapter 22.

6.3.5   *Audit of delivered performance data*

The delivered performance data, when computed for internal use within the service provider's or the network provider's organisation need not be audited externally. However, if these data are to be used for customers' benefit or any other body external to the supplier, it adds confidence to these recipients if the data are audited by an independent body. When SLAs or punitive measures are to be authorised for under achievement, it is desirable for performance figures to be audited.

Guidelines for auditing may be found in the following documents:

BS EN 45011: 1998 [2];
BS EN 45012: 1998 [3].

6.3.6   *Review of delivered performance and analysis of execution gap*

The delivered performance statistics ought to be reviewed if any of the following implications are pertinent:

- Assessment of delivered performance shows a variation from the planned levels of quality. This is the execution gap.
- Comparison against customers' perception of quality shows there is a substantial lack of correlation between customers' and delivered qualities.

The delivered performance is normally monitored by the network or service provider. The data are compared to the planned levels of quality. Where there is significant variation (gap) between delivered and planned quality, investigations must be initiated to identify the causes of these variations. Examples of possible causes are: (a) faulty measuring equipment, (b) adverse conditions affecting performance, (c) optimistic planning assumptions. It is important to identify the causes for the gap in order to carry out remedial actions and to restore the delivered performance to its planned levels whenever possible.

In the next chapter, customer perception of quality is compared against the delivered quality of service and the implications on the provider (analysis of the perception gap).

6.3.7   *Service and network providers' roles and responsibilities*

Where service providers are not network providers, it is necessary for the relationship between these bodies to be discussed and agreed amongst themselves. The service provider is expected to deal with all aspects of quality with the customer. In addition, the service provider has the responsibility to ensure that the network provider agrees and monitors specific performance parameters within the network to ensure that the correct end-to-end service quality is maintained. The range of parameters to be monitored and the elements to be measured must be agreed between the two parties for effective management of the delivered performance. Such agreement of mutual responsibilities should also be extended to international network and service providers when services are provided beyond the border of a country.

6.4   Summary

Management of delivered quality, even though the most prevalent among the service providers, requires better professionalism for optimum effectiveness. Scientific determination of relevant sample sizes, appropriate parameters (for optimum benefit to customers, regulators and the service providers) and specification of the most efficient monitoring systems are part of this process. Determination of the delivered performance leads to the next steps in the management of quality: the management of customers' perception of quality, analysis of the execution gap and establishing remedial actions.

6.5   References

1   OECD: 'Communication outlook' (Organisation for Economic Co-operation and Development, 2001)
2   BS EN 45011:1998 – General requirements for bodies operating certification systems
3   BS EN 45012:1998 – General requirements for bodies operating assessment and certification/registration of quality systems

Exercises

1   Identify all services offered by any one service provider whose delivered performance needs to be reported on a regular basis to the customers at large.
2   Identify all performance parameters to be reported on a selected (or all) services identified in Exercise (1) above.
3   For one main service break down the end-to-end performance that needs to be reported to
    (a) element performance collection – specifying the frequency of measurement, sample size and method of measurement,
    (b) operational performance set-up – specifying the frequency of data collection and an outline of how these may be used to calculate regular performance statistics.
4   Outline an auditing system for the processes instituted by the service provider to calculate end-to-end performance of basic telephony over circuit switched network.
5   Produce a flow chart showing the key steps in the publication and distribution of performance statistics for a service provider for residential customers.

Chapter 7

Customer's perception of Quality of Service

7.1 Introduction

A customer's perception of quality is a judgement made by the customer about the overall excellence, or otherwise of the delivered product or service. Customer's perception is the main criterion by which a service provider can assess and measure the value of the quality it delivers. Indeed, the service provider's decision making process would be incomplete without this information. However, the information contained within customer perception surveys is usually subjective and also likely to change over time with changes in the industry structure, environment, technology, competitive activity and good or bad media commentary. According to one study [1]

> 70% of customers dissatisfied with a service will go elsewhere, but only 5% will tell you they are unhappy... Dissatisfied customers tell an average of 10 people about their poor experiences while satisfied customers will tell only 5. It costs up to 5 times as much money to attract a new customer than to keep an existing one, but 95% of dissatisfied customers will stay loyal if they are handled properly.

These figures will normally vary with the type of service, from country to country, with competition, circumstances and time. Irrespective of the precise magnitude of such figures, it is clear that it is very costly to have dissatisfied customers. In more recent times internal studies by a service provider have shown that it is necessary for a customer to be 'delighted' rather than just be 'satisfied' to retain their custom. In times of active competition this aspect of provider–customer relationship ought to be studied, managed and reviewed by every service provider. Therefore, a service provider must analyse the credibility of perception ratings and constantly review customer's perception of their products and services and take any remedial action necessary to improve or correct the current perception. (See Figure 7.1.)

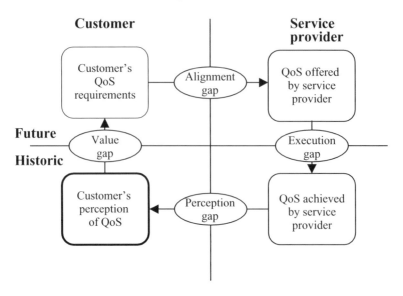

Figure 7.1   The quality cycle – customer's perception

7.2   Applications

The principal applications of customer's perception ratings on quality are

- as an indicator of market's perception (i.e. 70% of customers find 'call quality' as being good);
- to compare service providers quality (i.e. 70% of customers rate service provider A's 'call quality' as very good while only 30% of customers rate B's 'call quality' as good);
- to analyse perception gap – the gap or the difference between delivered and perceived quality (Section 7.3.3);
- to evaluate new network technology insertions or changes;
- to evaluate new service applications;
- to establish customer requirements for Quality of Service (QoS) parameter and referred ranges (e.g. the mean opinion score (MOS) mentioned in Chapter 11 for VoIP);
- to prioritise quality improvements for maximum benefit to both customers and the service provider.

By the successful application of perceived quality data and quality programmes coupled with good advertising, a service provider could considerably increase customer perception of their products and services and hence enhance their company image with only a marginal increase in cost. Service providers can also concentrate on parameters which customers rate as priority.

7.3   Management of customers' perception of quality

7.3.1   General

The key issues to be addressed in the management of perceived quality are

(a) the measurement of customer perception;
(b) analysis of the perception gap;
(c) modification of quality programmes.

  These are covered in more detail below.

7.3.2   Measurement of customer perception

The principal activities in the measurement of the customer perception may be grouped as follows:

(a) selection of service or group of services to be surveyed;
(b) choice of performance criteria to be surveyed;
(c) survey samples and size;
(d) questionnaire design and implementation;
(e) publication of results.

7.3.2.1   Selection of service or group of services

There are two categories of surveys. The first is to establish customer perception of services whose delivered quality has been monitored. The purpose of these surveys is to compare the perception against delivered quality. The parameters on which perception rating is to be gathered would be the same as those for delivered quality to enable comparisons to be made. The other category is to establish what the customer's perception is of any given facet of the business; whether it is quality perception of a specific parameter, specific service or indeed anything that requires focused study. The selection of a service or a group of services will reflect the aim of the survey. Indeed there are no boundaries to carry out a survey. Surveys of this nature would reflect a specific need to study customer's perception on quality of a service, group of services or any facet of business performance. The set of parameters on which customer perception is to be surveyed needs to be determined in light of the objective of the survey.

7.3.2.2   Choice of performance criteria

As mentioned in the section above the parameters selected for comparison of customer perception with delivered quality would be the same or similar. Only the wording of the questions need to be worked on to obtain the perception rating. On the other hand if a general study is the objective, the survey questions have to be designed for this purpose. The method proposed by Zeithaml *et al.* [2], could be adapted

Table 7.1   SERVQUAL questions and groupings

Groupings	Expectation score	Perception score
Tangibles	SERVQUAL questions 1–4. Example: Excellent companies will have modern-looking equipment.	SERVQUAL questions 1–4. Example: XYZ Co. has modern-looking equipment.
Reliability	SERVQUAL questions 5–9. Example: When excellent companies promise to do something by a certain time, they will do so.	SERVQUAL questions 5–9. Example: XYZ Co. promises to do something by a certain time, they do so.
Responsiveness	SERVQUAL questions 10–13. Example: Employees in excellent companies will never be too busy to respond to customers' requests.	SERVQUAL questions 10–13. Example: Employees in XYZ Co. are never too busy to respond to your requests.
Assurance	SERVQUAL questions 14–17. Example: Employees in excellent companies will have the knowledge to answer customers' questions.	SERVQUAL questions 14–17. Example: Employees in XYZ Co. have the knowledge to answer your questions.
Empathy	SERVQUAL questions 18–22. Example: The employees of excellent companies will understand the specific needs of their customers.	SERVQUAL questions 18–22. Example: Employees in XYZ Co. understand your specific needs.

for this purpose. The method identifies twenty-two performance criteria, which are of concern to customers. These have been grouped under five headings: reliability, responsiveness, tangibles, assurance and empathy as shown in Table 7.1.

They recommend that survey questions be asked in two stages. In the first, the customer's *importance* rating for each of the twenty-two criteria is ascertained to identify their expectations. In the second, the same questions are asked, slightly re-worded, to seek their rating for perceived performance. Before analysing the resulting data, customers are asked to distribute 100 points to five questions, representing the five groupings as shown in Table 7.1. The relative weights of the five groupings are established from the allocated points. The interview data is then prioritised for all parameters, together with gaps between expected and perceived quality. Such prioritisation, it is claimed, will enable the service provider to allocate resources in areas where they are most needed to improve quality. The approach by Zeithaml *et al.*, though very methodical, needs to be adapted for the telecommunications industry and to suit the relevant country culture.

7.3.2.3 Survey samples and sample size

To identify the customers (samples) and the number of such customers (sample size) the following criteria must be considered:

- Customers who have had recent quality related experience with the supplier.
- Customers who have had recent contact with the supplier.
- Customers chosen purely on a random basis (from the customer grouping identified for the survey, for example, the transport sector).

Customers who have had recent quality related experience in the area to be studied with the supplier (depending upon whether the experience was good or bad) are best suited for inclusion within the sample. However, the available sample size of such customers may be small but this would easily be offset by the freshness of their experience.

Alternatively, customers could be chosen who have had recent contact with the service provider, though not necessarily a quality experience. Care must be taken in choosing these customers, as they will not want to be continually researched, even if they perceive benefit from the research. The usefulness of the responses given by this sample may be of limited value due to the relevance of their recent experience. This method of selection is the next best and should only be chosen after the first category of customers has been exhausted and the sample size is not considered sufficient.

The third category of selection is from the total customer population of a particular customer grouping (i.e. all customers within the retail sector). This group will include customers from the above two categories and also those who have not had any experience with the supplier and will constitute by far the largest sample size. Choice of samples from this group may also be of limited value as their perception will be based upon company image and not past experience and this could be misleading if the results are applied to a particular product or service quality enhancement, or any conclusions are drawn.

The customer sample size for any survey will depend upon the customer grouping chosen. A sample size of approximately 30–50, would be sufficient to provide good results if all the customers within the sample have had a recent quality experience in the area to be studied. However, as we progress into the other two categories of customer groupings, the available sample size grows. If the range of responses vary significantly the sample size will need to be increased to provide statistically robust results. Customer responses to opinion rating should be studied both from a statistical viewpoint as well as from a psychological viewpoint.

7.3.2.4 Questionnaire design and implementation

The questionnaire should contain questions to reflect the objectives of the survey, as discussed in Sections 7.3.2.1 and 7.3.2.2. The principal guidelines indicated in Sections 4.4.3–4.4.5 for the capture of customer's quality requirements may be applied to the determination of customer's perception of quality. The question design and structure are critical to the successful management of this viewpoint of quality. Badly designed questionnaires will only reveal part of the 'whole truth' and could show the service provider to be either disproportionately good or bad. If, for example, the structure and content of the questionnaire tended to hide or diminish customer dissatisfaction with a product or service, the service provider will be lulled into a false sense of security with the belief that they are providing better quality than they in fact are. The worst realisation of this fact comes when the customer takes their custom elsewhere, with the consequential loss of revenue. The converse of this is when the structure and content of the questionnaire produce false results that show the product, service or service provider as under-performing in terms of customer satisfaction. The service provider will then tend to waste valuable resources in attempting to fix a problem, which does not exist. Either of these two scenarios will have the effect of reducing the service provider's profitability by loss of revenue or unnecessary increase in cost.

The design of the questionnaire will also depend upon the mode of implementation. The three main methods of questionnaire implementation are

- postal questionnaire;
- telephone interview;
- person-to-person interview.

This is covered in some detail in Section 4.4.3.

The perception ratings for each implementation methodology must have attached an associated confidence weighting so that the service provider and customer can have a clearer understanding of the data and its interpretation. The design of questionnaires and their implementation are specialised areas and for maximum benefit they should be designed and implemented by professionals with the relevant expertise who also have an understanding of the telecommunications industry.

There may be a need to carry out customer perception on a regional basis to investigate a specific problem or to assess the effect of some change in the procedure or the plant of the service provider affecting performance.

It is also worth bearing in mind that when surveys are carried out it may be desirable to change the order of questions in order not to let 'interest fatigue' creep into the voice of the interviewer. Should this happen the interviewee may give a disinterested reply thereby affecting the accuracy of the survey.

7.3.2.5   Publication of results

Only the results of the customer surveys specifically obtained for external use are suitable for publication. A good method of displaying these results is in a graphical form. A hypothetical example is shown in Figure 7.2. The information in Figure 7.2 shows the variation in customer perception of voice quality over time. From this information a service provider can determine the acceptable level of customer perception (e.g. 80% of customers think that voice quality is good or better). This type of information can also be used to trigger action by the service provider. The service provider can assign limits to this parameter and if the perception falls within those limits no action will be taken. However, if the perception falls outside those limits then action should be taken to correct the situation.

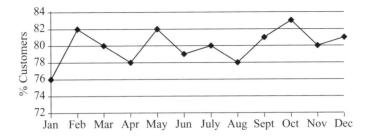

Figure 7.2   *Percentage of customers stating voice quality was good or better in a hypothetical customer perception survey*

7.3.3   Analysis of the perception gap

Perception gap is the 'gap' variation between customer's perception ratings and what one would expect from the level of delivered quality. One of the main activities in the management of customer's perception of quality is the analysis correlation of customer's ratings of Delivered Quality and Perceived Quality. In an ideal world there would be 1 : 1 correspondence between the delivered quality and customer's perception ratings. In reality this situation is only rarely achieved and it is necessary to carry out an analysis for the lack of 1 : 1 correlation. As the delivered quality data is objective, the main reason for the variation and, therefore, the gap is the shortcoming in the customer perception ratings. The principal shortcoming of any customer rating is its sensitivity to subjectivity.

In Section 3.2.4 it was mentioned that customers' awareness could affect their perception of quality. The variance ought to be carried out on a parameter-by-parameter basis. Where perceptions are obviously coloured or based on inaccurate information the customer's awareness needs to be raised to overcome this problem. If customers, for example, perceive that the service provider's billing formats do not meet their requirement but are unaware of the totality of the billing formats on offer and the customisation available within those standard formats, customers' perception can be raised significantly by a simple customer awareness programme, communicating the number of billing formats available and the customisation of those formats. Ignorance on tariff, unfounded assumptions, false expectations etc. can influence the stated perception rating of the customer. Meeting customers' communication needs may mean different service features to different people and some features may be superfluous. In other cases service features may not be adequate. Adverse recent publicity in the media can influence the respondent's perception of experienced quality of service. Personal preferences and prejudices could influence people's perception ratings.

The validity of the survey needs to be checked or reviewed, as some of the results or areas for improvement may not have been well defined within the original survey. This review should check the customer selection, sample size, questionnaire design and method of implementation. Audit programmes are likely to highlight any shortcomings with a perception study and regular audits should be applied to customer surveys.

Accurate pinpointing of the cause of variation ought to be attempted. Success in pinpointing the cause of the variation or the gap will be dependent upon the intimate knowledge of the market by the service provider. This leads to the next step – modification to the quality programme.

7.3.4 Modification to the quality programme

In quality management, any modifications to the activities influencing quality must be evaluated before implementation. If the perception rating on availability of a particular service has declined and the cost of improvement far outweighs the likely revenue increase, it may not be worthwhile for the service provider to carry out the modifications. However, other considerations may apply. It may be in the interests of the service provider to spend the necessary resources to improve their company profile on quality. Such profile improvement could have a beneficial effect upon the entire service provider's portfolio.

In Section 3.2.4, the relationship between advertising and customer's perception of quality indicated that advertising could be used to improve perception of a product, service or service provider image. However, care must be taken when using advertising to promote perception because the perceived quality could remain low or even deteriorate if the advertising promises too much.

Before committing valuable resources to a quality-improvement programme, it is advisable to gauge the customers' priority weighting for the quality parameter to be enhanced. For example, if a perception survey showed that customers were dissatisfied with 'voice call quality', but their priority weighting for this parameter was low, then the service provider would do better to use resources to improve a different parameter with a higher customer priority rating.

The improvements indicated from the perception ratings could also be submitted to the customers' requirements capture for the next round of management activities of the quality cycle.

7.4 Summary

Customer's perception of quality, expressed by opinion ratings, may be used for effective management of quality by the service providers. In the future more use could be made of customer perception surveys as part of the decision-making processes within a service provider's organisation. More work is also likely to be carried out in understanding the relationships between customer perceptions, expected quality, delivered quality, competition, technology, advertising, customer requirements and other variables. Improvements identified from perception survey ratings could be incorporated into the customers' requirements capture process (Figure 7.1) for inclusion in the next round of the management activities of the quality cycle, thus completing the quality cycle.

7.5 References

1 BROWN, T. 'Understanding BS 5750 and other quality systems' (section 3.6, Gower, ISBN 0- 566-07455-9)
2 ZEITHAML, V. A., PARASURAMAN, A. and BERRY, L. L.: 'Delivering quality service' (Free Press, New York, 1990)

Exercises

1 Design a questionnaire to administer to the business community a service most considered by them as an 'essential lifeline communication service'.
2 Carry out a gaps analysis of existing customer perception results with the delivered quality statistics and develop recommendations for closing any negative gaps.

Section III

Existing and emerging network and services

This section addresses the main QoS issues for legacy and emerging networks, and services. On the legacy network, the basic telephony QoS issues are identified together with guidelines on its management. After a brief description of the Internet architecture, the QoS issues of real-time (e.g. Voice over IP) and non-real-time applications (e.g. e-mail and some e-commerce) are addressed. Mobile and satellite communication systems are included in identifying their unique QoS issues. The principles of addressing QoS issues in this section may be applied to any service in telecommunications.

Chapter 8

Network evolution and its performance

8.1 Introduction

Given the impact of network performance on the QoS perceived by end customers and its role as a platform for the provision of services, it is necessary to understand how the network is likely to evolve in terms of technology and architecture. Telecommunications networks have been continuously evolving since their inception and exhibit the classical 'life cycles' generally attributed to consumer products. For the market life cycle, a new product will exhibit a growth of sales until market saturation after which sales decline because the market is fully penetrated or is overtaken by a better product. A telecommunication example is the Telex network service. Early growth is often influenced by the need to establish the network platform necessary to launch new network services, that is, technology push. Once critical mass is achieved by the uptake of new network services, then market pull dictates the rate of penetration of the service and is the dominant driver influencing the penetration of technology. The decline phase is determined by obsolescence where a more cost-effective technology with improved service potential becomes available. For obvious reasons all life cycles must overlap.

Application of the network life cycle analogy to the historic development of the network, its current phase of modernisation by digital technology and possible future evolution is shown in Figure 8.1. The timing of the life cycles are relevant to the BT network and will vary for networks of other telecommunications operators depending on market and political influences and technical competence, but primarily relies on the availability of capital to fund the necessary investment. It should also be noted that life cycles are becoming progressively shorter. Similar life cycles can be plotted for data networks, where the early X25 networks are being subsumed by the more efficient Frame Relay networks and broadband Switched Multimegabit Data Service (SMDS) networks. Mobile network life cycles have also progressed through the initial analogue first generation (1G) to digital (2G) and the upcoming broader band (3G) networks.

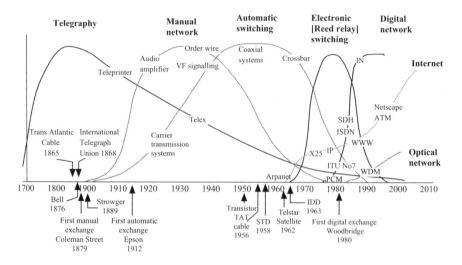

Figure 8.1   Network life cycles

In the earliest telegraphy life cycle [1], network performance was dominated by the resistance of telegraph circuits because signals were interrupted direct current and long distance telegraph circuits were therefore exorbitantly expensive due to the cost of high gauge copper cables. The advent of voice frequency telegraphy overcame this constraint but performance was then affected by loss, that is, the impedance of lines. QoS was determined by errors in a message and factors contributing to this were the skills of the operators as well as line errors.

The early manual network was characterised by increasingly efficient manual exchanges moving from Magneto exchanges with their cumbersome ringing through the local battery to central battery systems. Manual trunk working was streamlined by the use of a single order-wire for operators to request connection of the individual circuits between exchanges. High customer satisfaction was achieved by interaction with the human operator but once connection was established, the dominant performance parameter was loss, measured by the decibel (dB) power ratio. This era also saw the introduction of basic electronics with the valve amplifier, used to overcome the constraints of loss, and low capacity carrier systems made possible by negative feedback techniques. Noise and distortion then became noticeable to customers who were becoming increasingly familiar with the telephone.

The introduction of the analogue automatic network was the first major modernisation exercise driven by the need to reduce operating costs. However, customers were presented with a more complex interface to the network and the range of services enjoyed by customers was reduced from voice operated, intelligent services such as call transfer, follow me and call duration and charge advice. To this day, the customer is still presented with a user-unfriendly interface to the network and voice operation is a long-term objective. Major switching milestones included progressive use of common control to facilitate the introduction of services such as Subscriber

Trunk Dialling and International Direct Dialling to Stored Program Control, which marked the introduction of computer techniques to switching. The repertoire of signalling systems increased by the use of voice frequency signalling necessitated by the use of amplified lines. At this time, although the design of the Public Switched Telephone Network (PSTN) was dominated by the need to reduce loss to an acceptable level, a wide range of other performance parameters (described in Chapter 9) affected customer's perception of quality.

The most recent thrust in PSTN development has been its modernisation by the replacement of analogue switching and transmission equipment by their digital counterparts to create the integrated digital network. This comprises stored program controlled digital exchanges inter-connected by digital transmission equipment using common channel interprocessor signalling conforming to ITU-T No7 standards. Such a network has considerable cost savings arising, primarily, from the elimination of the per channel primary multiplexing equipment and signalling equipment at the transmission/exchange interface. It also establishes the basic infrastructure for the ubiquitous intelligent multiservices network. The customer benefits of the digital network have been expanded by extending the 64 Kbit/s channels and enhanced signalling capability to customer premises to form the integrated services digital network (ISDN), by Basic (2B + D i.e. 144 Kbit/s comprising two 64 Kbit/s channels plus a signalling channel of 16 Kbit/s) or Primary (30B + D) 2 Mbit/s access from ISPBXs. Centralised intelligence is penetrating the network in the form of the Intelligent Network (IN) used primarily to support advanced Freephone and VPN (Virtual Private Network) type services. The problem of loss has been overcome by loss-free digital technology in the core of the network but the access network (local loop) has been largely immune from modernisation, and line resistance and loss continue to restrict the maximum reach from a local exchange to the furthest customer to around 5 km, thus exerting a major constraint on network design by influencing the number of local exchanges required in a PSTN. However, QoS was vastly improved due to the low loss of the end-to-end connection and the reduction in call set-up time.

The provision of an integrated digital network with its stored programme control has led to a proliferation of network services as illustrated in Figure 8.2. Each service has its own unique set of QoS characteristics which are often related to network performance characteristics.

The Internet and Internet Protocol (IP) technologies are now beginning to exert a major influence on strategic thinking about how to meet the needs of the information society and is the start of the IP life cycle.

8.2 Drivers of change

The major driver of change is the rapid development of computer technology as illustrated in Figure 8.3. The well known Moore's Law says that chip capacity and processing power (MIPS, millions of instructions per second) roughly doubles every two years and, as a consequence, this is mirrored by the increase in modem speed, the increase in transmission speed and the increase in application requirements (one could

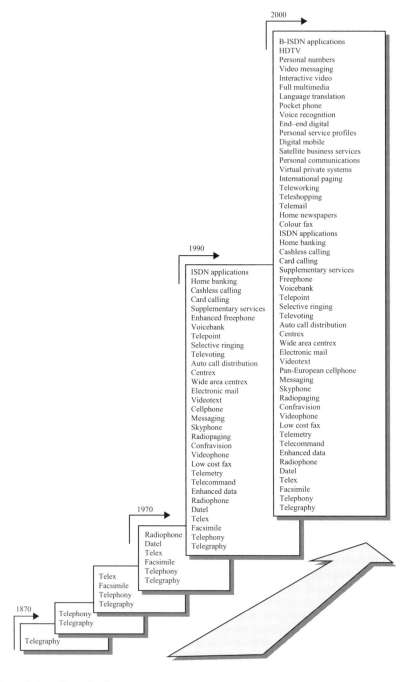

Figure 8.2   Growth of services

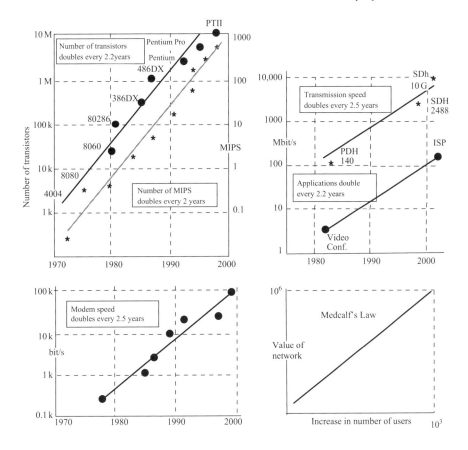

Figure 8.3 Technology trends

Source: Meijer & Geraads (2)

postulate a 'Gates extension to Moore's Law' which states that increased capacity will always be filled by the requirements of the next upgrade to Windows). Although unconnected, Medcalfe's law is relevant because it states that the network becomes more valuable as it reaches more users, 'an increase by a factor of a thousand in the number of users makes a network a million times more valuable.' Medcalfe also postulated that the earning value of equipment meeting his standard (Ethernet LAN) would increase by the square of the number of companies that would embrace this standard.

Disruptive technologies have a fundamental impact on the way in which people work and can significantly reduce the barriers of entry for new players and/or change the economics of service delivery. It can make the assets of existing players obsolete, cause loss of market share and put traditional margins under threat. The rapid growth of IP networks resulting from the Internet explosion could be considered as a disruptive technology. Local Multipoint Distribution System (LMDS) is another disruptive technology. As a broadband fixed wireless point to multipoint access system,

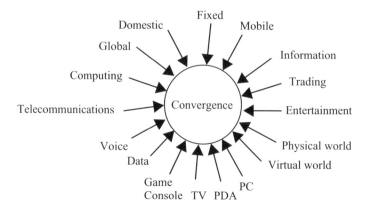

Figure 8.4   Convergence

operating in the 28 GHz region, it has the potential to deliver data rates of 50 Mbit/s down-stream and 10 Mbit/s upstream with a range of around 8 km. Also, its on-air statistical multiplexing makes efficient use of bandwidth and is suited to ATM and IP packet based protocols. It therefore breaks in the incumbent Telco's control of access to customers and facilitates cost efficient and flexible provision of broadband services. Perhaps the most disruptive technology from the telecommunications viewpoint is the Internet and its use of the ubiquitous IP protocols.

Stabilising technologies tend to protect existing assets while adapting them to future requirements. An example of such a technology is xDSL, which allows broadband services to be delivered to end customers over the existing copper loop (within certain constraints of reach and cable fill).

Driven by the Internet, the telecommunications world is being revolutionised by the many dimensions of convergence (Figure 8.4). It will open up new opportunities for Telcos, but will bring many new competitors into the market.

Telcos have a proliferation of solutions such as the PSTN, data networks (X.25, frame relay, SMDS), ATM and/or IP networks, analogue and digital mobile networks etc. These are complex overlay networks, difficult to manage and very expensive because of the lack of interworking between platform management systems. Thus migrating to a single multi-service network can provide numerous benefits. It can help solve the problem of network congestion whilst leveraging sunken investment, reduce operating costs and promote revenue growth from new multimedia services, whilst contributing to leadership in an increasingly competitive market.

8.3   Network architecture

The increasing complexity of modern networks requires careful consideration of architectural standards to ensure that the network elements interact to provide end-to-end service with the appropriate performance and that interconnectivity between

Overall architecture

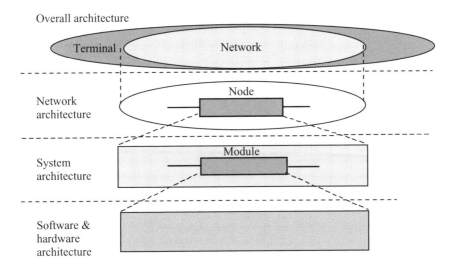

Figure 8.5 Network architecture

functional and global networks is achieved through appropriate gateways. Figure 8.5 illustrates how the architectural structure can be built up from the lowest hardware and software module to the overall system architecture. It is a very simplistic view of a multidimensional object, which embraces

- functional and physical separation, essential to ensure the flexibility of service-independent physical network platforms to enable network operators and customers to purchase equipment from a variety of vendors;
- specification of high or low functionality and type, that is, transport, control or management, embedded or external;
- geography and ownership.

An alternative view embracing these principles is shown in Figure 8.6.

Ideally, meeting customer requirements, which dictates service definition, should drive network architecture. This, in turn, enables the functional architecture to be specified in terms of the elements and overall network. The physical architecture comprises the network elements, their system architecture and overall topology of their interconnection. Finally, the physical architecture of the geographic location and size can be defined to meet market demand in the most cost effective manner, together with the staffing and support systems to give the required QoS.

The modern digital telecommunications network is analogous to a Rubik Cube. Its structure can be considered as a series of layers in all of the vertical and horizontal planes and, like the Rubik Cube, all dimensions of the planes have to be properly aligned to achieve a viable solution. The network comprises a bearer transmission network, which carries a number of separate functional networks each in turn carrying a number of network services, see Figure 8.7.

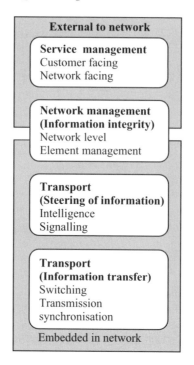

Figure 8.6   Network architecture

The principal functional networks are the PSTN, Private Circuit (PC) network, Telex network, Public Data network (PDN), Visual network used for video-conferencing and the mobile cellular radio network. These functional networks use a common transmission bearer network to interlink their switching centres and the local access network or Loop connects customers to them. A common physical layer provides the common infrastructure of ducts, buildings and power. The performance of the physical, network bearer and loop layers therefore impact on the functional networks and can affect the QoS supported by these networks. Access to the functional networks from the Operations and Maintenance Centres (OMCs), Network Management Centres (NMCs) and Data Collectors (DCs) for network and service management functions are by means of the Administration network. Gateways are provided to link some of these networks together to provide maximum transparency to customers to allow them, for example, to send messages over the public data network from terminals connected to the PSTN and, also, to link the functional networks to the international network, networks of Other Licensed Operators (OLOs) and third party Value Added Networks (VANs).

The layered model, shown in Figure 8.8, can best represent the architecture of the PSTN.

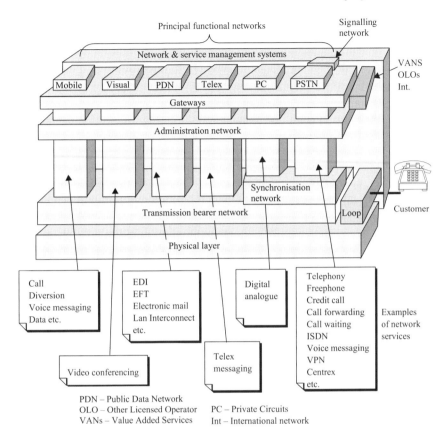

Figure 8.7   Physical structure of modern network

The layer beneath it supports each layer, as follows:

Physical layer – provides duct, buildings and power. Its impact on performance can be catastrophic in terms of, for example, building fires, power failures and excavation of duct routes by civil works contractors.

Transmission bearer layer – comprises the network of transmission links and flexibility centres (known as Repeater Stations in the UK) and the local loop that connects customers to the network. The transmission network can also be represented by the layered model shown in Figure 8.9, where the lowest 'Systems Layer' contains the point-to-point high capacity transmission (line or radio) systems; this supports the 'Capacity Layer' which contains the blocks of capacity, usually 34 Mbit/s in the BT network, which may traverse a number of transmission systems; the top 'Functional Layer' contains the various functional networks such as the PSTN, Telex, Data and Private Circuit networks. Each layer carries 'characterised' information in terms of bit rate and format and, conventionally, the higher layers are towards the user applications, whilst the lower layers are towards the Transmission media. The upper layers

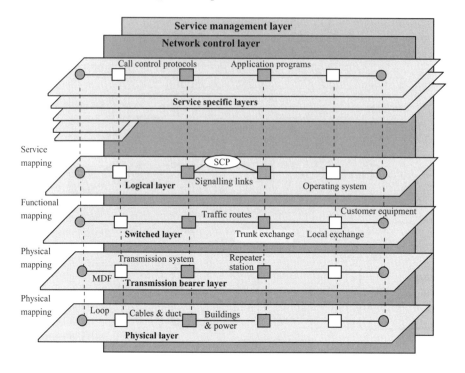

Figure 8.8   Architecture of PSTN

can be further classified into three groups namely; 'Circuits' which are normally end-to-end between users, 'Paths' which are blocks of multiplexed circuits, and 'Sections' which are specific to the transmission media.

For all layers, except the bottom, a link between two nodes in one layer, known as the 'Client Layer', will be carried by an end-to-end connection in the layer below, the 'Server Layer'. The process of transferring the characteristic information of a client layer on to a server layer is called 'adaptation'. The characteristic information in each layer has rate and format; the latter comprising the payload (the internal format of which the layer does not need to know) plus, in some cases, overhead information generated by the layer for use at the ends of the connection to ensure its integrity across the layer network.

Transmission impairments, such as, loss, error, noise and distortion can have a major impact on perceived QoS. Catastrophic failures can also occur when duct and cables are disturbed due to the operations of civil engineering contractors.

Switched layer – comprises the switching centres and their interconnecting traffic routes. The volatility of modern traffic requires congestion control, including Automatic Alternative Routing (AAR) and network traffic management, to minimise the effect on QoS.

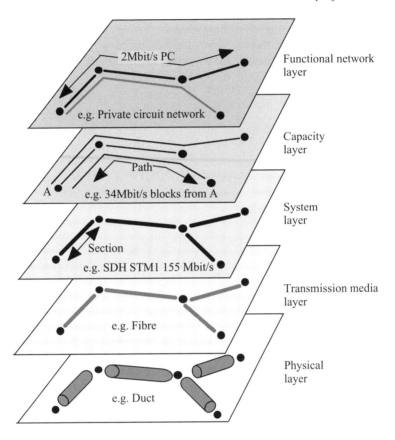

Figure 8.9   Layered architecture of core transmission network

Logical layer – the Integrated Digital Network (IDN) is analogous to a network of interlinked computers each having an operating system with specific application software packages to provide particular network services, that is, a distributed processing system. The logical layer describes the application software in the switching nodes in terms of generic and service specific attributes. Generic attributes are those common to all nodes, such as, switching a call between the input and output ports of an exchange. Service specific attributes are those necessary for the network to carry out a particular service. This layer also contains the generic functions of the IN platform that are embedded in the Service Control Point (SCP). The establishment of calls and services requiring specific features may require an interchange of information and instructions between exchange processors – this is carried out via the signalling network. The network control architecture is not well understood and can have a major impact on network integrity (see Chapter 9); for example, software problems in the signalling network have been responsible for some major network outages in the past.

Service specific layers – each service specific layer describes how the combination of generic and service specific attributes in particular nodes, together with messages between those nodes over the signalling network, create specific network services. For simple services such as local calls only the local exchange is involved, that is, the application software recognises the local number, routes the call to the called number and initiates charging. The design quality of the application software for individual services can have a major impact on their QoS. A particular problem can be 'feature interaction' where one service can have an undesirable impact on other services.

Network and service management layers – are orthogonal to the other layers to which they are connected for managing the network and the services it provides. The operational support software together with the competence of personnel staffing network and service management centres will clearly impact on QoS.

8.4 Architecture evolution

The development of the network architecture must be carried out with an understanding of the way the existing network is being evolved towards the future multi-bit-rate broadband network. To reduce the impact of competition from the information community and maximise new opportunities, the evolution of the telecommunications network will need to adapt to the new environment. Traditionally the Information Technology (IT) and Telephony world have been quite separate, but the increasing use of IT in business and the home in conjunction with communication, together with the emergence of networks, such as the Internet, that can transport both voice and data, has seen the convergence of the two worlds. The convergence will be speeded up as the information era is entered. But, as demonstrated in Figure 8.10, there are many differences between the two worlds.

However, the future network architecture begins to resemble that used by the IT industry where, for example, IBM, Compac etc. provide the hardware platforms which support a common (open) operating system (e.g. DOS); Microsoft Windows offers an open middleware platform which can be used by applications developers to efficiently develop and deliver and it also simplifies usage by common format and functions such as Open, Print, Save, Cut, Paste etc. It could be considered that the bit

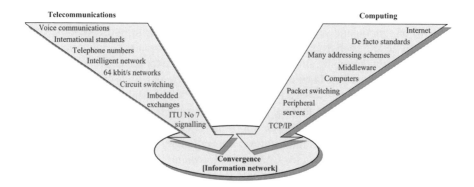

Figure 8.10   Telephony/IT convergence (Valdar [3])

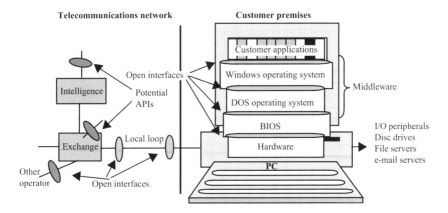

Figure 8.11   Open interfaces [Valdar]

transport network with its basic functions such as call control, redirect, calling line identity etc. is the equivalent of the hardware/DOS platform; the telecommunications middleware platform contains value add functions such as billing, authentication, network management etc. with open interfaces to the SCP and associated service creation and management environment. Hence, it is this middleware component that is likely to be the battleground between the Telecommunications and IT industries, given that the functionality can either be provided from within the network or at the periphery.

The difference between the IT and telecommunications approaches is illustrated in Figure 8.11 where a computer is connected to the PSTN.

It can be seen that the flexibility of the PC with its high functionality open interfaces has enabled thousands of innovative applications to be developed, all using well defined interfaces with good use of middleware features and rapid development of hardware that supports the old applications. Contrast that with the telecommunications network that has a few basic open interfaces with many of its services using service specific 'stove pipe' hardware and software platforms.

Network control can be considered as two layers, namely basic call and bandwidth control, and the middle ground for many of the functional groups such as navigation, access control, etc. which may be located in the network or at its periphery. Intelligent CPE can interact directly with services that interface with content. These functional elements can be considered as a layered architecture with open Application Programming Interfaces (APIs) between each layer, which allow any network operator or service provider to use any combination of layers in the stack. This architecture will be the same for national and global networks. The trend is, therefore, from the traditional 'Telco' world of application specific service logic, that is, stovepipes, to a new 'converged' world of application software that can be provided by any service provider, which uses general purpose reusable functions embedded in the network, that is, middleware. To allow the application software to use middleware requires an open interface, that is, APIs, as illustrated in Figure 8.12. Hence, the move is towards the computer world.

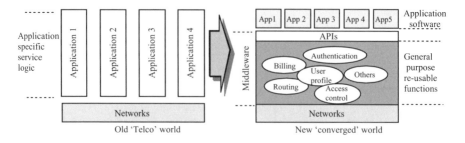

Figure 8.12   Convergence [Valdar]

The result of opening the network with APIs will result in the interworking of application software in the customer's terminal equipment with that in the network and also with that of independent and competitive service providers. It is equivalent to a distributed computing environment. This will drive the development of many innovative services; just as the opening of computer operating software by 'Windows' middleware has dramatically changed computing. Even now, almost any service provided by the embedded network can also be provided from outside of the network using the client/server technique with portable software, such as Java. However, contrast the present network, which is extremely robust against failures, with the fragility of the PC with its tendency to crash. This means that as the network evolves towards the structure adopted for PCs, it will be very important to take measures to preserve end-to-end integrity to ensure an acceptable QoS.

'The Parlay Group' [4], an industry consortium, consisting, amongst others, of BT, DGM&S Telecom, Microsoft, Nortel Networks, and Siemens, are developing an API specification that will accelerate the convergence of IP-based networks and the PSTN. This API, referred to as the Parlay API, will provide a common open interface into any kind of telecommunications network. The Parlay API will facilitate the seamless interworking of IP-based networks and voice networks while maintaining their integrity, performance and security. As such, it will stimulate a wide range of new services by giving IT applications developers controlled access to the functionality and intelligence of any kind of telecommunications network. It is a consortium that bridges the IT and telecommunications worlds and the group has established a WEB site at www.parlay.org. Work on developing APIs and interfaces within and external to switches to facilitate distribution of the functionality over multiple infrastructures is being carried out by the Multiservice Switching Forum (MSF) [5]. This is a forum comprising members from the IT industry, telecommunication operators and manufacturers [6].

The use of general-purpose functionality modules will enable a 'mix and match' approach to developing new services thus increasing the speed to market. Opening up the network by APIs will further stimulate development of new services by a multitude of independent service providers thus increasing traffic on the network and hence revenue, which will also accrue from interconnect charges from other licensed operators whose customers use services parented on the network. Opening

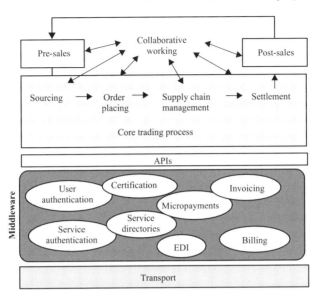

Figure 8.13   Facilitating e-commerce [Valdar]

the network by APIs is also fundamental to the introduction of e-commerce services. It will enable the distributed processing associated with trading in the information value chain to access general-purpose middleware in the network, as illustrated in Figure 8.13. Provision of such middleware by Telcos will enhance the value of their network thereby increasing revenue opportunities. In this case, network performance will have a major impact on the quality of the trading process.

The Trend towards adopting an IT view of the network together with the new environment of competition, regulation and information/e-trading services, requires a different view of architecture. This results in a number of architectural views as illustrated in Figure 8.14. The trading/services model provides a view of players who are involved in transactions that span the information value chain; they may influence the functionality required by the network. It could also represent the commercial model increasingly being used to offer international services. This view is vital in understanding the QoS requirements of the players themselves and also their impact on the QoS of end users. The techno/regulatory view identifies open technical interfaces that are specified by regulatory authorities. The logical and physical views represent the traditional view of network architecture.

An extension of the middleware concept is active networking where the network is no longer viewed as a passive mover of bits but rather as a more general computation engine where, in addition to carrying data, the packets traversing the network contain programs which invoke the switches to perform user-controlled operations supported by switches and Routers. This could be used to create, for example, 'spawning' networks, which is programmable networking capable of dynamically creating new network architectures on the fly. These networks are capable of spawning distinct

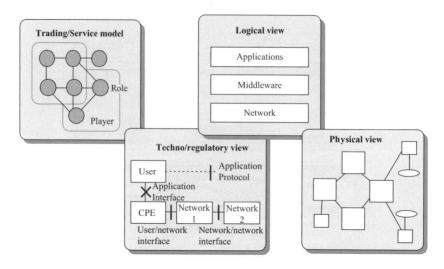

Figure 8.14   Architectural views [Valdar]

'child' virtual networks with their own transport, control and management systems. Such a child network operates on a subset of its parent's network resources supporting the controlled access to communities of users with specific requirements.

8.5   E-commerce

The rise of the Internet has caused a paradigm shift in the conduct of trading by the introduction of electronic commerce (e-commerce) that is, the selling/purchasing via an electronic communications medium. This can be via TV, fax, online networks and the Internet, particularly when facilitated by the ease of use made possible by the World Wide Web (WWW). The efficiency of trading is enhanced because much of the transaction processing is automated, for example, e-mail, online directories, trading support systems, customised services and goods, ordering and logistic support systems, settlement support, management information systems etc. It has also caused a paradigm shift from the traditional telecommunications function of providing a communication channel from a source to a destination to providing a channel between a seller and a buyer. Hence, networks will have the potential to provide applications that support e-commerce transactions and customer satisfaction will now be coloured by the whole purchasing/transaction process. These new requirements can best be appreciated by considering the value chain, shown in Figure 8.15, between end users and suppliers of information or goods.

The Information value chain starts with raw information that has the key attribute of timeliness but is not necessarily saleable if it lacks corroboration and context, via processing, to give it marketable contextual validity; it may then be stored pending

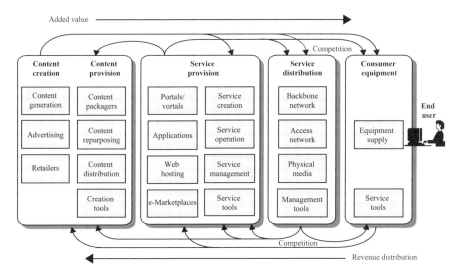

Figure 8.15   Information value chain

retrieval by the broker or end user. Information has no value unless it can be found, thus the search part of the chain is particularly important and must minimise the efforts of the end user. Supporting the chain are the various applications that facilitate manipulation of the information, for example, authoring tools for information creation and processing. The view of the information value chain presented in Figure 8.15 is an extension of the four-segment model of the information market adopted by Oftel in the UK, and the European Commission. Value is added down the chain towards the end user and the revenue is distributed up the chain, perhaps according to the value added by each stage. More information on the value chain and e-commerce is contained in Chapter 28. There will be considerable competition between actors populating the value chain in order to capture as much of the end user's revenue as possible.

There appears to be no accepted architecture that can be used for the design of information and e-commerce services and this chapter outlines some initial thoughts on how it might be developed. Spanning the value chain and orthogonal to it is a layered service architecture, shown in Figure 8.16.

The lowest layer is the *communications infrastructure* and represents that which is necessary to transport information to and from the end user. A number of architectures have been proposed by the various standards bodies and industry forums, but this layer is primarily the core and access elements of the service distribution part of the value chain. However, the layer also encompasses communication between the other elements in the value chain as well as customer private networks, for example LANs, WANs, Intranets, Extranets, PABXs etc. The performance of this layer includes traditional telecommunications parameters, outlined in Chapter 9, and affects the quality of information presented to the user in terms of, for example, the clarity of pictures and latency in retrieval of information.

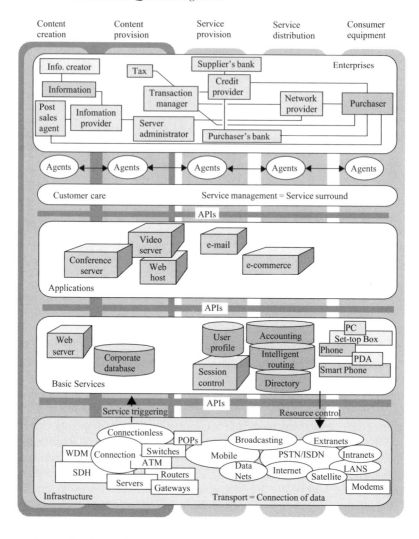

Figure 8.16   Service architecture

The *services layer* contains basic services and application support triggered, in some cases, by the communications layer for routing and bandwidth instructions etc. Basic services are generic to all services and this layer contains the applications and end user equipment necessary to find the broker/supplier/information and deliver/bill. Performance issues related to this layer encompass the accuracy of navigation and billing etc.

The *applications layer* contains the systems and hardware necessary to realise the information and e-commerce services that will be provided over the information infrastructure. The user's transaction experience will be heavily influenced by the distributed processing environment of this layer.

Customer care is provided by the layer containing the operational support systems for service management and processes to provide quality service surround. Network management across a multi-player, multi-domain, multi-system, multi-vendor transport layer and distributed processing system will cause unique problems. Distributed system management involves monitoring the activity of the system, making management decisions and performing actions to modify the behaviour of the system. The problem will be exacerbated when it is an active, programmable system where the system can be dynamically modified by potentially multiple users. This requires monitoring and visualisation of current state, audit trails and accountability to determine ownership of programs and network components, which may require assurance and certification. It is likely that agent technology will be used to manage this type of complex distributed system with 'policies' on who can use network resources; authorisation policies specify what actions an agent is permitted or forbidden to perform on a set of target objects, obligation policies then specify what actions must be performed by an agent on an object.

Service management is equally difficult and calls for intelligent systems to figure out what service centre should be accessed to deal with certain problems. There is a clear opportunity for outsourcing to take over such responsibilities.

The hypothetical *enterprises layer* contains the trading model (processes, agencies, resources etc.) necessary for e-commerce in the electronic market place. It is used to define the distributed processing requirements for the application layer. Understanding e-trading requires a sophisticated model, the enterprise model, in order to analyse the organisational environment in which the actors are involved, their requirements and roles, and how they relate to each other. The enterprise model occupies the enterprise layer, 'enterprise models depict a set of entities and identify the contractual and commercial relationships which exist between them' (Martin, Cobra). Considerable original work has been carried out on enterprise modelling by the European collaborative research programme known as COBRA [ACTS AC203].

Open APIs are used to interconnect the layers to provide flexibility for creation of innovative services by independent agencies. The end user interface is simplified by the use of collaborating electronic agents in the elements of the value chain.

Today's network has historically evolved into a series of 'stovepipes' where a mix of service and transport platforms have been introduced as new services and technology have been introduced. For example, consideration of the existing transport network illustrated in Figure 8.17 shows a variety of platforms and interfaces tailored to meet the performance requirements of individual services. The complexity of network and service management is high, creating expense and difficulty in meeting customer QoS requirements. As the transport network evolves it becomes progressively simpler, thus improving performance and QoS.

A possible view of the future network is illustrated in Figure 8.18. Since the QoS for the end-to-end transaction depends upon the distributed processing environment that spans the information value chain, it follows that the future network must

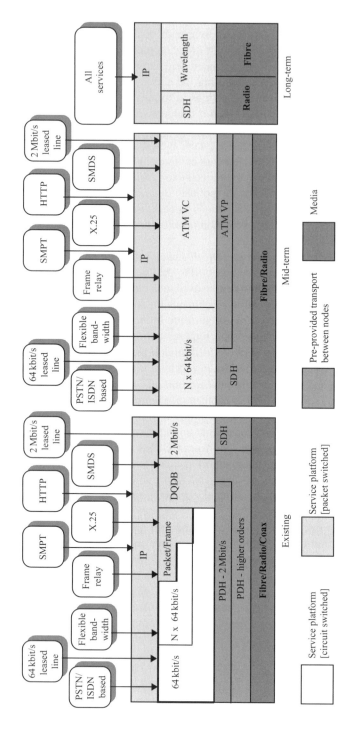

Figure 8.17   Today's network [Valdar/Ward]

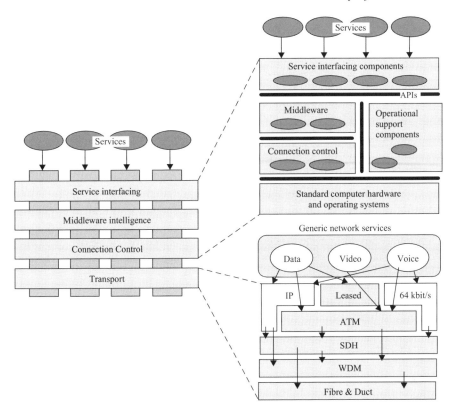

Figure 8.18   Future network [Valdar/Ward]

also encompass the full value chain. However, the extent to which application software is embedded in the telecommunications operators network will depend on many technical and operational considerations.

8.6   IT viewpoint

It should be recognised that although the telecommunication and IT worlds are converging, their views on infrastructure are different. For example, an alternative approach to the future structure may be to consider the functional blocks required for the information infrastructure. The functional approach has been developed by the Digital Audio-Visual Council (DAVIC) [7] which has defined about 250 functions and their open interfaces. Combinations of these functions, when put together in a specific way, can provide individual services. However, there are ten core functional groups, which are basic to the needs of information services. These are shown in Figure 8.19.

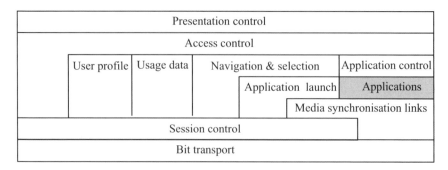

Figure 8.19   Principal DAVIC functional groups

Applications will invoke the core functions as required, or release or extend them with more specific functions (which may in turn invoke core functions). The core functions are

Bit transport – which provides the physical and logical links for the required bandwidth between the required points to be connected.

Session control – controls the bit transport functions, calling on them to establish or change a logical connection and determining the data rate and protocols to be used.

Access control – provides facilities to authenticate the user and verify access rights to the network, specific applications and related content, goods and services. Also provides verification of credit and payment.

Navigation, programme selection and choice – to enable the user to find and choose application or content, probably by the use of hierarchical menus.

Application launch – provides the facilities to run an application, obtaining and loading any necessary code (if not resident).

Media synchronisation links – provides links between objects, that is, sound segments, sub-titles, still and moving images and applications to achieve a multimedia presentation.

Application control – provides control of an application's behaviour, for example, pause, rewind etc. or content options/interactions.

Presentation control – for control of delivery and display of multimedia information, for example, sub-title activation, choice of language etc.

Usage data – the collection, storage and supply of data relating to user's consumption of material, resources and applications, for such things as payment, marketing and resource utilisation etc.

User profile – stores and utilises information about individual users and their behaviour to control access, assist navigation and correctly bill for services received.

Each of the functions has its own performance requirements which if not met will impact on the QoS seen by the end user.

8.7   Conclusions

The objective of this chapter is to present a network background to the subject of telecommunications QoS. A historic view of network evolution shows that the network evolves as a series of life cycles, each becoming progressively shorter. Current evolution is towards a multiservices IP/ATM network. However, the current expansion of the Internet is stimulating the growth of e-commerce and this requires the future network architecture to embrace the whole of the information value chain. Application software will now include all of that required by the distributed processing system needed for e-commerce transactions and QoS will be dependent on many more network performance parameters than hitherto. The chapter has presented a mainly telecommunications centric view of the future network architecture, but there are other views from the computing and information communities that must not be ignored.

Acknowledgement

Acknowledgement is made to Prof. Andrew Valdar of University College London for permission to use diagrams from his lecture handouts.

8.8   References

1   STANDAGE, T.: 'The Victorian Internet' (Berkley Books, 1998)
2   MEIJER, J. W. and GERAADS, J. M. G.: 'Convergence and Divergence in Business Communications,' 37th European Telecommunication Congress (FITCE '98) August 1998
3   VALDAR, A. R.: 'Next generation networks', International Conference on *InfoCommunication Trends*, Budapest, 2001, 11–12 October 2001.
4   Parlay.: web site www.parlay.org
5   MSF.: web site www.MSForum.org
6   WARD, R.: 'The Multiservice Switching Forum – An architectural framework for the 21st century' *The Journal of the British Telecommunications Engineers*, 2000, **1**(4), pp. 22–30
7   DAVIC.: web site www.davic.org

Chapter 9

Network performance engineering of legacy networks

9.1   Introduction

Network performance relates to the technical capability of the network in terms of the performance of particular combinations of items of equipment employed to deliver services. It is a major contribution to the QoS that customers experience when using the network. This chapter deals with network performance for legacy networks, that is, analogue and digital circuit switched networks. Performance aspects related to future ATM/IP networks and services are covered in other chapters, although the principles and some parameters are the same, for example, such networks often share a common transmission bearer network with legacy switched networks and hence their performance can be influenced by the transmission performance parameters.

Network performance is objectively measured by parameters meaningful to network operators and, in many cases, have a somewhat tenuous relationship to QoS perceived by customers. In some cases, several network performance parameters may contribute to the same impairment perceived by customers. Also, there are particular key parameters that fundamentally affect the quality of each service offered by the network, for example, error performance for data services and delay for voice services. These relationships are illustrated in Figure 9.1.

The relationship between QoS expectations and network performance parameters is shown in Figure 9.2. It should be noted that whereas performance parameters are well specified and measurable, QoS is a subjective awareness by customers that can vary depending on perceptions and expectations, and is very difficult to measure. Also, the term QoS is sometimes misused for network performance.

The dimensions of quality of service as seen by the customers and their relative importance are difficult to forecast and quantify. Unless QoS is specifically targeted in a Service Level Agreement (SLA), the non-technical segment of customers do not understand QoS parameters, hence targets are generally set in terms of past experience rather than theoretical considerations. Criteria include billing errors, waiting lists

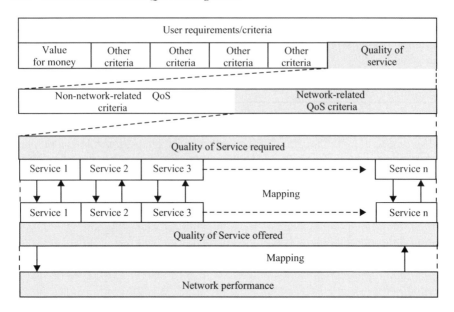

Figure 9.1   Quality/network performance relationship

for provision of service, wrong numbers, time to repair, poor transmission, fault incidence, failure of call completion and dial tone/post-dialling delay.

Performance impairments affecting switched services can be categorised as:

- *Transmission performance* – the ability to transport information between source and destination without distortion or undue loss and is mainly influenced by the performance of transmission systems.
- *Call processing performance* – this relates to the ability of a network to accept and interpret routing information from the customer and establish a connection to the required destination within prescribed response times. The main drivers are the performance of exchanges and signalling systems.
- *Availability performance* – this is broadly defined as the proportion of time for which satisfactory service is given.

Performance management is necessary to ensure that planned performance is achieved so that in a competitive situation the dominant Telco can retain, and perhaps gain market share by differentiating on the QoS delivered to its customers. Such a strategy requires good measurement of the QoS delivered as an essential feedback loop of quality management. Performance management must therefore:

- measure the QoS delivered to customers for each service offered. It must be recognised that this is just one of the four dimensions of service quality, namely quality delivered, quality offered, quality required and quality perceived. Also, the performance captured only represents those parameters that can be measured

Customer expectation	Network perfomance parameters
Negligible fault/no loss of service	Network availability
Minimal call connect failures	Connection establishment failure Resulting in no tone Misrouting of calls
Good transmission quality	Transmission loss Impulsive noise Psophometrically weighted noise Single frequency noise Loss distortion with frequency Group delay distortion Propagation delay Echo loss Stability loss Quantising distortion Sidetone loss Crosstalk Error performance Jitter & wander Slip rate
Connection establishment failure due to network connection	Grade of service
No cut-offs	Premature release of established connections
Rapid call set-up	Delay to dial tone connection establishment delay

Figure 9.2   Principal performance parameters

from the network and its operations and not the 'softer' dimensions of customer satisfaction;

- provide an overall check on the overall health of the network with data to assist the planning and operations process;
- review trends in the performance of individual equipment to assist purchasing and design decisions;
- measure efficiency, effectiveness and cost of key processes and workstrings;
- closely monitor QoS delivered to 'Other Licensed Operators' (OLOs) connected to the network, this is often a regulatory requirement.

Network performance parameters are established, generally taking account of ITU-T recommendations and economic considerations and are used for system design, network planning, specification and procurement of equipment and network operation and service specification. The parameters are often quoted in terms of 'design objectives' which are the values to be expected when equipment is operating in a defined environment and generally used in specification clauses in procurement documents. It is important to distinguish between design objectives and the various commissioning and maintenance values adopted operationally as illustrated in Figure 9.3.

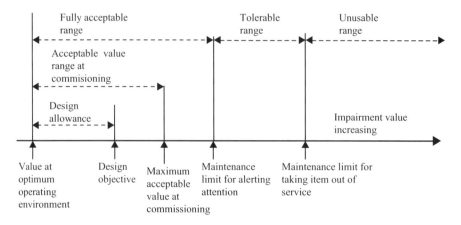

Figure 9.3   Performance objectives

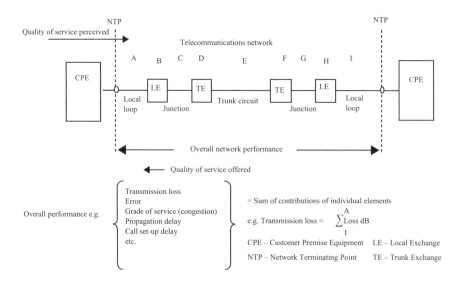

Figure 9.4   Apportionment of network performance

Overall network performance parameters are generally empirically derived and must be carefully apportioned over the contributing network elements to give the required target result. The apportionment is influenced by economic considerations, ITU-T recommendations [1] and the requirement to give the associated end-to-end QoS target to customers for international calls, calls between networks of other network operators and where calls terminate in customer private networks. The application of this principle for performance parameters is shown in Figure 9.4.

9.2   Performance parameters for transmission networks

9.2.1   Transmission plan

The network transmission plan is a set of network design guidelines recommended by ITU-T to ensure that end-to-end transmission is stable and acceptable to users. The recommendations lay down references against which national and international networks can be designed and cover such things as the:

- overall signal strength at various points in the connection;
- control of signal loss and the electrical stability of the connection;
- limits on acceptable signal propagation time;
- limit on acceptable noise disturbance;
- control of sidetone and echo;
- limits on acceptable signal distortion, crosstalk and interference.

9.2.2   Transmission loss

Transmission loss is defined as the ratio of the input power to output power expressed in decibels ($dB = 10 \log_{10} Pi/Po$). The loss across a voice connection comprises the loss in translating sound into electrical inputs to the network, the loss across the network(s) and that incurred at the receiver. With no loss the sound is too loud but too much loss results in an inaudible signal. To facilitate network design a system of 'Loudness Ratings' has been developed, namely:

- the Send Loudness Rating (SLR) is the signal loss expressed in dB (the standard unit of measurement, the 'decibel') when the electrical signal amplitude is compared to the original acoustic input, and
- the Receive Loudness Rating (RLR) of the receiver is the acoustic output compared to the electrical signal input.

Determination of these Loudness Ratings for specific telephone instruments (and associated local lines) used to be carried out by subjective comparison between these instruments and standard reference instruments [2]. Today, computerised electro-acoustic measurement equipment is used to objectively measure losses and calculate Loudness Ratings. The overall acoustic-to-acoustic loss between customers, or Overall Loudness Rating (OLR) is therefore the summation of SLRs and RLRs plus the network loss, that is,

$$OLR = SLR + \text{Network Loss} + RLR \text{ (dB)}$$

This is illustrated in Figure 9.5, which shows the reference points at which SLR and RLR apply.

The end-to-end transmission loss must be acceptable to customers on an international connection and the ITU-T G-series of recommendations specify the transmission performance of the national and international components of the links.

One of the features of digital transmission is that no loss is incurred over the digital path; however, connections with zero inter-exchange loss would result in calls being

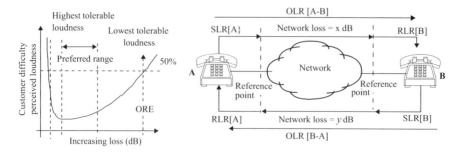

Figure 9.5   Derivation of overall reference equivalent

in a range approximately 6 dB louder than customers would prefer. Thus it is necessary to design the digital switched network to a 2-wire to 2-wire loss of about 6 dB
on all classes of connection, assuming line feeding and telephone instruments do not
change. With the Integrated Digital Network (IDN), the 2-wire local loop is converted
to 4-wire at the local exchange line termination unit and then into the digital format
with a net loss of 1 dB in the transmit direction; at the destination exchange the signal is
reconverted to analogue with a further 6 dB loss giving an overall loss of 7 dB between
reference points. Whereas a digital network gives superior transmission performance
to analogue, the inherent loss of the analogue/digital conversion in digital exchanges
can obviously degrade overall performance if they are introduced into an analogue
switched network. Likewise, there are undesirable losses introduced when interworking between analogue and digital networks, which constrains routing options.

Although the advent of digital networks has largely overcome the historical problems of transmission loss, it still represents an undesirable constraint on the design of
efficient network structures because of the losses in the local access network (local
loop). The two parameters that must be taken into account when planning the loop
are transmission performance and signalling resistance. Transmission performance
allows information to pass to the customer without extraneous noise and at an acceptable level. It is the attenuation of the line, expressed in decibels (dBs), measured at
1600 Hz and the limits are generally 10 dB for direct exchange lines and 8 dBs for
PBXs – the latter to allow for losses within the PBX.

Effective signalling from the customer equipment to the local exchange depends
largely on the resistance of the line. The maximum allowable signalling resistance
depends on the type of local exchange and customer equipment. Both the transmission
performance and signalling resistance are dependent on the length of circuit and
conductor gauge and can be expressed in terms of dB/km and Ohm/km for any type
of cable.

9.2.3   Stability

Insufficient loss around the 4-wire loop could result in positive feedback and the circuit oscillating in an uncontrolled manner – this is sometimes known as 'singing'. This
condition could occur if the balance of the hybrid transformer that converts the 2-wire

to 4-wire is disturbed thus causing a low loss between the 4-wire connections. It is therefore necessary to have 'stability loss' around the 4-wire loop to prevent instability [3]. Stability loss is defined as the lowest loss between equi-relative points at a 4-wire interface, from the receive to the send ports, measured at any single frequency in the band 0–4000 Hz. This measurement is applicable to all phases of the connection, for example, connection establishment, information transfer and release. A digital network has independent GO and RETURN paths, thus providing effective 4-wire working. Thus, the 6 dB loss provided on the single pair gives a stability margin of 12 dB.

9.2.4 Echo

Echos are caused by reflection of the speakers voice back from the distant receiving end due to an imperfect line balance at the hybrid that causes part of the signal energy transmitted in one direction to return in the other. The signal reflected to the speaker's end is known as *talker echo* and that reflected to the listener's end is called *listener echo*. The echo delay time is equal to the time taken for propagation over the connection and back again and is, therefore, related to the line length, the delay introduced by the switching systems, regeneration and signal processing. The obtrusive effect of echo on the customer increases with delay, ranging from a hollowness effect to large delay causing the speaker to interrupt his/her conversation and reducing the intelligibility of the received speech. Echo can be reduced by sufficient loss in the echo path. This is known as 'echo loss' and is defined as 'the loss between equi-relative level points at a 4-wire interface, from the receive to send ports measured as a weighted quantity in the frequency band 300–3400 Hz. The measurement is applicable during the information transfer phase of the connection'. The ITU-T recommends [4] that the mean value of echo loss presented by a national network should not be less than $[15 + n]$ where n is the number of circuits in the national chain. Within a small country, like the UK, the relatively high loss of the old analogue network gave satisfactory echo performance. But the reduced loss of digital networks together with the increased delay has required considerable care to avoid the expense of echo control. The 6 dB loss together with the loss across the 2/4 wire conversion (Hybrid) is usually sufficient to suppress echo but when digital networks are interconnected the additional delay could cause problems.

The ITU-T recommend that echo control devices should be incorporated if the one-way delay of the echo path exceeds 25 ms [5]. Echo control is required on very long distance calls, for example, inter-continental calls, and echo suppressors or echo cancellers carry this out. An echo suppressor is a voice-operated attenuator fitted on the path of the 4-wire circuit, which is operated by speech signals on the other path, hence interrupting the echo path. Unfortunately, if the connection comprises a number of circuits with echo suppressors connected in tandem, these can operate independently causing a lockout condition. Also the switching time of the echo suppressors is generally too slow for data. A more satisfactory echo control device is the echo canceller, which cancels the echo by subtracting a synthesised replica of it, derived from the received signal.

9.2.5   Delay

Propagation delay is an inevitable consequence of transmitting signals over distance and one-way delay is defined as the time taken by a signal applied at the input of an equipment (or connection) to reach the output of that equipment (or connection). Excessive delay not only incurs the risk of echo but also impairs communication. For speech, it leads to confusion because of the response delay, which can cause a speaker to repeat his sentence when the other party is responding. For data and signalling where the protocol requires a rapid response, retransmission may occur with a consequent breakdown of communication.

The problem is most acute on transmission via satellites in geostationary orbits where a one-way delay of 260 ms will occur leading to a pause between speech and response of about 0.5 s. The ITU-T recommends [6] that, except under exceptional circumstances, maximum one-way delay should not exceed 400 ms for an international connection. The one-way delay for connections originating and terminating in the UK is 23 ms. Connections that are routed within the UK network to, or from, an international gateway should not exceed a one-way delay of 12 ms. Within the digital network, delay is increased due to the switching mechanism (which writes and reads from buffer stores), multiplexing and other signal processing [7].

9.2.6   Error

An error is defined as a single bit inconsistency between transmitted and received signals. It is an important impairment in digital networks, particularly when they carry data services. Error parameters and target values that apply to the 64 kbit/s level of the digital hierarchy are specified by the ITU-T [8].

9.2.7   Slip

A synchronisation network is used to ensure that all digital exchange clocks are operating at the same average frequency. If not, the information rate of a signal received at an exchange would be different from the rate at which the exchange could process the information and retransmit it. This results in loss of information, if the input rate was faster, or repeated if it was slower. This process is known as 'controlled slip'. If the unit of information that is deleted or repeated is a complete frame (256 bits), it is known as frame slip. Slips will occur at each switching unit in the IDN at a rate proportional to the frequency difference between the incoming bit streams and exchange clocks and will cause error degradation to services, particularly data services. Uncontrolled slip is the unforeseen event resulting in the deletion or repetition of bit(s) resulting from a lapse in the timing process. Since it is indeterminate, it would normally lead to the loss and subsequent recovery of frame alignment with a potentially serious effect on the service carried. The maximum theoretical slip rate for a single plesiochronous inter-exchange digital link, where the clocks of the digital exchanges have a long-term frequency accuracy better than one part in 10^{11}, is one controlled slip every 70 days. ITU-T has produced recommendations for controlled slips in a hypothetical reference connection [9].

9.2.8 Jitter

Jitter is defined as short-term variations of the significant instants of a digital signal from their ideal positions in time. Jitter arises from the way that bit timing, associated with a digital signal, is extracted from the signal itself and on the 'justification' associated with higher order systems. Low frequency jitter can accumulate as the signal passes along a chain of systems. PCM-encoded speech is fairly tolerant of jitter but digital-encoded TV is much more sensitive to it. Jitter amplitudes are defined in terms of the nominal difference in time, expressed as unit intervals, between consecutive significant instants of the digital signal. Jitter (and wander) levels can be maintained below maximum limits by controlling the 'transfer characteristics' of jitter (or wander) between equipment input and output ports.

9.2.9 Wander

Wander is defined as the non-cumulative variations of the significant instants of a digital signal from their ideal positions in time due, for example, to temperature variations of the transmission media that, in turn, cause variations in propagation times. Typical variations, expressed in units of time, are 0.3–42 ns/km (monthly variation) and 0.6–77 ns/km (annual variation) for 140 Mbit/s system on optical fibre.

9.2.10 Sidetone

Telephones effectively contain a 2-wire/4-wire hybrid to isolate the transmitter from the receiver. Leakage between the transmitter and receiver is known as sidetone, and a small amount is necessary to make the telephone feel live to the user. An excessive level can make the talker speak too softly and also impair the received signal by excessive room noise transmitted as sidetone. Sidetone is defined as the proportion of the talker's speech that is fed back to his/her ear and is measured in terms of a 'Sidetone Masking Rating' (STMR), which is the weighted loss, in decibels, between acoustic interfaces at the transmitter and receiver. ITU-T recommends values of at least 12 dB [10].

9.2.11 Noise

The noise performance of a connection can best be described in terms of three separate parameters, namely,

- *Impulsive Noise* – occurs when a noise burst of a defined threshold power and duration is detected at the output of equipment. It is generally experienced as the 'clicks and bangs' one hears during a telephone conversation. The older switching and signalling systems together with poor joints in the local loop generally generate it.
- *Psophometrically weighted noise* – subjectively the effect is a relatively low level of background noise or 'hiss' which is present throughout the information transfer phase. It is measured as the total noise in the speech band of 300–3400 Hz with a filter, which has a weighting characteristic similar to the frequency response of

the ear. The noise is measured in terms of power and typical ITU-T objectives for analogue FDM systems are 3 pWOp/km [pW = picowatt and O indicates that the value is measured relative to the 0 dB reference point and p = psophometric noise] for line circuits and 200 pWOp for channel modulators.

- *Single frequency noise* – is defined as the noise power at a discrete frequency, generally measured in a 10–30 Hz bandwidth. This type of noise in the speech band can be very annoying and needs to be at least 10 dB lower than the psophometrically weighted noise to be masked by the background noise.

9.2.12   Quantisation distortion (quantisation noise)

The process of converting an analogue signal into a PCM digital signal involves a process of sampling the analogue waveform at regular time intervals and measuring its amplitude, then translating these amplitudes into a coded binary digital signal. Since the quantisation process turns a continuous signal into discrete voltage steps, it is not possible to reconstruct the original signal except as an approximate representation. Hence non-linear quantisation distortion is introduced which on voice communication appears as noise; therefore, it is often referred to as *quantisation noise*. The smaller the quantising levels the less will be the distortion, but this will require more digits in the code words representing the levels which increases the bandwidth required to transmit them. This type of distortion is expressed in terms of Quantisation Distortion Units (QDU). One PCM system using 8-bit coding with 256 quantizing levels is considered to produce 1 QDU, which is equal to a signal to distortion ratio of about 36 dB. When n such PCM systems are cascaded, then n QDUs are amassed. One 7-bit PCM system, with less quantisation levels, is considered to produce 3 QDUs, which equates to a 6 dB worsening of the signal to distortion ratio. ITU-T recommends a maximum of 14 QDU for an international link [11]. Given the dynamic range of speech, quantising distortion will be more apparent when soft speakers are communicating. Therefore, the distortion can be minimised by arranging for the lower signal amplitudes to have smaller quantising levels that is, companding. Two logarithmic quantising scales for 8-bit encoding have been defined by the ITU-T, namely: A law that is used in Europe and μ law used in America and Japan.

9.2.13   Loss distortion with frequency

This is defined as the logarithmic ratio of output voltage at a reference frequency divided by its value at any frequency in a specified band, with the input signal level constant. It is more loosely referred to as frequency response and observed as a variation of transmission loss with frequency.

9.2.14   Group delay distortion

When the time taken to traverse a system is a function of the frequency, signals at different frequencies arrive at different times and the variation in this delay is called group-delay distortion [12]. It is defined as the envelope delay measured at the output at a given frequency, compared with the minimum measured envelope delay at any frequency in a specified frequency band. The effect is also a differential shift in phase over the frequency band and this is inevitable when the circuit(s) contain transformers and coupling capacitors. Subjectively, the impact on speech transmission is negligible since the ear is insensitive to differences in phase. But the effect on a pulse shape is to spread it in time, hence if the signal waveshape is to be preserved (e.g. data and TV signals) then the distortion must be minimised.

9.2.15 Crosstalk

Crosstalk may be defined as the presence of unwanted signals coupled from a source other than the connection under consideration. The disturbing signal is sent over one channel and measured in the disturbed channel. For speech, even if low-level crosstalk is unintelligible, it can give the user an impression of lack of secrecy and it is therefore an undesirable reduction in quality of service. Crosstalk generally occurs in multi-pair cables in the local loop by capacitive coupling between the pairs. It can also be caused by non-linearity in Frequency Division Multiplexing (FDM) systems from which intermodulation products of frequencies in one channel may fall within the frequency bands of other channel(s). Crosstalk is primarily affected by

- the level of signal which if high is more likely to cause crosstalk into another channel;
- the frequency of interfering signal which is more likely to cause crosstalk as it increases;
- the balance of the circuits about earth, if each wire in a cable has equal capacitance to earth and to each other the crosstalk between pairs will be zero. Unbalance increases the risk of crosstalk as does balance to earth of the impedance terminating the line;
- the quality of the screening between circuits.

When crosstalk is transmitted over the disturbed channel in the same direction as its own signal, it is known as *Far End Crosstalk* (FEXT); when it is transmitted in the opposite direction to its own signal it is known as *Near End Crosstalk* (NEXT). However, crosstalk can usually be detected at both ends and NEXT and FEXT refer to the near and far end measurements. If the power of the crosstalk signal, measured at the near or far end of the disturbed channel, is $P1$ and the power of the disturbed channel's own signal is $P2$, then the

$$\text{Crosstalk ratio} = 10 \log_{10} P2/P1 \text{ (dB)}$$

The crosstalk ratio is usually measured when test signals of the same level are applied to both the disturbing and disturbed channels. The risk of crosstalk being audible is negligible if the crosstalk ratio exceeds 65 dB [12]. Crosstalk is most prevalent in the local loop due to capacitive coupling between parallel copper pairs. This coupling becomes greater as frequency increases, hence crosstalk becomes a major constraint when broadband systems, such as, ISDN and Digital Subscriber Loop (xDSL) are introduced in the local loop.

9.2.16   Local loop unbundling

Incumbent Telcos have a major competitive advantage in providing broadband services now that xDSL technologies allow high bit rates to be carried over the copper local loop. Regulators, for example, Oftel in the UK, therefore, require the incumbent to open up the metallic access network to enable others to provide services directly to end users, that is, local loop unbundling. The high frequencies used in DSL techniques can cause crosstalk interference between copper pairs in the same cable. These problems can be exacerbated in long lines, high power signals and where different vendors supply DSL equipment. To avoid such crosstalk problems (sometimes known as spectrum pollution), Oftel's cross-industry Network Interoperability Consultative Committee (NICC) DSL Task Group devised an Access Network Frequency Plan (ANFP), which defined the limits on power, frequency and location of DSL systems. The power and frequency limits are defined in a family of Power Spectral Density (PSD) masks, each designating the maximum power that can be transmitted at any relevant frequency [13].

There are four masks (Figure 9.6) each of which applies at a specific location in the loop, showing the line spectrum that is available to a DSL modem at that point. The current ANFP version only deals with Asymmetric Digital Subscriber Loop (ADSL). Also, it is not permitted to deploy ADSL in reverse configuration, that is, with the high bit-rate channel running from the customer.

The quality of the local loop, in terms of resistance, insertion loss, crosstalk and noise, determines the data rate that can be transmitted. But the quality of lines in

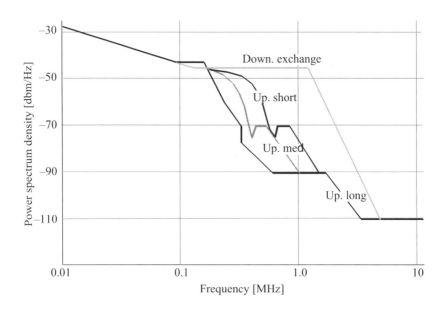

Figure 9.6   ANFP mask definition

an access network is very diverse, each pair being different. Hence there is a wide range of bit-rates that can be carried from 8 Mbit/s for short high quality to 2.5 Mbit/s for typical lengths to much lower where distances are long and quality low. This poses particular problems when deciding what ADSL bit-rate should be contractually offered as a service. There are also, regulatory problems when a loop is unbundled to allow competitors to lease pairs for their own high-speed access. The regulatory position is to ensure non-discrimination so a minimum loop quality needs to be set such that a reasonable proportion of pairs are suitable for unbundling. If the standard is too low, a large proportion will be available but there will be no guarantee that a high-speed service can be offered. If too high, a lower proportion of pairs will be available resulting in a low penetration of the service. Also, setting such a standard does not take account of future degradation due to, for example, spectrum pollution from other high-speed systems using pairs in the same cable.

9.3   Call processing performance

Performance parameters can be classified in terms of their impact on the processing of calls [7], that is, access to network (or call establishment) phase, or disengagement phase, information transfer phase and availability. Most parameters described so far have been mainly concerned with the information transfer phase and availability. Those affecting the call processing phases are

- *Delay to dial tone* – the time span from the instant a valid seizure signal is received by the network [local exchange] to the instant dial tone is returned by the network.
- *Removal of dial tone* – the ability of the network to remove dial tone on receipt of the first dialled digit. Failure to remove dial tone results in an unavailability condition.
- *Connection establishment delay* – the time span from the instant the last digit of a valid destination address is received by the network (originating local exchange) to the instant ring tone or the network returns number engaged tone.
- *Connection establishment failure due to network congestion* – Following the input of valid address information by the customer, connection to the required destination is not established due to network congestion.
- *Connection establishment failure due to no tone* – following the receipt of valid address information, connection to the required destination is not established and no tones or announcements are returned from the network.
- *Misrouting of calls* – the probability of the network connecting the call to a destination other than that requested.
- *Incorrect network tones or announcements* – the probability of the network returning a tone/announcement which is incorrect for the circumstances applicable.
- *Premature release of established connections* – the probability of the network releasing an established connection when the conditions for release have not been met.

- *Delay to connection release* – the time span from the instant the connection release request is received by the network to the instant all circuits in the connection are released and are ready to be used in establishing other connections.
- *Connection release failure* – the probability of the network not releasing an established connection after the conditions for release have been met.

9.4   Network congestion

A common network cause of poor QoS is traffic congestion. A network is of dimension to meet the peak demand, that is, the busy hour, with a small proportion of calls rejected due to insufficient equipment, because it is uneconomical to cater for the greatest possible traffic demand that could arise. This congestion [14] is known as the Grade of Service (GoS) β, defined as:

$$\beta = \frac{\text{traffic lost}}{\text{traffic offered}} = \text{probability of congestion}$$

Whilst modern circuit switched networks are resilient against small, unexpected surges of traffic because of the employment of routing techniques such as automatic alternative routing and dynamic routing, traffic is becoming ever more volatile due to, for example, media driven events such as television or radio phone-ins. The impact of such events can be minimised by network traffic management techniques such as call-gapping which restricts calls to particular destinations from entering the network, for example, 1 in n calls or 1 call every t seconds is allowed on to the network. Severe traffic congestion can cause problems in common channel (ITU-T No 7) signalling networks due to the high volumes of initial address messages generated by the high calling rates generated.

The dramatic growth of Internet traffic is a particular problem where access to Internet Service Providers (ISPs) is obtained via the PSTN. This arises due to the long holding time associated with 'surfing the net' over a network designed to meet a demand of short holding-time telephone calls, for example, one hour-long Internet session is the equivalent of the traffic generated by 20 three-minute duration telephone calls. Additionally, circuit switched networks where a dedicated connection from source to destination is held for the duration of each call, are very inefficient for Internet calls where the bursty nature of the data transported means that the connection is not used for a relatively large proportion of the time. Telecommunication Operators (TOs) are beginning to stream such traffic on to a more efficient IP network, where network resources are only used when information is transported, rather than installing additional circuit switched resources, which are becoming obsolete.

9.5   Network interconnect apportionment

When a connection spans more than one network, end-to-end performance targets must be apportioned over the interconnected networks. Figure 9.7 illustrates the

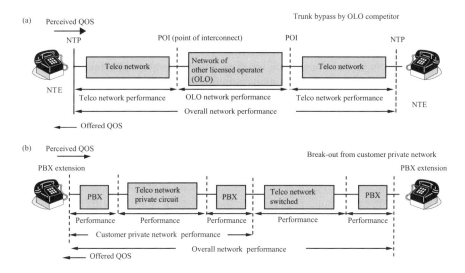

Figure 9.7   Examples of connections affecting network performance and QoS

increasing complexity of performance engineering when dealing with interconnected networks and the difficulties of apportioning a number of performance parameters over different networks to give a satisfactory end-to-end quality of service.

9.6   Availability

This parameter describes how much of the time a system or network will be available and operating, measured as the ratio of uptime to total time, as a percentage. From a customer perspective, unavailability is a more meaningful measure since it represents the time during which a failure is experienced, generally expressed as annual downtime. Some customers, particularly data users, are more concerned at the frequency of service disruption rather than the total downtime. Overall unavailability is heavily influenced by the reliability of equipment, which is generally measured in terms of Mean Time Between Failures (MTBF measured in hours or years), Mean Time to Repair/Replace/Restore (MTTR) and, sometimes, Mean Time to Failure (MTTF) which describes the uptime, or Mean Down Time (MDT). MTTR and MDT are usually measured in hours. For components and small assemblies, whose mean lives are too long to be meaningful, Failure Rate, usually measured as Failure in 10^{-9} Item Hours (FIT) is used.

In theory, the MTBF of an item of equipment can be calculated from the predicted reliability of its components [15]. The calculations become difficult due to design measures taken to improve reliability, such as redundancy. Such predictions are carried out by manufacturers to provide evidence of compliance to required MTBF specifications, but they are unreliable and generally do not match achieved MTBFs. Achieved MTBFs should be statistically observed over a sufficiently large sample.

Typical values are 100 km fibre system = 3 years and 2–8 multiplex equipment = 30 years. MTTRs can often be inferred from performance targets, for example, the percentage of faults cleared within x hours. They can be customer specific depending on the level of service care contracted, for example, a private circuit 'Reduced Charges Scheme' may guarantee that customers are rebated up to 100% of rental charges if six faults during a year are not cleared within five hours.

It can also be assumed that a normal statistical distribution of faults obtains in which case the tail customers of the 'bell' get an abnormally bad service, hence 'tail management' is an important aspect of performance management.

The impact of unavailability on customers may be expressed as

- *Unacceptability* – where the user deems the service effectively unavailable due to the poor performance, for example, bit error rate that is encountered. However, this is user specific since error on speech is not serious but can have a major impact on a data connection.
- *Partial inaccessibility* – where the customer is unable to access parts of the network due to, say, congestion or isolation of a local exchange or isolation from another part of the network.
- *Total inaccessibility* – occurs when the customer is completely isolated from the network. This could impact on one customer for a short break, for example, ITU-T define a complete failure on a 64 kbit/s connection when the error ratio exceeds 1×10^{-3} in each second for more than 10 consecutive seconds. More serious examples impact on groups of customers, the larger the group the more serious the isolation on a group of customers, and since business and society is becoming ever more dependent on telecommunications, the impact of service outages become more severe. In particular, total network failures, sometimes referred to as 'brownouts', can have a major economic impact on a country. Moreover, the more technology sophisticated the network becomes the more fragile it is to such occurrences. Hence, it is essential that future network designs are resilient and self-healing. Prof. Peter Cochrane (BT) has postulated that network disasters are analogous to earthquakes and can be categorised in terms of the Richter scale [16]. Total network information capacity outage (loss of traffic) in customer affected times is thus

$$D = \log_{10}(NT)$$

where N = number of customer circuits affected and T = total downtime. The qualitative ranges of media, regulatory and governmental reaction relative to this disaster scale are illustrated in Figure 9.8. However, the problem is exacerbated when networks are interconnected.

Such major failures are often classed as *major service failures* and the mean time between them is known as MTBmsF. They are sometimes defined as a failure that results in either exchange isolation (that is, the loss of all incoming or outgoing service or standard tone(s) for 30 s or more), or restricted service from the loss of 30% or more circuits in a route, loss of access to 50% of exchange traffic carrying capacity or loss of service to 50% or 500 customers.

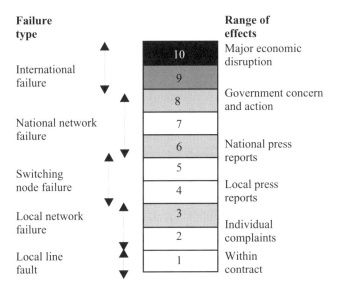

Figure 9.8   Network outage disaster scale

9.7   Network integrity

Network integrity can be broadly defined as 'the ability of a network to retain its specified attributes in terms of performance and functionality'. Clearly, isolated incidents that only impact on a small number of customers should not be classed as a breach of network integrity. It is, therefore, generally accepted that the definition should focus on major outages or performance degradation affecting large numbers of customers. The exception is where loss of an important, say, IN service is experienced where a relatively small number of major business customers subscribe to it. In the USA, the Federal Communications Commission (FCC) defines an outage as 'a significant degradation in the ability of a customer to establish and maintain a channel of communications as a result of failure in a carrier's network'.

The widespread use of processor controlled technology in the networks with high capability common channel signalling and 'intelligent' platforms create an increasingly unstable equilibrium where comparatively minor perturbations can cause severe network outages. The vulnerability of these new technologies is exemplified by the severity of the signalling related failures (sometimes known as 'brownouts') that afflicted several network carriers in the USA during 1990/91. For example, in January 1990, over half of the AT&T network (in the USA) failed for a period of around 9 hours due to a software problem, causing severe disruption to long distance services [17]. The outage was due to a single-bit 'soft-glitch' cascading through the SS7 network (ITU-T recommended Signalling System No 7), which was subsequently found to be a single AND instead of an OR condition in several millions of lines of software code [18]. In June and July of 1996, there were major outages that affected millions of customers of Bell Atlantic and Pacific Bell (USA) caused by a faulty software upgrade (supplied by DSC Communications Corporation) to SS7 Signal Transfer Points (STPs)[19].

These incidents prompted the US Congress to severely criticise the FCC for not being proactive in protecting the communications infrastructure [17]. They concluded that

- The public switched networks are increasingly vulnerable to failure: and the consequences for consumers and businesses and for human safety are devastating.
- The problem of network reliability will become increasingly acute as the telecommunications market becomes more competitive.
- No Federal agency or Industry organisation is taking the steps necessary to ensure the reliability of the US telecommunications network.

Subsequently, the FCC established the Network Reliability Council and also gave the remit to Alliance for Telecommunications Industry Solutions (ATIS) to produce monitoring and measurement proposals [20]. The FCC has imposed a reporting regime which requires them to be notified of all outages of at least thirty minutes and affecting 30,000 or more customers, plus outages affecting emergency services and key facilities.

The potential problems due to network complexity are exacerbated when networks are interconnected and this is becoming a major regulatory issue since such policy is aimed at allowing unrestricted interconnect between the networks of established network operators and competing operators/service providers. An example is the European Union Open Network Provision (ONP) policy. The European Commission has recognised the risks of this policy to network integrity and has commissioned work to determine how the risk can be minimised. In the UK, the regulator (OFTEL), has established the Network Interoperability Co-ordination Committee (NICC) as the consultative forum to advise the regulator in relation to optional and essential interfaces considered important in the progressive deregulation of telecommunications systems and services.

9.7.1   Network integrity definition

Since the definition of network integrity is 'the ability of a network to *retain* its specified attributes in terms of performance and functionality', it considers integrity in terms of robustness, invulnerability and incorruptibility. In this sense, the degree of integrity of a system is inversely related to the risk of failure, that is, the higher the degree of integrity, the easier it is for the system to maintain its attributes, and hence the less likely it is that unexpected perturbations can result in failure. Figure 9.9 [21] shows that the degree of integrity can be divided into a collection of states of integrity. At one end there is 100% integrity, which means that the system is absolutely robust and will not be affected by unexpected problems. At the opposite end, there is 0% integrity, meaning that any perturbation will cause failure. Between those two limits, the network can be at different states of integrity.

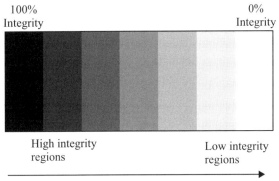

Figure 9.9   Integrity definition

The different bands between 0% and 100% integrity should be represented as a continuous gradient. However, this is not feasible in practice. Due to the extreme complexity of the problem, the parameters introduced to measure integrity must take discrete values, and hence the need to introduce real boundaries between bands with different values of integrity. The separation between these bands will not always be strict and there may be overlaps. Breach of network integrity and failure are related concepts, but network integrity goes beyond failure identification. Loss of integrity means moving from one band to another band of lower integrity, but this does not necessarily imply failure. Breach of integrity would imply failure only in one of the following cases:

- if the system was initially in a band of very low integrity, because any minor perturbation would put it in a situation of failure; hence, it is important to operate in a high integrity band;
- if the loss of integrity is very high, causing the system to move from a band of acceptable integrity to the region near 0%; in some cases loss of integrity cannot be avoided, for example, there may always be unforeseen problems which, depending on the state of the system might have a chain reaction effect, leading to serious failure; but it is desirable to minimise such instances by identifying possible sources of failure and designing them out, or monitoring the relevant parameters to take avoiding action.

9.7.2   Network control

Modern networks are software controlled and the services they provide are established by the interaction of application programs in appropriate network elements via messaging over the signalling network. Interactions also occur between support systems and embedded network intelligence for the purpose of service and network management. In the future, it is increasingly likely that information technology (IT) applications in customer terminal equipment for multimedia services will require heavy interaction with network control. Should invalid messages be exchanged between application programs then miss-operation can disrupt the normal operation of the network. In extreme cases, such a malfunction can cause invalid messages to propagate through the network, closing it down. Hardware faults, software 'bugs', operational errors, deliberate sabotage or incompatibilities in signalling and/or software can cause such problems. When constituent parts of the networks of different TOs and service providers (SPs) are interconnected the probability of such perturbations occurring rises causing threats to network integrity. Moreover, the behaviour of such interconnected software resources is non-linear and possibly chaotic and hence almost impossible to predict. Network integrity is therefore a control-engineering problem and one of the more important aspects of modern Performance Engineering.

There is an analogy to the personal computer (PC) on one's desk; almost everyone has experienced a 'crash' on a PC due to a faulty application program or unusual set of circumstances. But it is quite different dealing with a PC problem compared to that where a whole network has crashed. A major concern is the possibility that 'hackers' might introduce viruses into the network control systems, as they do with PCs [18].

A network is considered as a collection of distributed entities performing different tasks and interacting with each other to provide the required service functionality. The behaviour of such a system can be represented as an extended finite state machine [21]. The behaviour path is a set of states, and the transitions between them, are illustrated in Figure 9.10. A path of correct behaviour consists of a sequence of acceptable states. Under adverse conditions, the system might leave this path and move to an erroneous state. Some errors will not produce a major deviation from the path of correct behaviour and the system remains in the 'shaded' area of acceptable behaviour. Such minor deviations can be classed as Type 1 when they return to the correct path, Type 2 where they are detected and cause the process to stop and Type 3

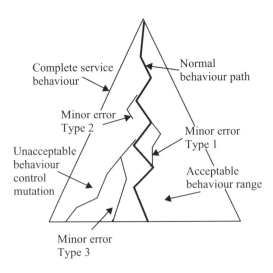

Figure 9.10   State transitions

that terminate in a final state outside the normal path but still within the safe region. Conversely, major errors or combinations of errors can cause 'control mutations' which take the sequence outside of the region of safe behaviour and result in failures.

The worst problems are likely to arise when a chain of errors occur, hence the need to identify possible sequences of error states that will progressively take the system away from normal behaviour. The correlation between different error states, that is, the probability that one error state leads to another, forms a large and complex decision tree. Regardless of the extreme improbability of entering a dangerous control sequence, identification of such sequences is extremely important. This is because a system operating for a significant length of time will explore a considerable fraction of all the possible sequences. Hence, it will eventually enter this dangerous region with catastrophic consequences. In summary, unlikely events are inevitable, provided enough time elapses, and if such unlikely events lead to a threat to integrity, the threat will manifest itself. The key point is to rapidly identify them and take the appropriate actions before the system falls over.

9.7.3   Network interconnect

The risk of breach of network integrity is increased when networks are interconnected and this is becoming a major issue now that regulatory policy is allowing many competitive operators and service providers to interconnect to the networks of established TOs. Hence, it is necessary to consider network integrity in the context of network interconnect. In the future environment of highly sophisticated Intelligent Network services, an increasing number of Service Providers (SPs) will also require access to fixed and mobile networks to offer specialised services. It can also be expected that SPs will offer value-added services from their own service platform connected to the transport networks of other operators to gain access to customers and route calls. Independent information providers will supply information services, particularly in the multimedia era. Hence, in an open network environment, customers will be able to gain services from a variety of interconnected constituent parts of TOs' and SPs' networks and service platforms, the points of interconnect being determined by a variety of technical and economic factors.

The most common form of interconnect is related to the Plain Old Telephone Service (POTS) where interconnection of Public Switched Telephone Networks (PSTNs) and mobile cellular radio networks to PSTNs are already successfully operational using ITU-T No 7 signalling. In the near future, the range of interconnected services will probably be extended by services requiring the transmission of Calling Line Identification (CLI) signals between networks. Intelligent services will follow with, for example, regulatory requirements for number portability between networks and the, so-called, CS 1 services such as virtual private internetworking. Likewise, the interconnection of SDH networks will require particular attention to be paid to control overhead aspects.

As increasingly sophisticated services are operated between networks, so signalling activity and interface complexity increases. It can, therefore, be concluded that there is an increasing risk of network integrity being compromised by problems arising from interconnected networks and equipment connected to the network. There is a broad correlation between the complexity of an interconnect, in terms of such aspects as signalling activity and the number and range of network elements that are accessed to provide a particular service, and the degree of risk to integrity, for example, POTS would be low complexity but IN based services, such as VPN, require a significant volume of transaction signalling to various databases. Clearly, conditions of interconnect will be influenced by risk and hence complexity. Thus the more complex services will require more demanding interconnect conditions in terms of, for example, testing, screening etc.

9.7.4   Measurement of failure impact

A broad definition of integrity is 'the ability of a network to retain its specified attributes in terms of performance and functionality' but there is a need to define network integrity in a manner that will allow network failure due to loss of integrity to be identified and preferably measured. The requirement to provide a quantitative definition arises because of

- the need to identify when a breach of network integrity has occurred;
- the need to understand the magnitude of the loss of integrity in terms of the impact on network customers;
- the need to construct meaningful SLAs or contracts between network operators for network interconnects, or impose regulatory conditions that can be arbitrated in quantitative terms.

The main problem in deriving an index is to quantify the impact on customers connected to the network(s). Questions that need to be addressed include

- Does a local exchange outage of x hours duration affect all customers connected to the exchange, given that only a proportion of them will attempt to make a call during that period?
- Does loss of connectivity to the rest of the network merely affect those who attempt to make calls during the outage period and what is the difference between partial and total isolation?
- What is the difference between a total outage and severe degradation of service where, say, loss of network synchronisation causes slip with severe error performance that has a different impact on speech than data?
- Is the impact on business customers greater than that on residential customers?
- What is the impact of loss of interconnect between networks? Is it different for fixed–fixed, fixed–mobile, PSTN–data etc.?
- How do you determine the impact on customers of different services? Does the tariff reflect the value of the service to the customers?
- How do you equate the difference between a large highly penetrated network and a small low penetrated network?

Three possible measures of network integrity have been identified, but none adequately answer the above questions. They are the 'Cochrane "Richter" Scale', the McDonald 'User Lost Erlang (ULE)' and the Alliance for Telecommunications Industry Solutions (ATIS) 'Outage Index'.

9.7.4.1 'Cochrane' Richter scale

Prof. Peter Cochrane (BT/UCL) has postulated that network disasters are analogous to earthquakes and can be categorised in terms of a Richter type scale, see Figure 9.8. The advantage is that the measure is easy to comprehend even though it has a nominal numerical value. It is impossible to derive an absolute value without having to specify outage in huge numerical values and the logarithmic scale (like the Richter scale for earthquakes) reduces the numbers to manageable sizes. However, for failures in the core of a network it would be necessary to convert, say, lost calls into affected customers as is done for the ATIS index (Section 9.7.4.3). It would also need modification to deal with outages of important individual services with low traffic volumes.

9.7.4.2 User lost erlang

The parameter called ULE is defined [22] as

$$ULE = \log_{10}(E \times H). \text{ for } E \times H \text{ greater than 1, where}$$

E = estimated average user traffic lost during the time of the outage in erlangs taken from historical records

H = outage in hours

Thus 1 ULE = 10, 2 ULE = 100 and so on. Hence, the unit is logarithmic (like the Cochrane Richter scale), represents the societal impact in terms of calls affected and the outage duration, and is easy to measure. It can also be translated into a measure of the estimated number of customers affected by dividing the erlangs by average calling rate in erlangs per customer line. The definition does not provide a measure of the geographical spread of the outage, which, for a small network could be extensive even though the value of ULE is modest.

Although loss of network integrity is commonly assumed to be the result of unavailability it can also be the result of degradation of other performance parameters that affect large numbers of customers, for example, poor error performance or the unavailability of a service or services. In such circumstances the ULE parameter could still be used as a measure of lost integrity by redefining E as 'the estimated average user traffic lost or of unacceptable quality of service during the time of the service disruption, in erlangs' and H 'the service disruption in hours'. The use of erlangs reflects the actual loss of usage rather than the absolute number of customers; it is also easy to calculate and comprehend and handles customer and traffic outages. However, it would need modification to deal with outages of important individual services with low traffic volumes.

9.7.4.3 ATIS index

The FCC in their February 1992 Report and Order challenged the telecommunications industry to develop a scientific method for quantifying outages. The Network Reliability Steering Committee passed the remit to the T1 Standards committee (both

are committees of the ATIS). The ATIS T1A1 Working Group on Network Survivability Performance proposed that the index [23] should be based on a combination
of service(s) affected, duration and magnitude (customers affected). 'In particular, it
should

- reflect the relative importance of outages for different services;
- be able to be aggregated to allow comparisons over time; and
- reflect small and large outages similarly to their perception by the public'.

An important objective of the exercise is to enable the summation of individual
outage index values, over time periods and relate it to the FCC reporting requirements
(<30 minutes, $\geq 30,000$ customers), for comparison purposes. Aggregation should
reflect the importance of outages to customers such that the aggregate index for
multiple small outages is less than the index for one large outage where its duration
(or magnitude) equals the sum of the small outages. Likewise, the aggregate index for
multiple large outages should be greater than the index for one very large outage over
the same measurement timescale (one year). The T1A1 have introduced a general
framework for quantifying service outage from a user's perspective. The parameters
of this framework are

- the Unservabilty (U) of some or all of the services affected by the failure, in terms
  of units of usage, for example, calls;
- the Duration (D) for which the outage exists;
- the Extent (E) in terms of the geographical area, population affected, traffic
  volumes, and customer traffic patterns, in which the unservability exceeds a given
  threshold.

The service outage can be categorised by sets of values for which the (U, D, E) triple
qualify for particular values of outage. This is illustrated in Figure 9.11 where the
triples are categorised in three regions of minor, major and catastrophic.

The index of an outage is the sum of the service outage index values for each
service affected. The service outage index of each service affected is the product of
the service weight (Ws), duration weight (Wd) and magnitude weight (Wm). Thus,
the outage index $I(O)$ for an outage O has the following form:

$$I(O) = \sum_{j=1}^{N} \text{Ws}(j), \text{Wd}(j), \text{Wm}(j)$$

where $j = 1, \ldots, N$ are the services.

The method is simple to understand and operate, is based on the FCC reporting
requirements and is weighted to be used to aggregate outages over a time period.
However, the assumptions are coarse and based on broad averages, for example,
a caller makes two re-attempts when blocking occurs, there is no distinction between
business and residential customers. It is currently used for a limited range of 'POTS'
services and does not differentiate between residential and business customers who
have different perspectives of the importance of telecommunications.

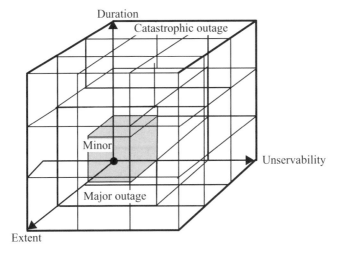

Figure 9.11   (U,D,E) qualifying regions

9.7.5   Risk assessment

Ideally the assessment of risk should be met by an assessment based on a model, which would assess the level of risk to integrity and the threshold criteria would determine its acceptability. Such models do not exist but a pragmatic and broad-brush methodology is illustrated in Figure 9.12 [20].

9.7.6   Development of new services

The process, from inception to launch of a new service starts with the feasibility study on which are based the requirements specification of the overall system. The architectural design sets the parameters within which the detailed design is carried out and the different components of the architecture and their interoperability are considered. This is followed by the detailed unit design and coding. Each of these processes requires a different level of testing. At the most detailed level, testing must ensure that the individual coding works. Then, interaction testing is necessary to ensure that when two or more components are put together they perform as expected. When the overall system design is tested the acceptance testing is carried out not only to ensure that the system works but that it also conforms to the requirements. The launch phase requires testing in an operational environment as well as trialing the in-service support processes.

At the customer requirements definition stage, consideration should be given to how the service will be used by customers, in particular the potential for undesirable feature interactions with other services. Early in the design stage, modelling should be carried out to assess the risk of feature interaction, that is, the undesirable interaction of services, and integrity violation, the objective being to remove problems at the early stages of the design cycle rather than the time consuming and costly method

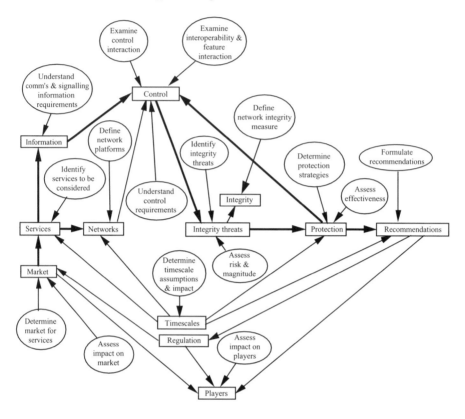

Figure 9.12   Risk assessment process

of identifying problems when the development is completed. At present, services are more often designed in isolation, without taking any account of existing services, and therefore may give rise to interworking problems.

Testing should be carried out at appropriate stages in the development and acceptance testing should embrace not only testing of the completed design on laboratory models but also intra-network field testing of the installed equipment and/or software. Where the service is to be inter-networked via the networks of other operators, comprehensive non-intrusive testing should be carried out between captive models followed by intrusive testing of the inter-networked services, in the operational environment, before customers use it.

Due to the high complexity of the services, it is not possible to test everything. Testing must focus on those areas where problems are more likely to arise, whereas other aspects that can be assumed to be non-dangerous will not be included in the testing. The level of testing carried out must also be subjected to risk assessment. Since it is clear that these limitations will exist, it is of paramount importance to understand what their consequences are. For this reason, a risk assessment methodology that helps to assess the probability, impact and consequences of actions must be developed and integrated into the development of systems. Risk analysis should be applied at the different levels of systems development, in order to determine the amount of testing that needs to be performed for different levels of risk. Increasing the amount of testing decreases the risks, but increases the costs and delays, so a balanced approach must be taken. The risks should be analysed, categorised and documented.

Part of this risk assessment activity would consist in keeping records of any constraints, decisions taken during the development of the systems and an assessment of the effects that these limitations would have in adverse conditions. Such an activity would

- provide a better understanding of the systems operations;
- make the information accessible to different people, instead of being only in the heads of those involved in the design process;
- help in identifying weak points in the systems and where problems are likely to arise;
- help in the diagnosis of failure and in a more rapid identification of the actions required to fix the problems;
- help in evaluating consequences of failure, for example, in terms of lost calls, damage to equipment, cost etc.

9.7.7   Integrity protection

There are a number of ways to protect the integrity of networks when they are interconnected. Some of the more common ways are

- *Gateway screening* – comprising software, usually embedded in gateway exchanges at the points of interconnect which contain a 'mask' of legitimate messages and therefore reject all others as invalid. Many operators use this for 'POTS' interconnect.
- '*Firewalls*' – the interconnection of more complex services employing, for example, transaction signalling may require more comprehensive protection, such as double firewalls. Typically this may be in the form of a Signalling Relay Point which, in addition to containing a mask of valid messages, can block messages to invalid signalling destinations (point codes), change signalling routing to avoid network 'hot-spots' and act as a fuse when unexpected signalling surges arise to prevent them from propagating into the network.
- *Mediation devices* – Whereas screening and firewall devices are used for blocking unwanted messages, mediation devices modify the content or structure of messages to achieve compatibility between source and destination. Such devices will probably be required for interworking of complex IN based services where interconnection is, say, at the SCP level.

9.7.8   Interconnect testing

In order to achieve seamless service operation with high integrity protection, a sequence of testing programmes must be carried out which encompass stand-alone

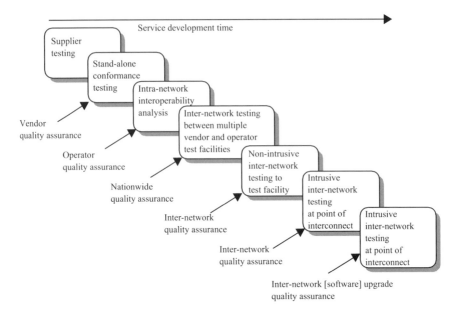

Figure 9.13   Testing sequence

technical analysis and auditing of individual systems, multi-supplier interoperability, as well as inter-network interoperability. The depth of testing will obviously depend on the complexity of the service and interconnect. Responsibility for these tests should be shared between the equipment vendors and the service providers/network operators (as shown in Figure 9.13).

The design for the equipment is usually in accordance with industry standards and the individual specifications of the client operator.

- The first step in the technical analyses is a thorough design review of the specifications. This concentrates on potential weaknesses, such as overload control algorithms, which have been identified previously. It usually becomes prohibitively expensive to correct a fault not detected at the design stage.
- The second stage involves stand-alone conformance testing on each system to verify the correct implementation of the specifications. This includes a check on the generic requirements and SS7 standards, capacity and performance, security, hardware and software quality, and, in particular, the feature and service capabilities. Any corrective actions can thus be performed prior to a product release.
- The third stage of testing considers interoperability necessary within the operators network, often between multi-vendor systems, for transparent service operation. Such tests are usually carried out within the operators test facility before intrusive tests are carried out in the network.
- Ideally, if network interconnection is likely for inter-networked services, the next stage of testing should be cooperative, between the interconnected test facilities

of a number of vendors and operators. For such multi-vendor/operator situations, diversity in the interpretation and implementation of the specifications can, in practice, lead to operational difficulties. Such testing could take weeks rather than days.

- Stage five arises when an application has been made for a specific interconnect, say, from a new entrant to an established PTO. If the switch were unfamiliar to the PTO, it would be reasonable to insist on interconnect testing to the PTO's test facility. In addition to straightforward interoperability tests, stress testing would probably be carried out to observe the reaction to such things as overload and fault recovery. Such tests and analysis could take 4–5 weeks.

- At the time of interconnect, intrusive tests should be carried out at each point of interconnect between the two 'live' networks. Such tests would typically comprise electrical and alarm tests together with test calls and SS7 (ITU-T No 7) signalling tests on each type of route (e.g. emergency, assistance, egress, ingress etc.) with diagnostic testers to check the message sequences. Typically, this might take 3–4 days for an experienced operator. Such tests may be reduced if other operators interconnecting to the PTO have used similar equipment without problems.

- If a new software generic is introduced or additional capacity/new routes added to the interconnect a reduced set of intrusive tests should be carried out.

Interconnect testing can be time consuming and expensive. Unlike Customer Premise Equipment (CPE) type approval, which is carried out by neutral bodies, interconnect testing is dictated by the dominant TOs. It is conceivable that TOs could impose over-elaborate testing that would discourage or delay other operators or SPs from connecting to their networks. In the USA, the telecommunications industry organisation, ATIS have produced standard and agreed test scripts, although their use is not mandatory.

In the future, it is conceivable that internetworked services may require the collaboration of resources in more than two networks. Hence, even though bilateral interconnect testing is satisfactory, the use of three or more networks to provide an inter-networked service may give rise to integrity violation in one or more networks.

9.7.9   *Software upgrades*

Digital exchanges and signalling transfer points are subject to regular changes in data and upgrades to software design. Hence, interconnected networks are in a state of constant change, which may invalidate previous interconnect testing. Furthermore, although standard interfaces may be assumed, in multi-vendor interconnected networks each manufacturer may interpret standards differently. Experience in the USA has shown that corruption of data and changes to software are a significant cause of intra-network outage, particularly for ITU-T No 7 signalling networks; it can be assumed that there are similar intra-network risks. Indeed, some operators insist on a reduced test programme being carried out when significant software updates are made at the point of interconnect.

In a multiple provider interconnected networks environment, it may be assumed that interconnect interfaces should be defined by standardised protocols, such as

ITU-T No 7. This implies that the individual software elements within the networks can be designed by different manufacturers and administered by different network operators. But, conformance to standards can still lead to incompatibility because each manufacturer may interpret protocol aspects differently. Given the continuous upgrade of software, somewhere in interconnected networks, compatibility between generic levels of software across the networks becomes difficult to maintain.

9.8   Conclusions

This chapter has briefly described the main network performance parameters which have a significant impact on customer satisfaction and, hence, in a competitive environment, market share. The increasing change to networks together with the introduction of new services makes performance engineering a complex task and it is necessary to issue detailed guidance to all concerned with the network and services it supports.

The most severe degradation that customers can experience is loss of access to the network due to its failure, the greater the network outage the larger the impact on customers. The modern digital network is, in effect, a very large distributed processing system with multiple varieties of exchange processor with a multiplicity of software operating systems and service applications. The complexity of such a system is high and rising with the growth of services and network interconnect between operators. Although standards bodies specify interfaces, it is not possible to produce a detailed unambiguous specification. Hence, application of standards depends on the interpretation of equipment suppliers. Therefore, the risk of incompatibilities is high and the probability of software driven network outages is large. Furthermore, events of the past have shown that even minor errors in software design can cause major service-affecting network problems when unexpected combinations of circumstances or perturbations occur. There is also a long history of 'hacking' into computer networks and the malicious insertion of viruses that can cascade through computer systems rendering them inoperable; there is no reason to assume that telecommunications networks are immune from such attacks.

It must be recognised that modern networks are fragile and that the design of networks and systems to maximise the preservation of network integrity is of paramount importance and will have an increasing influence on customer perceived QoS; particularly as networks become more complex and inter-networked services become more sophisticated.

9.9   References

1   ITU-T Recommendation I.350: 'General aspects of QoS and network performance in digital networks including ISDNs' (International Telecommunications Union, 1993)
2   McLINTOCK, R. W.: 'Transmission performance', in FLOOD, J. E., and COCHRANE, P. (Eds): 'Transmission systems' (Peter Peregrinus, 1991)

3   FLOOD, J. E.: 'Telecommunication switching, traffic and networks' (Prentice Hall, 1995, chap. 2)

4   ITU-T Recommendation G.122: 'Influence of national systems on stability, talker echo in international connections' (International Telecommunications Union, 1993)

5   ITU-T Recommendation G.131: 'Control of talker echo' (International Telecommunications Union, 1996)

6   ITU-T Recommendation G.114: 'One-way transmission time' (International Telecommunications Union, 1996)

7   COOK, G. J.: 'Network Performance', in J. E. FLOOD, (Ed.): 'Telecommunications Networks' (IEE, 1997)

8   ITU-T Recommendation G.821: 'Error performance of an international digital connection operating at a bit rate below the primary rate and forming part of an integrated services digital network ' (International Telecommunications Union, 1996)

9   ITU-T Recommendation G.822: 'Controlled slip rate objectives on an international digital connection' (International Telecommunications Union, 1988)

10 ITU-T Recommendation G.121: 'Loudness ratings of national systems' (International Telecommunications Union, 1993)

11 ITU-T Recommendation G.113: 'Transmission impairments' (International Telecommunications Union, 1996)

12 FLOOD, J. E.: 'Transmission Principles', in J. E. FLOOD, and P. COCHRANE, (Eds): 'Transmission Systems' (Peter Peregrinus, 1991)

13 CAMERON, A. *et al.* 'Local Loop Unbundling – an account of Key Operational Challenges Faced by BT' *The Journal of British Telecommunication Engineers*, 2001, **2** (1), p. 12

14 SONGHURST, D. J.: 'Teletraffic engineering' in J. E. FLOOD, (Ed.): 'Telecommunications Networks' (IEE, 1997)

15 GHANBARI, M. *et al.* 'Introduction to reliability' in 'Principles of performance engineering for telecommunications and information systems' (IEE, 1997, chap. 9)

16 COCHRANE, P. and HEATLEY, D. J.: 'System and network reliability' (Chapman and Hall, chap. 11)

17 US Congress: 'Asleep at the switch? Federal Communications Commission efforts to ensure reliability of the public telephone network'. Fourteenth report by the Commission on Government Operations, House Report 102–420, US Government Printing Office, 1991

18 WOHLSTETTER, J. C.: 'Gigabits, gateways and gatekeepers: reliability, technology and policy', in LEHR (Ed.): 'Quality and reliability of telecommunications infrastructure' (Lawrence Erlbaum Associates, 1995)

19 Common Carrier Bureau: 'Preliminary report on network outages', FCC 1991 (there was no final report)

20 WARD, K. E. *et al.*: 'Network integrity in an open network provision (ONP) environment'. Final report of study for the Commission European. University College London, 1994

21   MONTON, V., WARD, K. E., WILBY, M. and MASON, R.: 'Risk assessment methodology for network integrity', *BT Tech. J.*, 1997, **15** (1)

22   MACDONALD, J. C.: 'Public network integrity – avoiding a crisis in trust', *IEEE J. Sel. Areas in Comm. Mag.*, 1994, **12**, pp. 5–12

23   ATIS: 'Technical report on analysis of FCC-reportable service outage data'. Alliance for Telecommunications Solutions, 1994, document T1A1.2/94-001R3

Chapter 10

Internet, Internet services and Quality of Service framework

10.1 Introduction

There are more than 150 million hosts on the global Internet. The number of Internet users is over 513 million, worldwide [1]. The Internet has been doubling annually since 1988. The size of the global Internet is estimated to exceed the size of the global telephone network by 2006. It is estimated that e-commerce on the Internet will reach somewhere between $1.8 T and $3.2 T U.S. dollars by 2003 [2]. The Internet needs to provide service quality that end users require. End users expect the Internet to be accessible, reliable, secure and fast all the time – irrespective of who owns it and who runs it. It is critical for Internet Service Providers (ISPs) to provide QoS meeting end users' needs.

Quality of service continues to be the number one concern of the users of Internet-based services. Recent surveys and focus group studies indicate QoS to be the most serious limiting factor in Electronic-Commerce/Electronic-Business (EC/EB). Users consistently rank quality, reliability and security as 'very important'. Internet businesses lose billions of dollars each year because of slow and failed web services. As e-business/e-commerce grows exponentially so does the quality, reliability and security concerns of both service users as well as service providers. With stock, real state and many other online commodities now selling for thousands of dollars, fast and reliable Internet exchanges are becoming increasingly important.

This part of the book focuses on the QoS aspects of Internet-based telecommunication services. It develops a systematic structure from which service providers can measure and monitor the QoS of Internet and Web applications as being experienced by the end users and conduct near real-time root cause analysis of any quality degradation.

This chapter provides a high-level architecture of the global Internet, reviews ingredients of the World Wide Web (WWW), Internet access technologies, real-time and non-real-time Internet services, and a generic framework for identification and definition of parameters influencing users' perception of quality. Chapter 11 addresses the QoS of real-time applications including VoIP, facsimile and media streaming. Chapter 12 applies the methodology to non-real-time applications of electronic mail (e-mail), and web hosting/e-commence/e-business applications.

10.2   Overview

The end-to-end performance of Internet applications is intertwined with the performance of a collection of systems/subsystems that a user's transaction touches during an Internet service application. These include desktop, browser, access infrastructure, ISPs, backbones, peering, application and content servers, databases, gateways and gatekeepers, and other hardware/software that run these systems. Section 3 provides a short description of components of the global Internet and WWW supporting the end-to-end Internet services and applications. Additionally, users' expectations of service quality of Internet applications varies from application to application. The QoS requirements of real-time interactive applications such as voice is different from the requirement of a non-real-time application such as e-commerce. Section 4 describes and classifies real-time and non-real-time Internet-based applications. Finally, a systematic approach is needed to identify, define and monitor performance parameters impacting the end-to-end QoS. This chapter describes methodological foundations necessary for quantifying and monitoring the QoS of each of the Internet services. Section 5 introduces the QoS framework of accessibility, continuity, and fulfillment and demonstrates how it can be used to monitor the QoS as being experienced by end users.

10.3   The Internet

This section provides a short description of components of the global Internet supporting the end-to-end Internet services and applications. It focuses on components that impact the end-to-end quality of Internet services and applications as being experienced by users.

10.3.1   The Internet infrastructure

The Internet is a conglomeration of thousands of interconnected networks. It is a medium for information exchange between computers all around the world. Consumers use the Internet to exchange e-mail, pay bills, buy and sell stocks, shop and conduct research. Businesses use the Internet to sell, place orders, train employees, and conduct web conferences/meetings and provide customer service. Voice, video, music and fax are transmitted over the Internet. A high-level Internet infrastructure is shown in Figure 10.1.

10.3.1.1   Backbone

The interconnected networks communicate with each other over a suite of standardised protocols called Transmission Control Protocol/Internet Protocol (TCP/IP).

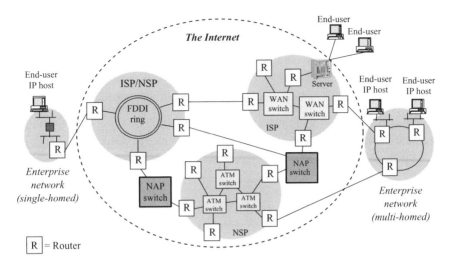

Figure 10.1   A high-level architecture of internet infrastructure

TCP/IP breaks up the data into 'envelopes' called 'packets'. Packets are the fundamental unit of transmission in any Internet application. The networks transmit Internet traffic, packets, over thousands of interconnected backbones. Backbones, each operated under distinct administrative control, consist of high-speed routers and links that transmit traffic at gigabit speeds. The different carriers that operate Internet backbones, exchange traffic with each other at Metropolitan Area Exchanges (MAEs) and Network Access Points (NAPs) also known as *peering*.

10.3.1.2   Peering

Different Internet networks exchange data at NAPs called 'peering' sites. NAPs enable users and sites on different networks to send and receive data to and from each other. These NAPs are distributed worldwide in major metropolitan cities including Washington DC, Chicago, San Francisco, Dallas, London, Amsterdam and Frankfurt. Peering sites are operated by commercial organisations. ISPs and Network Service Providers (NSPs) lease ports at peering sites and connect to their routers. Different peering arrangements are made between different network operators, meeting certain levels of service quality, and exchange speed.

Network service providers own and operate high-speed links that exchange data at peering sites and make up the Internet. At the peering sites, routers transfer messages between backbones owned and operated by many network service providers. To prevent traffic congestion at major exchange centres, network service providers have arranged private peering sites. Private peering sites are direct exchange places in which carriers can agree on many aspects of the data exchange including the quality and reliability related parameters of delay, service level and amount of the data to be transferred. The QoS can be monitored more closely at the private peering exchanges.

10.3.2   The World Wide Web

The WWW is a vehicle for multimedia presentation of information in terms of audio, video and text. It enables users to send and hear sound, see video, colours, and graphical representation. The basic idea is that a Personal Computer (PC) should be able to find information without needing to know a particular computer language. As long as they use WWW browsers (such as Netscape or Internet Explorer), all PCs are compatible with the Web. The advent of the WWW (1989) and browsers (1993) created the point-and-click environment by which users can easily navigate their way from computer to computer on the Internet and access information resources.

10.3.2.1   Browser

A browser is a program that requests and receives information from the web and displays it to the user. Browsers installed on PCs provide simple interactive interfaces for the users to access and navigate the web. Browsers are easy to use and have facilitated public use of the Internet.

10.3.2.2   Client–server

The WWW is based on the client–server model. The client is a service-requesting computer, usually a PC, which generates requests, receives and reads information from the service-providing computers (servers) within the Web. Clients utilise browsers to communicate with servers through the Internet and create graphical interfaces to access and navigate the WWW.

Depending on their function, servers may be classified into 'web server', 'application server' and 'database server'. The web server is the service-providing computer on which the web information is located. The server provides service to clients. It receives, processes and responds to the clients' (users') requests. The server locates the requested information and copies it to the network connection to be sent to the client. The server needs to meet the requirement of the application it is serving. The application server works in conjunction with the web server. It may process the user's orders, and check the availability of desired goods in an e-commerce application, for example. The database server provides access to a collection of data stored in a database. A database server holds information and provides this information to the application server when requested.

10.3.3   Internet access

Individuals, homes, small-office users, corporations and institutions connect to the Internet via many types of telecommunication services. These include analog dial-up lines, Integrated Services Digital Network (ISDN), Digital Subscriber Line (DSL) services, Cable Modem (CM), T1, T3, and most recently wireless access services. Dial-up, ISDN, and DSL access utilise the existing Public Switched Telephone Network (PSTN) infrastructure. CM access uses the cable TV network or community area TV (CATV) infrastructure. T1 and T3 access are private (dedicated or leased) high-speed connection lines.

Home or small-office consumers typically connect to the Internet using a web browser running on a PC that connects through a modem to the local ISP. On the other end of the line, ISPs aggregate traffic from many users and send it to the Internet backbones over high-speed lines. Corporate consumers use browsers running on computers in a local area network that connects to the Internet through routers and purchased or leased permanent T1/T3 links. In any event, the dial-up or dedicated connections to the Internet via NSPs or ISPs facilitate access to the information resources of the worldwide Internet consisting of thousands of separate and independent networks and millions of individual computers. The connection speed depends on the access technology used and it has an essential impact on the user's expectation of the performance as well as the user's perception of an 'effective' (or 'defective') Internet experience.

10.3.3.1   Dial-up access

Dial-up access utilises the PSTN's existing infrastructure in analog (copper) loops. Dial-up users program their PCs to dial the local telephone number associated with their ISP. The call is routed to the local telephone company's central office, which in turn sends the call to the central office connected to the subscriber's ISP equipment. The connection to the local central office is through two modems between which an analog 'voice call' has been established. ISPs have banks of modems in remote access server devices that handle calls received from the telephone company. The ISP has direct high-speed links to a regional data centre that is connected to the Internet. Today's low-speed dial-up phone line modem connections provide speeds up to 64 kb/s (kilo bits per second) and constitute the majority of homes and small businesses Internet access connections. Dial-up access imposes challenging performance problems because of their variable and asymmetric bit rate.

10.3.3.2   ISDN access

ISDN utilises a technology known as DSL that provides a higher speed and a more stable connection between the user and the PSTN over the local loops. A simplified view of the ISDN connection between a user and local exchange is the network termination equipment that provides ISDN modem functionality at either end of the traditional analog local loop. The digital link between network terminations is always on. ISDN provides point-to-point connectivity in terms of one or more 64 kb/s connections. Typical ISDN Internet access provides speeds up to 128 kb/s.

10.3.3.3   DSL access

A new DSL technology, generically called xDSL, offers higher-speed Internet access to homes and businesses. A popular version of the DSL-based access technology is asymmetric DSL (ADSL). A distinguishing feature of the ADSL service from the DSL-based ISDN is higher speed. ADSL is capable of delivering up to 9 million bit/s downstream (to the subscriber site – downloading) and up to 1.5 million bit/s upstream (from the subscriber site – uploading). The actual speeds depend on the quality of the loop over which the ADSL modems operate – longer or older loops lead to lower speeds. The number of high-speed subscriber lines increased from 1.8 million to 3.3 million users in the USA by the end of the 2001 (Yankee Group).

10.3.3.4   Cable modem access

In recent years the cable industry has been working to leverage the cable TV network infrastructure for residential and business high-speed Internet access. A cable TV network, initially designed for TV signals, utilises fibre optics from the 'head end' to the neighbourhood and coaxial cable for the last hop into the customer premises. The Hybrid Fibre Coax (HFC) technology provides an alternative high-speed Internet access. Using cable modems, subscribers connect their PCs to the cable TV network for high-speed full-time (always on) Internet access. Cable modem technology provides high-speed data transfer rates of up to 10 million bit/s depending on the number of users and the particular cable modem configuration. Cable companies are expected to increase their subscriber base from 3.7 million in 2000 to 7 million by year's end 2001 in the USA, according to the Yankee Group.

The high-speed DSL access or CM access are generally called broadband services. According to the New York Times Yankee Groups [3], by January 2002 some 10.7 million of the USA households will have a broadband service, or about 16% of all households online.

10.3.3.5   Dedicated access

Large and medium businesses with a high volume of traffic use dedicated, private lines to access the Internet. The ISP arranges for the dedicated T1/T3 line to run from the customer site to the ISP point of presence (PoP). T1 and T3 speeds are 1.54 million bit/s and 44 million bit/s, respectively.

10.4   Internet services and applications

The end-to-end users' experience of QoS of a given Internet application depends on the service quality of a number of distinct networks, and their associated hardware/software, systems, and subsystems that make up an end-to-end Internet application. Furthermore, different operators often administer the Internet distinct networks. An end-to-end Internet application, therefore, is impacted by a number of Autonomous Systems (AS) in the Internet. Figure 10.2 shows a high-level architecture of these autonomous systems [4]. The challenge is to ensure that all these systems and subsystems function together in harmony and create an 'effective' user's experience.

The user's experience of an Internet application varies from application to application. To formulate the concept of QoS from a user's experience there is a need to understand and consider the user's application. Understanding characteristics of applications is an important aspect of identification, definition and formulation of QoS of Internet services. In this section we provide a brief list and description of current services and applications. For the purpose of QoS, the Internet services may be classified into two general categories of 'real-time' and 'non-real-time' applications.

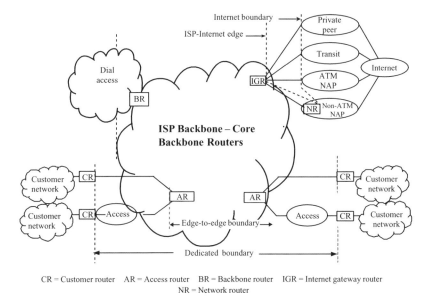

CR = Customer router   AR = Access router   BR = Backbone router   IGR = Internet gateway router
NR = Network router

Figure 10.2   A high-level architecture of Internet autonomous systems

10.4.1   Real-time applications

As was indicated earlier, the Internet is a packet-based network. An Internet application digitises the information into bits, forms a collection of bits into an IP Packet,
and transmits the IP Packet from source to destination(s) over the global Internet
network. Packets, therefore, are the fundamental units of information transmission in
any Internet application.

   Real-time applications are those applications in which a distinct notion of time-
lines and ordering is associated with their packet delivery. Packets must reach the
destination within a bounded time period and within bounded variations. Examples
include applications such as audio, video, media streaming, real-time fax and broadcasting. The total end-to-end time period within which a packet gets delivered from
a source to a destination is called one-way 'delay' and its variations are called delay
variations or 'jitter'. Real-time applications are sensitive to delay and jitter. The degree
of sensitivity depends on the type of the application. Real-time applications are further classified into two categories – 'interactive' (intolerant) and 'non-interactive'
(tolerant) [5]. The differences between different categories of applications boil down
to the differences in how applications react to packet delay and jitter.

10.4.1.1   Interactive (intolerant)

Two-way conversational applications such as voice and video are called real-time
interactive applications. In these applications audio/video signals are packetised at
the source and transported through the network to the receiver. The temporal ordering
as well as the end-to-end latency of these packets, as they cross the network, has

a fundamental impact on service quality as being experienced by the two end users. These applications are intolerant to the jitter and packet delay. End-to-end delays and jitter beyond certain limits could degrade the signal to an unacceptable quality.

10.4.1.2   Non-interactive (tolerant)

Real-time applications such as streaming audio/video (web-based broadcasting) are essentially one-way information transfers and, contrary to the interactive applications, users are not overly concerned if they see or hear events a few milliseconds (or sometimes seconds) after their occurrence. They are tolerant to delay and jitter. The requirement for the total latency and jitter for tolerant applications is less stringent than intolerant applications. Given levels of latency and jitter that may lead to an unacceptable interactive audio may be acceptable for a non-interactive audio streaming.

10.4.2   *Non-real-time (elastic) applications*

Non-real-time, or elastic, applications wait for the packets to arrive before the application processes the data. In contrast to the real-time applications that drop delayed packets and do not retransmit lost packets, elastic applications retransmit lost or delayed packets until they get to the destination. Examples of elastic applications include all other Internet services such as electronic mail (e-mail), IP-based fax, File Transfer Protocol (FTP), and web-browsing/e-commerce related applications. In short, elastic encompasses applications that are less sensitive to delay, jitter and packet loss. They process packets immediately after their arrival.

10.4.3   *Real-time versus non-real-time*

The fundamental distinction between the two applications lies in flow-control and the error-handling mechanism. In real-time applications flow-control is self-paced, in that the rate at which packets are sent is determined by the characteristics of the sending application. The network then attempts to minimise the latency and packet loss based on conditions imposed by the sender application. On the other hand, elastic applications use an adaptive flow-control algorithm in which the packet flow rate is based on the sender's view of the available network capacity and available space at the receiver's end – at that point in time.

   With respect to error handling, a real-time application has a timeliness factor associated with it. Packets must be delivered within a certain elapsed time frame and within application specified sequencing. Otherwise, the packet becomes irrelevant and can be discarded. Packets in error, out of sequence packets, and packets with latency beyond the application requirement must be discarded. A non-real-time application uses error-detection and retransmission recovery techniques to recover the transmission error.

10.5   Quality of Service assessment framework

This section introduces a framework for understanding end-user's expectations of Internet services. The framework will assist service providers to identify, define and measure parameters associated with each of the Internet-based services – at each of the stages of a user's experience. Similarities are demonstrated between the Quality of Service Framework (QSF) introduced by the telecommunication community for the PSTN-based services and that of the Accessibility, Continuity, Fulfilment framework introduced in recent years by the computer/communication community for Internet-based services. The evolution of the service reliability from a 'service provider centric' approach to a 'user centric' approach will be shown.

Prior to the Bell System divestiture of 1984, quality and reliability of telecommunications services were determined largely from the service provider's perspective. Divestiture and competition changed the service quality and reliability paradigm from 'service provider's view' to 'user's/customer's view'. The attention was focused on defining and measuring user's perspective of quality and reliability of communications services. In 1988, Richters and Dvorak of AT&T Labs introduced a QSF for defining the quality and reliability of communications services [6]. The QSF consisted of a matrix of 'rows' and 'columns'. Rows were made up of communication functions performed by the users when using a service, and columns were made up of quality and reliability criteria perceived by users. Communication functions were classified into three major categories of: Connection Establishment, User Information Transfer and Connection Release. Quality and reliability criteria included Speed, Accuracy, Availability, Reliability, Security and Simplicity. The International Telecommunications Unit (ITU) later adopted the framework. Specific QSF for each of the services including voice and facsimile were developed.

A similar framework can be developed to define, measure and monitor the end users' experience of quality and reliability of Internet-based services. Users would like to be able to *access* the desired application, initiate a transaction, *continue* using the application with no interruption for a desired duration of time, and *fulfil* the initiated transaction at a desired quality. Thus, for the Internet-based applications, user's perspective of quality and reliability is characterised by the *Accessibility (A)*, *Continuity (C)*, and *Fulfilment (F)* framework defined below:

Accessibility

Accessibility is the ability of a user to access the application and initiate a transaction. For instance, the first step in an e-mail application consists of accessing mail server(s). For a dial-up user, accessibility consists of two successive stages – first, the user must be able to access the dial platform and, second, the user must access the appropriate mail server (given the accessibility to the dial platform has been successful).

Continuity

Continuity is the ability to deliver the service with no interruption, given that accessibility has been successful and a transaction has been initiated. For the e-mail application, for instance, if the message gets sent or received completely with a proper connection closure, it is considered a successful transaction. If not, it is considered a failed transaction.

Fulfilment

Fulfilment is the ability to deliver the service meeting the user's quality expectations. In the case of e-mail, for instance, the quality of a transaction can be principally expressed in terms of throughput or speed. Once an access has been established for an e-mail application, the time that it takes to complete sending or receiving e-mail determines the fulfilment. E-mail transactions taking a longer time than what the user expects will be considered a 'defective' transaction.

The following chapters demonstrate the application of the framework to major Internet services including e-mail and web browsing.

10.6   Practical steps for measuring QoS

In practice, there are several essential tasks that need to be addressed to arrive at a programme capable of measuring QoS. Listed below is the set of essential tasks common to all Internet-based services:

Task 1 – Specify the Internet service

Specify the Internet service for which QoS needs to be measured. Specifically, these services could be any of the real-time services such as voice and video or non-real-time applications such as e-mail, web browsing, streaming, fax over IP and so forth.

Task 2 – Define end users' transaction

This is a unit of study to be observed and QoS parameters need to be measured. For instance, for e-mail service, a transaction would consist of the 'GetMail' experience – that is getting a mail from a mail server. For a fax over IP service, a transaction could consist of sending a set of fax pages. For a web browsing service a transaction could consist of the experience of downloading the main page of a web site.

Task 3 – Define QoS parameters

Using the QoS framework, define accessibility, continuity and fulfilment parameters impacting the quality of the specified transaction for each of the corresponding applications. The end result of this step would be a complete list of parameters classified within the accessibility, continuity and fulfilment framework. Parameters could be of the dichotomous type such as success or failure of accessing the application, or continuous, such as the response time.

Task 4 – Decide on measurement approach

There are two distinct measurement approaches called 'active' (intrusive) and 'passive' (non-intrusive):

- *Active measurement*: In an active measurement, a transaction request is initiated with the sole purpose of measuring the performance. An active approach emulates the users' experience and provides an end-to-end perspective of QoS. It generates the transaction needed to make the measurement and allows a direct observation and analysis. An active approach necessitates a carefully designed sampling plan including a representative transaction, geographically balanced sampling distribution and appropriate sampling frequency.

• *Passive measurement*: In a passive measurement, performance data are collected from network elements including routers, server log files and switches. An independent device passively monitors and collects performance data as the traffic traverses within the network. Since the measurement is made within a network it would be impossible to extract a complete end-to-end user's experience. Also, studies of competitor capabilities are not possible.

Task 5 – Sampling plan and data collection

This task involves many activities including the desired confidence level and accuracy and validity of the collected data [7].

Task 6 – Trend analysis

In practice many parameters, each representing a different segment of an end-to-end QoS will be measured. Trend analysis of the parameters over time requires advanced statistical techniques that need to be specified and formulated at this stage. Six Sigma [8] and Statistical Process Control [9] are appropriate techniques for trend analysis.

Task 7 – Set thresholds

Set the threshold against which the QoS parameters need to be controlled. Different thresholds could be set for a given parameter. A threshold could be set to meet the operations requirements, another threshold could be set based on the historical data, yet another threshold could be set to meet a Service Level Agreement (SLA). The QoS trend will be plotted against the specified thresholds. When a threshold is exceeded a corrective action takes place and a root cause analysis is conducted.

Task 8 – Detect, diagnose and resolve

A continuous monitoring and analysis of the QoS parameters enables service providers, network operators and system/equipment/product vendors to detect, diagnose and resolve failures impacting users' transactions. A QoS monitoring programme is essential in meeting end-users' expectations of service quality of the emerging Internet-based applications.

10.7 Summary

Methodological foundations necessary for quantifying and monitoring the QoS of each of the Internet services was described. A short description of the components of the global Internet and WWW supporting the end-to-end Internet services and applications was provided. Real-time and non-real-time Internet-based applications were classified. Distinctions between the QoS requirements of real-time interactive applications such as voice and non-real-time applications such as e-commerce were demonstrated. The Accessibility, Continuity, and Fulfilment framework was introduced and its application to monitor the QoS, as being experienced by end users, was demonstrated. Finally, a systematic approach to identify, define and monitor performance parameters impacting the end-to-end QoS was provided.

10.8   References

1   Internet Engineering Task Force IETF, RFC 2026, 1 January 2002
2   CERF, V.: 'The Internet is for everyone,', IETF, RFC 2026, 1 January 2002
3   HAFNER, K.: 'The Internet's invisible hand,' The New York Times, 10 January 2002
4   KOGAN, Y., MAGULURI, G. and RAMACHANDRAN, G.: 'Defect per Million and IP Backbone Availability', Committee T1 Contribution, September 2001
5   BARDEN, R., CLARK, D. and SHENKER, S.: 'Integrated services in the Internet architecture: An overview,' IETF RFC 1633, June 1994
6   RICHTERS, J. S. and DVORAK, C. A.: 'A framework for defining the quality of communications services,' *IEEE Communication Magazine*, October 1988
7   BRUSH, G. G.: 'How to choose proper sample size', American Society for Quality Control, Volume 12
8   HARRY, M. J. and LAWSON, J. R.: 'Six Sigma Producibility Analysis and Process Characterization' (Six Sigma Research Institute and Motorola University Press, Addison-Wesley, 1992)
9   GRANT, E. and LEAVENWORTH, R.: 'Statistical quality control' (McGraw-Hill, 1998)

Chapter 11

Quality of Service for real-time Internet applications

11.1 Introduction

Harvard professor Clayton Christensen wrote a book called *The Innovator's Dilemma* [1] in which he refers to a concept called 'disruptive technologies.' According to Professor Christensen, a disruptive technology is one that is so profound that it makes not only quantitative changes in price and performance but equally dramatic changes in qualitative ways. Voice over IP (VoIP) is an example of that in the telecommunications industry. VoIP is significant not just because some say it will afford cheaper service, which has yet to be proven, but because it promises new services through the convergence of voice and data infrastructures.

The evolution of VoIP parallels the technology adoption of the personal computer (PC) and the Internet. Hobbyists, motivated by the lure of free long distance telephone calls, drove the initial interest in VoIP. In the United States the Internet was usually accessed via a local phone call in the United States. These calls were insensitive to usage and so were considered 'free!' Packetisation and depacketisation were performed in the PC and initially the capabilities were PC-to-PC only. The call quality was inferior to that of the public network, but that was an understood trade-off. Some people, such as college students and international expatriate communities, were willing to accept this trade-off.

However, the industry is clearly moving in the direction of VoIP and call quality is being improved rapidly to meet customer demands. When traffic is routed over the public 'unmanaged' Internet, any connection may experience poor quality, depending on routing and network congestion. With managed IP networks, however, such as corporate intranets, the infrastructure and traffic is more controllable, so quality can improve.

Despite potential cost savings and improving call quality, the greatest motivator may lie in the ability to merge traditional voice and data services. An early example of this potential is in web-enabled call centres where a customer browsing a web site can click on an icon to speak to a customer service representative. On-line merchants have found that this can greatly improve their on-line sales. So, when VoIP services are discussed, it usually means more than just a voice call.

Based on an analysis of leading service providers, IDC estimates the total 2001 worldwide retail IP telephony market at 20.6 billion minutes, including both domestic long distance and international long distance minutes of use. IDC projects that this total will grow to 59.6 billion minutes in 2002, and to 745.8 billion minutes by year end 2007 [2]. While the trend is upward, the numbers are still small when compared to the trillions of minutes of traditional voice phone calls worldwide.

What is deterring a more rapid growth of VoIP services is the concern many customers have over quality and reliability. Along with the current inadequacies of the Internet's infrastructure, the lack of interoperability among current proprietary products makes building and operating a network more difficult. For example, differences in the voice compression algorithms of codecs, as well as their different software designs is a problem. Since few enterprise companies have deployed IP telephony for their businesses, the market is wide open for service providers who can deliver quality services to customers. This chapter will describe the QoS issues to be addressed when offering voice, fax and streaming media applications to customers.

11.2   Voice over IP networks

11.2.1   Architecture

Most VoIP services offered today are really a hybrid offering, where parts of the call are carried over circuit switching equipment and parts over an IP network. A pure IP call is one where the packetisation of the voice signal takes place in the customer premise equipment which is then connected to the internet (see Figure 11.1).

The first calls made this way used a PC software package, which digitised the voice signal. The quality was poor due to delay and early codec designs, but the price was right – free. There was no additional charge to internet accounts that already existed.

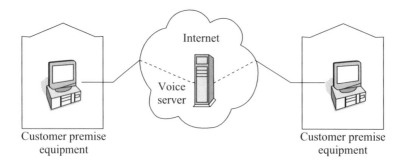

Figure 11.1   Voice service over an IP network

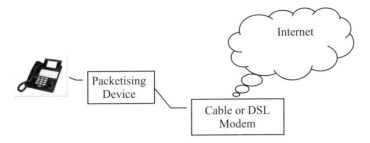

Figure 11.2   Telephone calls over the Internet

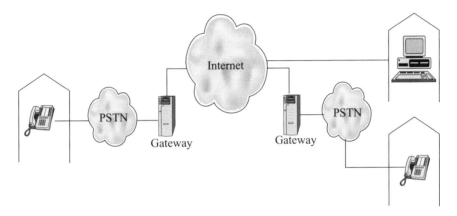

Figure 11.3   Voice service over a hybrid network

Since there was no real signalling involved, the service was quite rudimentary. It was used mostly by hobbyists and individuals who otherwise would not have made the call.

In mid-2001, Net2Phone, a worldwide telecommunications service provider formed in 1996, announced a new service allowing customers to make telephone calls over the Internet using their regular hand-sets. This service required customers to purchase a device which is placed between the regular phone and a cable or DSL modem. Cisco and others have announced similar devices (see Figure 11.2). These services are still very limited in their application, since only a small, but growing, number of consumers have cable modems or DSL access.

The more typical connection today is one involving a hybrid of circuit switching and IP technology.

Figure 11.3 shows a typical hybrid architecture. These hybrid architectures introduce a complex set of performance problems. Packet network technology can introduce delay, distortion from voice compression techniques and problems due to packet loss and packet replacement strategies. These factors, individually and in combination, can produce significant degradation unless properly managed.

A network management strategy used today is to over-engineer packet networks so that voice compression is not necessary, queuing delays are minimal and congestion is not an issue. This over-engineering strategy, however, is expensive which then motivates other QoS strategies such as packet network admission control, traffic conditioning, service differentiation and priority queues. Strategies such as MPLS, Diff Serv and IntServe are expected to achieve significant traffic flow gains. These strategies are beyond the scope of this book, but are described in many texts covering the Internet. It is anticipated that IP networks can thus achieve performance levels approaching today's circuit switched network quality. However, even with all these advanced strategies, the issues of delay, packet loss and compression will still be significant for voice QoS.

Figure 11.4 shows in more detail how a voice call would be delivered across an IP network. The voice call is routed to an access VoIP gateway which contains a voice coder to digitise the voice signal and create packets. It can use voice activity detection (VAD)/silence suppression to avoid sending empty packets which would contain only silence. Signal classification can also take place at the access VoIP gateway to distinguish between voice and non-voice signals, such as fax. The call is then routed across the IP network where packets can be lost and delay variation can be introduced as different packets take different routes. The packets are processed at an egress VoIP gateway which contains a de-jittering buffer. The de-jittering buffer smoothes out the delay between the packet arrival rates so that voice packets can be delivered at a constant rate. If a packet is lost, a loss-concealment algorithm must be used. Comfort noise can also be introduced at the egress gateway to provide a constant background noise to compensate for the empty 'silence' packets which were not sent. Echo control is also needed for these calls. Choices made in the management of these attributes can lead to different levels of QoS. A more extensive description of these quality impairments and their impact is given in Section 11.2.3.

11.2.2   QoS measurement

11.2.2.1   Classes of service

Since Internet-based applications are many, varied and new, such as e-mail, a method is needed to group these applications into categories or classes so their quality of service needs can be met. The ITU-T proposes grouping IP telecommunications transactions into six unique classes defined according to the desired performance QoS objectives. According to this recommendation, 'a packet flow is the traffic associated with a given connection or connectionless stream having the same source host, destination host, class of service, and session identification.' Detailed parameter specifics and explanations for each class are described in Tables 1 and 2 in Recommendation Y.1541.

Briefly, the classes are as follows:

Class 0:   Real-time, highly interactive applications, sensitive to jitter. Mean delay upper bound is 100 ms, delay variance is less than 50 ms, and packet loss ratio is less than 10^{-3}. Application examples include VoIP, Video Teleconference (VTC).

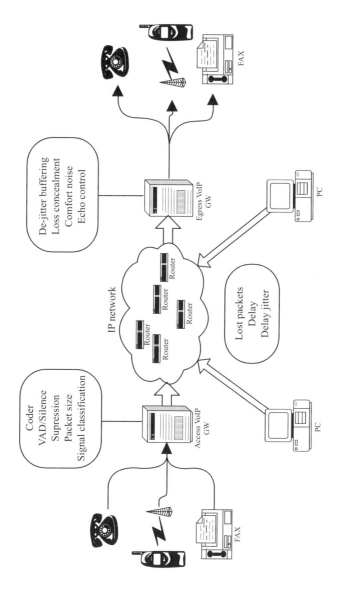

Figure 11.4 Major QoS/VoIP issues

Class 1:   Real-time, interactive applications, sensitive to jitter. Mean delay upper bound is 400 ms, delay variance is less than 50 ms, and loss ratio is less than 10^{-3}. Application examples include VoIP, VTC.

Class 2:   Highly interactive transaction data. Mean delay upper bound is 100 ms, delay variance is unspecified, and packet loss ratio is less than 10^{-3}. Application examples include signalling.

Class 3:   Interactive transaction data. Mean delay upper bound is 400 ms, delay variance is unspecified, and loss ratio is less than 10^{-3}. Application examples include signalling.

Class 4:   Low loss only applications. Mean delay upper bound is 1 s, delay variance is unspecified, and loss ratio is less than 10^{-3}. Application examples include short transactions, bulk data, video streaming.

Class 5:   Unspecified applications with unspecified mean delay, delay variance and loss ratio. Application examples include traditional applications of default IP networks.

These classes enable service providers to make agreements between themselves and also to create Service Level Agreements (SLAs) with their customers. A summary of this categorisation is given in Table 11.1.

Table 11.1   Classes of service

Class	Mean delay upper bound	Delay variance	Loss ratio	Application examples
Class 0 Real-time, highly interactive applications sensitive to jitter	100 ms	<50 ms	$<10^{-3}$	Voice over IP, Video Teleconference
Class 1 Real-time, interactive applications sensitive to jitter	400 ms	<50 ms	$<10^{-3}$	Voice over IP, Video Teleconference
Class 2 Highly interactive transaction data	100 ms	Unspecified	$<10^{-3}$	Signalling
Class 3 Interactive transaction data	400 ms	Unspecified	$<10^{-3}$	Signalling
Class 4 Low loss only applications	1 s	Unspecified	$<10^{-3}$	Short transactions, bulk data, video streaming
Class 5 Unspecified applications	Unspecified	Unspecified	Unspecified	Default IP networks

11.2.2.2   Quality of Service matrix

Users who make voice calls over a hybrid VoIP network use the same criteria to measure performance as they do for calls over a strictly circuit-switched network. Table 11.2 shows the QoS matrix for voice calls over a hybrid network.

While the quality criteria remain the same for voice calls over a hybrid network, some criteria will clearly change for a pure end-to-end IP call. A good example is the parameter 'number of digits dialed.' This parameter is related to the quality criteria of 'simplicity' for the communications function of connection establishment.

As indicated in Chapter 10, the computer/communication community for Internet-based services has been using the framework of accessibility, continuity and fulfilment to describe the technical quality attributes for Internet-based services. Accessibility is the ability of a user to access the application and initiate a transaction. For voice calls, accessibility is equivalent to connection establishment. It measures whether the call was established or whether it was blocked. Continuity is the ability to deliver the service with no interruption, given that the accessibility has been successful and a transaction has been initiated. For voice calls continuity measures the ability of the call to stay up for a specified length of time, usually the length of an average call. Fulfilment is the ability to deliver the service meeting the user's quality expectations. For voice calls it is equivalent to user information transfer. It measures the transmission quality of the call, such as the clarity of the connection and the ability to clearly understand the other party.

The telecommunications industry has been measuring the fulfilment attribute for voice performance in terms of customer ratings of speech quality. This type of measurement, called subjective testing, involves a sample of customers listening to and judging the quality of speech provided by a new technology or service prior to its introduction. This allows the service provider to ensure that the quality is acceptable before it is introduced to the network. Voice quality experts worldwide have developed agreed-to standards for the implementation of such testing, which are given in ITU-T Recommendation P.800 [3].

In these tests customers rate call quality on a five point scale. The five point scale is

5 = excellent
4 = good
3 = fair
2 = poor
1 = bad

The arithmetic average of these ratings is called the Mean Opinion Score (MOS). Because MOS scores are subjective, MOS ratings for a technology or service being evaluated are typically compared with a well established reference. For voice service over an IP network for example, the voice quality ratings would be compared to those of the current circuit-switched networks. Several charts in the next section will give the results of subjective testing for calls with certain impairments in terms of MOS.

While subjective testing is perhaps the most accurate prediction of customer opinion, it can be an expensive and time consuming task. For this reason subjective testing has also been used to generate data from which 'grade of service' models have been

Table 11.2   QoS matrix for voice over hybrid network

Communications function	Quality criteria: All parameters						
	Speed	Accuracy	Availability	Reliability	Security	Simplicity	Flexibility
Technical sales and planning	Response time	% Correct information	Hours staff can be accessed	% Optimal information	Confidentiality	Ease of contact	Options and alternatives
Provisioning	Time to deliver	% Correct	Hours staff can be accessed	% Optimal	Confidentiality	Ease of contact	Options and alternatives
Technical support	Time to respond	Document quality Knowledge level	Hours staff can be accessed		Confidentiality	Ease of contact	Options
Repair	Time to repair	% Correct	Hours staff can be accessed	% Robust	Confidentiality	Ease of contact	Options
Technical quality							
• Connection establishment	Dial tone delay post-dialing delay	% Wrong Number (Due to network)	% Blocked % No network response	% Outage (Due to network failures)	% Bridged connections	Number of digits dialed Understandable announce- ments	Number of alternate routes
• User info transfer	Propagation delay	Transmission quality	Dropouts	% Cut offs	Intelligible crosstalk		
• Connection release	Time to release	% Correct					
Billing	% Late	% Correct	Frequency		%Fraud	Understandable	Alternate programmes
Network/service mgmt. by customer							

Application: Voice; Service: Hybrid network offering

built. These models can be used to predict customer opinion when a new architecture or technology introduces impairments similar to those that have been previously judged in subjective tests. The most popular model of this kind today is the E-model [4]. This E-model supersedes the previously standardised grade of service model for circuit-switched networks which can be used to predict customer acceptance given objectively measured network impairments.

11.2.2.3   Measurement approaches

In general, there are two approaches to measuring the end-to-end QoS being delivered to a customer. There is the active, or what is sometimes called the intrusive approach, and the passive, or the non-intrusive approach. Each has advantages and disadvantages which will be discussed. It is recommended that both be used to fully understand the QoS that is being offered to the customer.

Intrusive (active) measurements

The intrusive measurement approach involves creating and measuring the customer experience from an end-to-end perspective, mimicking the customer experience as closely as possible.

For a voice call, test units can be placed in the network behind a traditional 2-wire customer interface to the network. These could be in people's homes or offices.

A host network computer directs the test units to initiate calls to each other over the network at a predefined time. For voice, these remote units can measure accessibility in terms of the probability of establishing the call and the time taken to establish the call. For continuity the units would measure the probability that the call stayed connected for a specified length of time. The inverse is the probability of premature disconnects or cut-offs.

For fulfilment for a purely circuit-switched call, the units would measure loss, noise and perhaps echo. A grade of service model would then be used to predict customer satisfaction.

For IP voice calls carried across a hybrid network, a different approach is required for intrusive testing. In this case a segment of speech can be sent across the network and stored. This received speech file can then be taken to the laboratory where experts or customers listen to these speech segments and give a customer opinion rating. Comparative, analytical methods are also being explored.

There are many advantages of using the intrusive testing approach. Perhaps the most important is that it closely mimics what the customer experiences. This approach is also useful for competitive characterisations. After subscribing to a service, as a customer would, any carrier can see the QoS that is being delivered to the customer. Intrusive network testing can be used to identify network weaknesses. Calls in a particular geographical area may be poor when compared to another area, or calls over certain network equipment may be poor. This type of testing can be used to establish benchmarks against which any network changes can be compared. This allows a network provider to monitor improvements to their network service. Lastly, this technique can be used to verify that network equipment is performing as expected over time.

A disadvantage to the intrusive testing approach is that it places an additional load on network resources. It also requires paying for test calls, which in a competitive characterisation mode can be expensive. There are also extensive logistics, site preparation, and equipment maintenance and calibration issues to be addressed.

Non-intrusive (passive) measurements

The complement to the intrusive testing technique is what is called the non-intrusive measurement of quality. In this methodology, in a circuit switched network, equipment is bridged onto both directions of the transmission of the call at a common point. Because this method is bridged on via network equipment, it does not allow the measurement of accessibility or continuity, but it can measure fulfilment. For a voice call, parameters like speech level, noise, echo path loss and echo path delay would be measured for both directions of transmission.

In an IP network, passive measurements come from gateways and routers which measure parameters such as packet loss. Off-line data processing analyses these values to determine QoS.

The advantage of non-intrusive testing is that the tests do not add any additional load to the network. These tests can run automatically since the network is running and it frees field personnel for other tasks. There is no per call test costs since it is monitoring traffic already on the network. A key advantage is that it does give an extensive cross section of customer calls and also the ability to classify call type. This approach also naturally includes all customer equipment since it is monitoring calls already on the network.

The disadvantage of the non-intrusive approach, of course, is that it is done inside the network offering the service so it is not really getting an end-to-end view of what the customer is experiencing. Competitive characterisations cannot be done with this methodology because access is needed to the internal portions of the network. Another disadvantage is that it can only monitor the traffic that is on the network. It cannot add any additional calls to the network to stress test any network segment.

11.2.3   *Factors affecting speech quality*

The new packet networks are designed to carry both voice and data services. Voice, however, has unique requirements which must be met. Because voice requires a continuous signal, voice service is very sensitive to packet delay and packet loss. For example, if voice packets are lost, the system must provide something for the customer to hear. To minimise cost, many carriers have considered voice compression techniques and other ways to maximise network utilisation. These techniques can have a major impact on voice quality. This section will describe the impact of network technology and design decisions on the end-to-end quality of voice the customer hears.

11.2.3.1   Delay

In general, connections that accumulate more than approximately 15 ms of one-way delay become susceptible to the impairment of talker echo [5]. Because echo is such a severe quality impairment, the industry has essentially solved the problem of echo by the proper deployment of network echo cancellers. In the IP packet network environment, echo cancellers are positioned at the edges of the network, actually on the time division multiplexing (TDM) side of the IP gateway. So positioned, these cancellers must deal with the access network delay, not the greater IP network delay. Figure 11.5 shows where echo control would be placed in a typical connection.

In early implementations, VoIP efforts were plagued by echo problems which were generally traceable to design flaws. The echo canceller functionality is part of the gateways' audio processing system (speech coding, echo cancellation, possibly VAD/silence suppression and noise generation) and is implemented with DSP technology. Because echo cancellation is complex, the DSP design task is both difficult and processor intensive. Basic design flaws such as poor double-talk detection algorithms or the allocation of less processing power than necessary have been routine. The harsh reality of echo impairment has interrupted VoIP introduction until either design improvements were achieved or more DSP resources were allocated to the task. It is worth noting that the necessary improvements were achieved as a product of extensive performance testing, against ITU-T Recommendation G.168 and, in many instances, against proprietary and speech-based tests. The underlying tests not only allow for performance predictions, but can also point the way to achieving performance improvements.

Connection delay can cause impairments other than echo if it gets large enough. Delay starts impeding the 'naturalness' of the back-and-forth conversation of the parties on a call, when the one-way connection delay exceeds approximately 150 ms [6]. This problem is often referred to as the problem of absolute, or echo-free delay. In the traditional circuit-switched network environment such long delays are associated with transoceanic connections and connections carried over satellite. On these calls, the customer experience is often described as the parties talking over each other. It is caused when the normal pauses in back-and-forth exchanges are extended by the additional connection delay. When the parties are strangers, the long 'pauses' are sometimes misinterpreted as the behaviour of a 'slow' thinker or of a person who is carefully, maybe too carefully, considering his or her replies. In any case, the ability to have a phone conversation that resembles face-to-face conversations, is increasingly lost as one-way connection delays stretch past the 150 ms threshold.

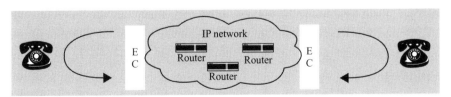

Figure 11.5   Echo control

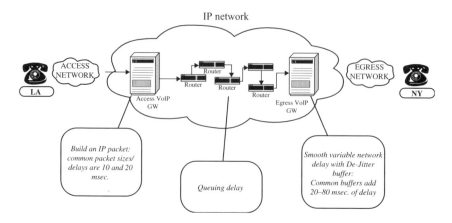

Figure 11.6   Sources of VoIP delay

Delays in the traditional circuit-switched environment have typically been well below the 150 ms threshold. However, the situation has changed dramatically with the evolution from circuit-switched to IP-based network transport. Figure 11.6 shows the multiple sources of delay in a VoIP network.

To encapsulate the digitised voice signal in a packet, the voice stream must be delayed as the packet is built. Common packet sizes are 10 and 20 ms. The time to build a packet must be added to the baseline delay introduced by the connection length (propagation delay) and traditional signal processing activities (e.g. echo cancellation).

It is also necessary to use a buffer at the end of the packet network to accommodate variable delay, or delay jitter, introduced by the network. This de-jitter buffer acts to smooth out delay variations that are present in packet networks. This is particularly important for voice, where speech must be delivered at a constant rate. Such a buffer is also necessary to give the gateway the ability to re-order packets that have arrived out of order. Without such a buffer, all packets that arrive at the terminating packet network gateway at other than the expected time must be discarded, resulting in potentially unacceptable packet loss rates. Typical delay-jitter buffers hold two to four packets and thus introduce an additional 20–40 ms of delay for 10 ms sized packets and 40–80 ms for 20 ms sized packets.

In addition to the delay added by the packet creation and buffering processes, there are router-associated delays imposed on packet delivery, and possibly additional delay associated with such compression techniques as low bit rate speech coding and/or voice activity detection with silence suppression, if these techniques are in use. Given these factors, one-way delays in the packet network environment can easily exceed 100 ms.

While this is still below the usual requirement of 150 ms, problems arise when a call originates and/or terminates on a digital cellular service. One digital cellular link adds approximately 90 ms to the connection delay. A cellular-to-wireline call carried over a packet network thus accumulates almost 200 ms of one-way delay. Where both parties are using their cell phones the connection delay approaches 300 ms. Other access/egress technologies such as DSL and cable telephony also add nontrivial delay to the overall connection when using forward error correction schemes to increase their service reach.

11.2.3.2   Packet loss

The packets carrying the voice information along with the packet header information are sometimes 'lost,' or more accurately, discarded. There are two major causes of packet loss on backbone IP networks. One is congestion, where the network routers/switches are offered more packets than their queues can accommodate. The other is untimely delivery of the packet to the egress gateway, such that the time of packet delivery is outside the delay-smoothing capacity of the gateway's de-jitter buffer. In either case the packet is not processed normally but rather discarded. Thus, the information in the packet is not passed on to the customer.

The effect of packet loss on the customer depends on at least four factors:

1   percent packet loss, or the rate of loss, that is, the percent of transmitted packets lost 0.01%, 0.1%, 1%, 10%?
2   packet loss distribution, or what is the pattern of packet loss? Are the packets being lost in a random fashion over time, or are 'bursts' of consecutive packets lost?
3   packet size, since larger packets contain more voice information; and,
4   packet loss concealment strategy, or the strategy used to 'fill in' or conceal the lost packet.

In early 'whitepapers' covering VoIP the threshold of 5% packet loss was adopted as that point where packet loss becomes a serious performance problem. The support for this 5% threshold was not clearly spelt out and recent studies suggest a more conservative threshold. (It is interesting to note that the application of facsimile transfer requires relatively stringent network packet loss performance, and that the desire of IP network providers to carry Fax traffic has required packet loss performance levels of well below 1%. For more detail on IP performance issues for facsimile service see Section 11.3.) Establishing a hard threshold for packet loss is complicated by the related issue of the background packet loss pattern.

Voice quality studies reinforce the intuition about packet loss pattern: When lost packets occur together, in a burst, the audio effect for the customer is significantly worse than when packets are lost one at a time. This is because bursty packet loss results in a larger segment of speech being lost or distorted, causing an impairment that is much more noticeable to customers. Furthermore, the larger the segment lost, the harder it is to cover up or conceal. The actual pattern of lost packets seen on any actual network is variable. Depending on the traffic load and the moment-to-moment health of network elements, both random and bursty patterns can be found. The prudent network manager should assume the worst, that packet loss will be bursty often enough that plans for controlling the percent of packet loss on a network need to assume this (bursty) pattern.

If, for whatever reason, a scheduled packet is not available for 'play out' at its scheduled time, a substitute packet is played. What kind of substitute packet is played? An obvious strategy is to simply play out a packet length's worth of 'silence.' Where lost packets are rare and occur randomly in time, this strategy of playing silence has little if any quality effect. But when the overall packet loss percentage approaches 1%, especially where the packets are being lost in bursts, silence substitution is associated with enough distortion of the speech signal to significantly reduce customer opinion of transmission quality.

To improve matters, alternate strategies have been developed that attempt to conceal the fact that a packet has been lost. Concealment here is in terms of the effect on the customer. These strategies attempt to offer the customer something other than silence, something that makes the audio stream sound less interrupted and choppy. Relatively simple packet loss concealment strategies re-play the last valid packet to fill the gap created by a single lost packet. If more than one packet in sequence is lost, either this same last-good-packet is continually replayed (maybe at an attenuated level), or noise is played out, or the silence strategy is then adopted. Of course, these concealment algorithms cease once valid packets become available. In general, these simple strategies improve matters relative to doing nothing (i.e. playing silence), but it has been observed that the strategy of repeatedly re-playing the last good packet can, depending on its duration, intensity and frequency of occurrence, introduce a 'squeal' that is unacceptable.

More sophisticated techniques for concealing packet loss, techniques that play out an estimate of the speech information lost, have been developed and are incorporated into the recommendation for certain low-bit-rate coders. This is not true for the G.711 coder, but work is going on in the standards arena to gather industry acceptance for some loss concealment techniques that can be used with G.711 [7]. Results to date indicate that these more sophisticated techniques do very well at hiding lost G.711-based packets. These techniques work well enough to allow the packet network provider to accommodate some increased level of packet loss over what can be tolerated for voice with the simpler techniques.

Figure 11.7 shows the results of an experiment measuring customer MOS as a function of random packet loss for three lost packet concealment strategies, one of which is an AT&T experimental algorithm.

11.2.3.3 Compression strategies and effects

Currently some packet networks carry full-rate coded speech, that is, speech coded with the 64 kbps G.711 algorithm used in the circuit-switched network. Other packet networks, in order to save bandwidth, make use of one of the available low-bit-rate codecs such as 8 kbps G.729 or G.723.1 that can operate at as low as 5.3 kbps. In addition, some packet networks employ silence suppression where periods of non-speech (or silence) are detected by a voice activity detector and are suppressed, or not transmitted, through the network. With this strategy the bandwidth that would otherwise be used to carry packets of silence information through the network can be reassigned to another 'connection' where active speech is present.

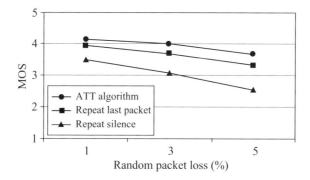

Figure 11.7   Loss concealment effects

A separate area of scientific research has grown out of the desire to code speech at rates less than 64 kbps. An early need for such low-bit-rate coding schemes was on transoceanic voice paths where capacity was expensive and, therefore, a precious commodity. The need has exploded with the evolution to digital cellular networks where, again, the available bandwidth is limited.

In the design of these low-bit-rate coders alternate strategies for reducing the coding rate are evaluated against the goal of maintaining the highest possible sound quality. The coders that have emerged, and are in use, have some amount of distortion associated with them. However, for some of these coders, for example the G.729 codec, the quality reduction they bring is small enough that the industry has generally accepted them as 'carrier grade' or 'toll quality.'

When it comes to choosing a particular speech coder, the packet network provider must decide whether the quality reduction introduced by that coder will combine with other sources of quality degradation, such as network packet loss, to result in an overall quality reduction that is too severe. Again, the decision is one of trade-offs. The choice of coder can affect how tightly packet loss needs to be controlled. Also, since there is a delay penalty associated with low-bit-rate coders, the choice of a particular coder may be driven by the available delay budget.

Another issue is whether the choice of coder will limit service options. While a particular coder might introduce an acceptable level of quality reduction when it is used once on an end-to-end basis, the quality penalty accumulates with multiple, or tandem, encodings. The speech signal would receive multiple encodings by the same codec on a connection where a teleconferencing bridge is in use (because the audio mixing function requires that the non-wave form, low-bit-rate coded speech be transcoded to another form for mixing and then the mixed signal re-coded with the low-bit-rate coder for play out from the bridge). Multiple encodings, but of different codecs, would also occur if digital wireless customers are carried over a backbone packet network that employs compression.

Figure 11.8 shows the mean opinion score for various coders as a function of packet loss and end-to-end one-way delay. Clearly the full bit rate coders, G.7.11, provide the best QoS.

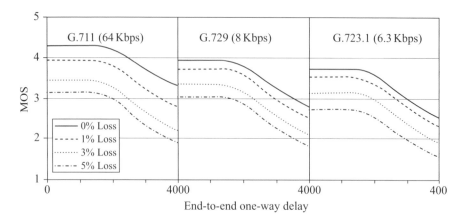

Figure 11.8   Coder, packet loss and delay effects

11.2.3.4   Silence suppression

The capacity constraints associated with transoceanic links spurred the design of the compression technique called silence suppression. In general, the speakers on a connection are only actively speaking less than 50% of the time. Also, any connection has some amount of 'dead air' as the speaker pauses to take a breath. These two factors result in the transmitting circuit being idle on average about 60% of the time. By designing a voice activity detector that can identify this idle time, the transmitting network element can share existing bandwidth among multiple speakers by 'stuffing' speech information from one speaker into the bandwidth that would otherwise be used for transmitting silence from another speaker.

This technique can also be used on packet networks. There are problems, however. One problem is that the up-front voice detector can falsely declare a period of speech as one of silence, especially at the transition point where a silence period ends and speech begins. This results in clipping of the speech stream that can make the speech sound distorted and can reduce intelligibility by clipping off important bits of the speech signal. This tendency for the voice activity detector to clip the speech stream can be significantly reduced, and thus controlled within acceptable limits, by careful design.

Unfortunately, there is a second potential impairment associated with silence suppression that keeps this compression strategy out of popular use. This impairment is that of noise-pumping. Parties on a connection are often located in noisy environments and this background noise gets picked up by the transmitting telephone and sent through the network to the listening party. Under normal conditions, where silence suppression is not in effect, the listening party hears one of two states, speech + noise or noise alone. With silence suppression in use, the listening party hears either speech+noise or quiet. Under just about any background noise condition, this cycling of the transmission of speech + noise and no-noise (quiet), called noise-pumping, is unnatural and causes customers to lower their opinions of the network

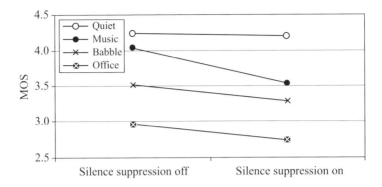

Figure 11.9   Effects of silence suppression on MOS

transmission quality. This quality penalty is especially acute when the background noise is something identifiable, such as music in the background.

Figure 11.9 shows customers' mean opinion scores as a function of different background noises with silence suppression activated or not activated. Obviously, when the background is quiet there is not much difference in mean opinion score ratings. However, when music is played, people are clearly expecting to hear the next part of the music. Activating silence suppression causes a large degradation in peoples' opinions of those calls.

One strategy to counter this noise-pumping impairment is to have the decoder at the end of the connection play out 'comfort' noise instead of silence. More sophisticated implementations of this strategy do a spectral analysis of the background noise and try to shape the play-out noise to better match the original. This strategy has only limited benefit given the challenge of matching noises that range from the simple, such as the steady whine of a computer terminal in operation, to the complex, such as the lyrics to a favourite song. There is work underway to do a better job at reproducing the noise signal for the listening party; to date, this work is still confined to the laboratory.

How a network provider chooses to build a network consisting of different speech coding techniques, the use or not of silence suppression/voice activity detection, packet sizes, signal classification, de-jittering buffer sizes, loss concealment strategies, and echo control, can obviously lead to different qualities or grades of voice service.

Table 11.3 shows a hypothetical differentiation in grades of voice service going from gold to bronze when a network provider has made different choices among delay, voice activity detection, packet loss and coder bit rate. This is simply an illustration of what is possible on these networks because of these new technologies.

11.2.4   Summary

While the VoIP market is small today, it is definitely the way of the future and growing. QoS is the number one concern among the customer base. People expect the QoS that they have become accustomed to on today's circuit-switched networks. This section

Table 11.3   Multiple grades of voice service

Level	Delay	VAD/SS	% Packet loss	Coder
Gold	<150 ms	NO	<0.1%	G.711
Silver	<400 ms	NO	<1.0%	G.729
Bronze	>400 ms	YES	<5.0%	G.723.1

has shown that QoS is measurable and manageable in these new networks. It encompasses the performance of the underlying network as well as the VoIP technologies in the network, and is the key to customer acceptance of VoIP.

11.3   Fax over IP networks

11.3.1   Introduction

The dramatic growth of IP networks has introduced new methods of transporting fax images. Traditionally, two fax machines were connected to the PSTN via analogue loops and communicated directly with each other through a dedicated, circuit-switched path. Now for reasons of economics and convenience, an increasing number of faxes are being transmitted over hybrid PSTN–IP networks and in some cases, through completely IP networks.

Users with access to the Internet can now send and/or receive faxes using e-mail or web applications. These services can provide businesses with substantial savings by reducing or eliminating the per minute charges incurred on the circuit-switched network. New and existing network providers have also reduced the costs of long distance calls by building IP networks which interconnect with the PSTN and provide long haul transport both domestically and internationally.

Although there has been significant growth in the number of faxes transmitted or received directly over the Internet, there is an embedded base of more than 70 million fax machines connected to the PSTN [8], which is continuing to increase every year. A substantial majority of all fax traffic still originates and/or terminates at these PSTN-based machines. Customers whose fax machines work well on today's circuit-switched network, expect their fax service to continue to work well as networks evolve to IP.

11.3.2   Architecture

Figure 11.10 provides a diagram of a hybrid PSTN–IP network. As it shows, faxes can be transmitted or received directly to or from a PC or other devices directly connected to an IP network. However, IP telephony gateways also make it possible to transmit faxes from fax machines connected to the PSTN over IP networks.

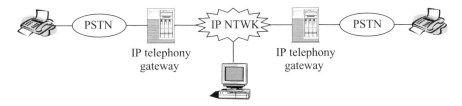

Figure 11.10   Fax over a hybrid circuit switched & IP network

Users who send and receive faxes directly over the Internet typically subscribe to a specialized IP fax service. To send a fax they use software on their PC to transmit the fax image to a server associated with the service. This could be done using e-mail or via a web application. If the fax is being sent to a fax machine connected to the PSTN, the telephone number of the fax machine is included in the e-mail address used or is entered as part of the web application. The server completes the fax transaction by making a normal telephone call over the PSTN to the receiving fax machine and transmitting the fax image using the standard fax protocols given in ITU-T Recommendation T.30.

Faxes sent to a user directly connected to the Internet reverse this technique. As part of the IP fax service, the users are provided with their own unique telephone number for receiving IP faxes. When a fax machine connected to the PSTN dials this number, it is routed to one of the IP fax service's servers. The server acts like a receiving fax machine and stores the received fax images as files. When the fax call is completed, the server sends the fax as an attachment to an e-mail to the user.

One new development depicted in Figure 11.10 is the use of an IP network to transport faxes between two fax machines connected to the PSTN. In this situation, the IP Telephony Gateways packetise and depacketise the information transmitted by the fax machines. These gateways typically use one of two techniques to convert the fax call to packets. The first is to treat the fax call like a voice call and put the pulse code modulation (PCM) information the gateway receives from the PSTN directly into packets. The second technique gateways can use is to demodulate the information in the fax call before packetising it and then remodulate the signal at the terminating gateway before it is sent back into the PSTN. This technique is called fax relay. ITU-T Recommendation T.38 provides guidelines for packetising V.21 protocol information and the demodulated fax image data. Using this technique, the gateways minimise the bandwidth required to transmit the fax call and can also treat voice traffic differently, which allows the use of low-bit-rate codecs and silence suppression for voice, as discussed earlier in this chapter.

11.3.3   Measurement of quality of service

Users who transmit faxes over hybrid PSTN–IP networks use the same criteria to measure performance as they do for faxes transmitted over connections which just use the PSTN.

The majority of the time, the category having the largest impact on overall customer satisfaction with fax is User Information Transfer and specifically, the speed, accuracy and reliability of it. As Table 11.4 shows, the performance of fax in these areas is measured through the following parameters:

- Time it takes to complete a fax transaction;
- Speed the fax machines use to send each fax page;
- Percentage of pages that are transmitted without any errors;
- Percentage of pages which have severe errors (usually requiring retransmission of the page);
- Percentage of fax transactions successfully completed.

The key parameters impacting fax performance are the transaction completion percentage and the percentage of severely errored pages. These parameters measure how often users have to set up additional calls to retransmit fax pages. Newer fax machines have features like error correction and the ability to store fax images, so that the machine can make multiple attempts to transmit a fax without user intervention. Currently, about 90% of fax machines being sold have the capability to use error correction techniques. Error correction has significantly reduced the percentage of errored and severely errored pages that occur. However, since fax machines use retransmission to correct errored information, error correction does increase the transaction time making it a more important parameter. Poor connections that cause many errors can significantly increase the transaction time and can noticeably increase the cost of sending faxes. This increase in transaction time can negate any savings users may gain from using networks which charge cheaper per minute rates, but offer lower transmission quality.

11.3.4   Fax QoS

As was mentioned earlier in this book, the terms accessibility, continuity and fulfilment are currently being used to describe the technical quality of various services.

The accessibility of fax services has two distinct components. The first component is the ability to establish a connection between two fax machines. This is impacted by the availability of facilities or bandwidth in the networks handling the call and is measured by the call completion rate. The second component of accessibility is the ability of the fax machines to use all of their features. In packet networks that demodulate and remodulate fax signals, the packet gateways can modify the fax protocol messages to limit the features that the fax machines use. This can reduce the transmission speed used by the fax machines from 33.6 to 14.4 or 9.6 kbps. This restriction is typically due to limitations of the demodulation/remodulation software in the gateways. The gateways can also restrict the use of nonstandard features that some fax machines use to optimise fax transmissions. There is currently no standardised metric for measuring cases where the network limits the capabilities used in a fax call.

The continuity of fax services measures the ability to complete the full fax transaction. This is primarily measured by the percentage of completed transactions. Incomplete transactions can be caused by call cutoffs, transient network problems during the fax call, severe distortion from noise or echo, or lost packets. The T.30 protocol used by fax machines was designed to be very robust to most network

Table 11.4   *Quality matrix: Real time fax over IP*

	Speed	Accuracy	Availability	Reliability	Security	Simplicity	Flexibility
Technical sales and planning	Response time	% Correct info	Hours can be accessed	% Optimal information	Confidentiality	Ease of contact	Options and alternatives
Provisioning	Time to deliver	% Correct	Hours can be accessed		Confidentiality	Ease of contact	Options and alternatives
Technical support	Time to respond	% Correct info	Hours can be accessed	% Accurate information	Confidentiality	Ease of contact	Options
Repair	Time to repair trouble	% Correct	Hours can be accessed	% Robust	Confidentiality	Ease of contact	Options
Technical quality							
• Connection establishment	PDD	% Wrong number	% Blocked, % No response	Outage due to network failure	% Bridged connection	Ease of digits dialed	Alternate routes, Options
• User info transfer	Transaction time, % pages at each speed	%Error free pages, % severely errored pages	Drop outs	% Completed transactions	Probability of fax packets being monitored		
• Connection release	Call clearing time	% Incorrectly cleared					
Billing	% Late	% Correct	Frequency		Fraud	Understandable	Delivery methods
Network/service mgmt. by customer							

problems. When errors occur in protocol messages or messages are lost, most fax machines request a retransmission of the last message. Most fax machines also are capable of downspeeding to as low as 2.4 kbps to accommodate connections with poor transmission quality.

The fulfilment of fax services measures the quality of the transmitted images and the time it takes to complete the fax transaction. The quality of fax images is measured by two parameters: the percentage of error free pages and the percentage of severely errored pages. Pages that have severe errors, typically, must be retransmitted. Errors in fax transmissions can be caused by high levels of distortion, transient events like impulse noise or gain hits, or lost packets. The introduction of error correction techniques into most fax machines has significantly reduced the incidence of errored fax pages. However, error correction requires the fax machines to retransmit any frames of information with errors which can significantly increase the transaction time of fax calls. The transaction time is also significantly affected by the transmission speed the fax machines use. The transmission speed that fax machines can achieve is closely correlated with the transmission quality of the connection they are using.

11.3.5   Factors affecting IP fax performance

The introduction of an IP network into a connection primarily affects the parameters discussed above through impairments in the IP network and the design of the Internet telephony gateway. IP networks add three impairments that can affect fax performance: lost packets, delay and delay variation. The loss of packets can have a major impact on fax calls transmitted over hybrid PSTN–IP networks. Because the fax information is being transmitted in real time between the two fax machines, lost packets cannot be retransmitted and they cause errors in protocol messages or fax image data. Low levels of nonbursty lost packets will force most fax machines to send protocol messages multiple times if they are corrupted. However, some fax machines will disconnect the call if a V.21 protocol message contains a silent period caused by a lost packet. If error correction is not used, lost packets that occur when image data is being transmitted will cause errored pages. Bursts of lost packets will cause severely errored pages that will usually require retransmission. High levels of lost packets could cause fax calls to disconnect before completing and if error correction is used, will significantly increase transaction times.

11.3.5.1   Delay

IP networks also add significant amounts of delay to connections. Most fax machines are normally robust to delay though and are not affected by the range of delays added by an IP network, unless echo is also present on the connection. Since Internet telephony gateways use Echo Cancellers (EC) to eliminate echo, delay should not be a problem for fax performance. However, if the gateway's ECs cannot sufficiently cancel echoes, the echoes of protocol messages can cause fax machines to terminate connections.

Delay variation can be a serious problem in IP networks. If the gateways do not have de-jitter buffers that can guarantee that packets will be received by the time they need to be transmitted to the receiving fax machine, then delay variation will cause an increase in lost packets. Depending on the fax machines being used, this could cause fax transaction failures, errors in fax images, or noticeable increases in transaction times.

Currently, IP networks treat fax packets the same as all other packets. This means that fax packets which require very low lost packet rates are treated the same as packets for other applications, like e-mail or web browsing, which do not require as stringent lost packet rates to have acceptable performance. If IP networks implement QoS indicators within packets which give higher priority to applications like fax, noticeable improvements in fax performance could be achieved. As IP networks evolve, this is an important area for guaranteeing that fax performance will be acceptable.

11.3.5.2   Signal classification

A number of decisions made in the design of the gateway to handle fax calls can have an impact on performance. If the gateways do not use fax relay to demodulate and remodulate the fax signals, an initial design decision is whether to treat fax calls the same as voice calls. If the gateways are using G.711 codecs without silence suppression for voice calls, then fax calls can be treated the same. However, if voice calls are compressed using low-bit-rate codecs like G.726 or G.729A or if silence suppression is used, then fax calls will have to be treated differently. Using a low-bit-rate codec with fax will reduce the speed at which pages are sent. Silence suppression can cause portions of the fax signals to be clipped causing transaction failures. If fax calls are treated differently from voice calls, then the gateways will require signal classifiers to identify fax calls and give them special treatment. Poorly designed signal classifiers can cause fax transaction failures if fax calls are misclassified at either the start of the call or during the call.

A signal classifier is also needed if the gateways use fax relay to demodulate the fax signal and use techniques like those given in ITU-T Recommendation T.38 to transport fax packets across the IP network. If the gateways do use fax relay, they can also modify the information contained in the fax signals. Some of the V.21 protocol messages used in a fax call exchange information about the capabilities of the fax machines. These include the maximum speeds at which the fax machines can transmit images, the resolution level the machines support and are using, whether the machines support error correction mode and whether the machines support any nonstandard features which could be used to improve performance. If the gateways do not support some of the features the two fax machines have, it can modify the protocol messages sent by the two machines to eliminate features it cannot support. This could significantly reduce the speed used during the fax call and could eliminate the use of error correction during the call. Both these changes could substantially impact the performance perceived by the user.

11.3.5.3   De-jitter buffer

The design of the de-jitter buffer used by the gateways can also have a substantial impact on fax performance. If the buffer size is too small, delay variations in the IP network could significantly increase lost packets. However, increasing the size of the buffer directly increases the end-to-end delay of calls which can degrade other applications like voice. Some gateways are now implementing dynamic de-jitter buffers which change size in reaction to the delay variation occurring during a call. This can optimise the buffer size to minimise the delay added to calls and the number of packets lost because they arrive too late to be transmitted. However, because changing the size of the buffer during a call means that information either has to be added or deleted to the actual signal being transmitted, every change to the buffer size could create errors in a fax signal. These errors could require retransmission of protocol messages or image data or if error correction is not being used will create errors in the fax images being transmitted. If the size of the buffer changes frequently, the errors added to the fax signal could cause the transaction to fail.

11.3.5.4   Packet Loss Concealment

A last design decision for the gateway that impacts fax performance is the Packet Loss Concealment (PLC) algorithm used by the gateway. If a packet is lost in the IP network or arrives too late to be played by the terminating gateway, the gateway must replace the packet with something. Typical PLC algorithms replace lost packets by playing silence, repeating the last good packet, or using information from previous good packets to estimate what would have been in the lost packet. Various combinations of these algorithms have been implemented in Internet telephony gateways. Fax machines are particularly sensitive to silence periods during fax signals. Some machines will terminate a call if silence is detected during a protocol message or image data. Most fax machines will not terminate calls if lost packets are replaced with a signal that resembles the modulated signal it was expecting. However, replacing lost packets with previous packets or estimates based on previous packets creates errors in the signal and will require retransmission of protocol message or image data and could cause errors in fax images.

11.3.6   QoS measurement results

An analysis of fax performance over the IP network of a start up company offering cheap rates for voice and fax calls provided some interesting results. The service required callers to dial into one of the company's Internet telephony gateways which were located in various regions around the USA. After connecting to the gateway, the caller went through an authorisation process and then dialed the PSTN telephone number for their destination. The company indicated they supported fax service, but provided no quality guarantees. Testing between various domestic sites showed that the IP-based service had a transaction completion rate of only 85% compared to 99% for calls between the same sites over the PSTN. Only 30% of the pages transmitted over the IP network were error free compared to 99.5% over the PSTN. One of the most interesting results was the speeds the fax machines achieved over the IP network. The gateways in the network actually modified the fax protocol messages that indicated that the fax machines could transmit at 14.4 kbps. The gateways changed these messages to indicate that only 9.6 kbps was supported. Even with this change, the fax machines had to drop their speeds to lower rates than 9.6 kbps on 25% of the test calls made over the IP network. In comparison, every test page transmitted over the PSTN was sent at the maximum speed of 14.4 kbps.

Although these results are from a very early implementation of an IP network supporting fax calls, they underline the impact of the Internet Telephony Gateway's design and the performance of the IP network on fax performance.

11.3.7   Summary

Although only a small percentage of fax traffic is currently routed over IP networks, it is growing rapidly like VoIP traffic. Customers using IP fax services expect QoS levels near those achieved over the PSTN and use the same parameters to measure fax QoS. The QoS of IP fax services is closely related to the performance of the IP network used to transport fax packets and the design of the Internet telephony gateways used to packetise fax calls. If the IP network has excessive lost packets, fax transactions could fail or contain severe errors which will require retransmission of fax pages. The use of fax relay, low-bit-rate codecs, silence suppression, PLC algorithms, and dynamic de-jitter buffers in Internet telephony gateways can all significantly impact fax QoS.

11.4   Streaming media

11.4.1   Introduction

The advances of technology, communication protocols and commercialisation of the Internet has made many communication applications both possible and popular using the Internet. Initial Internet applications started out as access to text, such as e-mail and bulletin boards, which later progressed to text plus image access in the form of web pages. The progress has continued to other applications such as streaming audio and video, which require a larger and more consistent bandwidth, and which are more sensitive to network impairments.

Streaming Media services are defined in this context as rich audio/video content 'played' over the Internet by streaming packets from the host location of the content, as opposed to downloaded content played from a user's hard drive [9].

Initially, streaming video clips lasting for a few minutes became popular at web sites used for reporting news and for advertising purposes. Later, streaming video started being used for broadcasting events lasting hours. Some companies such as AT&T have started testing a 'Video on Demand' Internet streaming video service aimed at 'Pay-per-view' providers looking for another avenue to distribute their programmes, such as sports events and concerts, internationally.

Even though the quality of video viewed on the Internet today is considerably inferior to that of television, the growth of streaming media (i.e. video accompanied with audio) has been phenomenal because of ease of access. The NetAid concert in October 1999 set a world record for the largest Internet broadcast event for a single day – 2.5 million streams [10]. The BBC, in the previous year estimated that its streaming audience was growing by 100% every 4 months [11]. In addition, businesses often use streaming videos to enhance their enterprise communications with employees around the world.

Similar to the situation with voice services offered over IP networks, users cite a poor viewing experience as a leading reason why they are not using the service more often. But as the technology evolves to make higher and more consistent bandwidth available to the users including the 'last mile' connection to the users' computers, the viewing experience is expected to improve for the users resulting in streaming media services becoming more popular.

This section provides an overview of the working of streaming media and parameters that relate to its QoS.

11.4.2 The architecture of streaming media

Video streaming is a new, evolving, and changing technology. Industry wide standards have not yet been established. Many products currently available on the market are proprietary and may not be compatible with others [12]. A high level description of how a typical streaming media application works is provided below (see Figure 11.11).

11.4.2.1 The client

Once connected to the Internet, the user accesses a streaming video by selecting the appropriate URL by clicking on the hyperlink. This action directs the user to a server

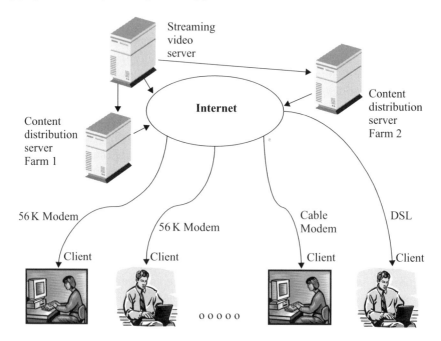

Figure 11.11   Streaming media architecture

where the requested video resides. The server then asks the user questions about the connection speed. The choices include modem speed, such as 28 or 56 kbps, or broadband, such as 110 or 300 kbps etc. There is also a choice of type of streaming video player, such as Real Player, Windows Media Player, QuickTime etc. The higher the connection speed, the better the resolution, and the viewing window can be correspondingly bigger.

After the selections of the connection speed and the player, a video viewing window opens on the screen and the user's machine, the client, tries to establish a connection to the video server over the Internet. After a few seconds the connection gets established and the buffering process begins, where, over the next several seconds, the player builds up a buffer of streamed packets before it starts to play the content. The function of a buffer is to maintain a steady stream of packets during the play session and minimise interruptions that may otherwise occur due to network impairments. The player determines the buffer size based on the bandwidth available at the time the connection is established. The buffer size is optimised for viewing quality. The aim is to determine the smallest buffer size for the video to play without having to rebuffer. If the buffer size is too large, the start up time is longer. If it is too small, the buffer could run out of packets and will need rebuffering, thereby, affecting the viewing quality.

Once the buffer is full, the player starts and the user can view and listen to the streamed video. If the stream of packets slows down due to heavy network traffic or other reasons and the buffer is emptied, the player stops and waits for rebuffering before continuing with the playing.

11.4.2.2   The streaming media server

The streaming application content resides in a file on one or more servers connected to the Internet. In case of live events, the content is generated near real time and made available on the servers. Each server farm can provide thousands of simultaneous video streams. Depending on the need, one or more server farms are used for streaming. Each 'server farm' consists of multiple servers.

The original content is created and placed on the streaming server by following these key steps:

1   The video is recorded in analogue or digital format.
2   The content is converted into a digital file (avi, mpeg, mov, etc.) and captured on the hard disk using a video capture card (e.g. Osprey, Winnov, Intel).
3   The digital file is prepared using digital video editing software, and/or video switching equipment.
4   The digital video file is compressed and encoded for the Internet (tools include RealSystem Producer, Windows Media Encoder etc.) and placed on the server for distribution.

For a given video clip, there are multiple file versions encoded and placed on the server for distribution to match with the users' connection speed and streaming player. Companies like RealNetworks [13] and Microsoft Windows Media [14] now offer encoders with features that create a single streaming file to accommodate multiple connection speeds. In this case, the server determines a user's connection speed based on the established connection and streams the packets accordingly. During the course of streaming video viewing, if the connection speed changes due to network conditions, the server dynamically adjusts the stream to optimise the viewing experience. Because the streaming audio/video performance is sensitive to bandwidth fluctuations, the streaming server is connected to the Internet with a constant high bandwidth connection. Because the audio and video applications are more sensitive to delay variation than an occasional packet loss, the packets are streamed using the User Datagram Protocol (UDP).

11.4.3 Quality of Service measurement

Just as streaming media technology and services are evolving, so are the methodology and algorithms for monitoring the performance of these services. To measure the QoS from the end user perspective, it is important to identify which parameters affect the viewing experience.

To ascertain which parameters affect the end user perception of service quality, subjective testing is used. Streamed videos created over a wide range of possible options and delivered over a wide range of possible network conditions are assessed by a number of test subjects representing a typical target audience. The test subjects rate each rendered video on a base scale, such as 1–5, or 1–10. Based on the subjective ratings and the associated objective measurements of performance parameters, a model or an algorithm can be developed to translate the objective measurements into an equivalent subjective rating. The model thus developed is called the Grade of Service model. The model gives an idea of which parameters affect the viewers' perception of service quality and to what extent. These models are still being developed.

11.4.3.1 Service performance parameters

The end user experience is a function of both encoding the streaming media and its rendering. Thus parameters related to both encoding and rendering are captured for service monitoring. The key areas that determine performance and the associated parameters are as follows.

- *Startup time*: This is the time it takes from pressing the play button until the video clip begins to play. Startup time equals the total time required to initiate the streaming connection (includes DNS look up and time to first byte), Redirection Time and Initial Buffering.

  - *Initial connection*: This is the time it takes to establish a connection between the streaming server and the streaming client (or player).
  - *Redirection time*: This is the time it takes to transfer the request to the final server that renders the video.

- *Initial buffering time*: This is the time it takes to start viewing and hearing a streaming media clip from the time the data starts arriving in the buffer of the client.
- *Video quality*: This covers the visual performance during the play session. It is determined by the parameters described below. The first two are related to service design, while others reflect network impairments.

  - encoded bit rate of the video stream when it was created;
  - encoded frame rate of the video stream when it was created;
  - frames that were lost or dropped during the play session;
  - packets lost or arrived late during the play session;
  - video bandwidth observed during the play session – its average and the associated variation;
  - number of times rebuffering occurred and the time it took for rebuffering.

- *Audio quality*: This part covers the audio performance during the play session. It is determined by the parameters described below. The first two are related to service design, while others reflect network impairments.

  - number of audio channels;
  - encoded bit rate per channel;
  - packets lost or arrived late during the play session;
  - audio bandwidth observed during the play session – average and the associated variation;
  - number of times rebuffering occurred and the time it took for rebuffering.

Players such as Real Player and Windows Media Player can provide measurements for most of the above-mentioned parameters.

Performance monitoring companies such as Keynote Systems, Inc. [15] measure and report streaming media performance from the end user perspective.

11.4.3.2   Technical quality in the QoS matrix

The QoS matrix includes the service quality related to many functions such as sales, provisioning, technical quality, billing, repair and technical support. Considering the free availability of streaming media applications, the technical quality aspect is the most pertinent at present and hence that is covered here. As the services evolve to include 'Pay per View' and subscription based services, other functions will also gain importance and are shown in the QoS matrix in Table 11.5.

From the end user perspective, service performance for streaming media can be categorised into the following three areas.

- *Accessibility:* This is the ability of a user to access the application and initiate a transaction. For streaming media, it would mean the ability of a user to establish a connection with the streaming server.
- *Continuity:* This is the ability to experience the service with no interruption, given that accessibility has been successful and a transaction has been initiated. In the case of streaming media, if the streamed file gets played until the end of

Table 11.5   QoS matrix for streaming media

	Speed	Accuracy	Availability	Reliability	Security	Simplicity	Flexibility
Technical sales and planning	Response time	% Correct info	Hours can be accessed	% Optimal information	Confidentiality	Ease of contact	Options and alternatives
Provisioning	Time to deliver	% Correct	Hours can be accessed		Confidentiality	Ease of contact	Options and alternatives
Technical quality							
• Accessibility (Connection establishment)	Connection time	% Correct site connected	% Site not found, % Request timed out	No connection due to network failure		Ease, # of client information questions	
• Fulfillment (User information transfer)	Packet delay	Packet loss, bandwidth, Frame rate, Bit rate		Loss of signal, # of Rebufferings			
• Continuity (Connection release)				% Incomplete rendering, Loss of connection			
Billing							
• Delivery	% Late	% Correct	Frequency		Fraud	Understandable	Delivery methods
• Inquiry	Response time	% Correct	Hours can be accessed	% Accurate information		Ease of contact	Delivery methods
Repair	Time to repair trouble	% Correct	Hours can be accessed	% Robust	Confidentiality	Ease of contact	Options
Technical support	Time to respond	% Correct info	Hours can be accessed		Confidentiality	Ease of contact	Options

file is reached, or until the user stops the playing, it is considered a successful transaction. If not, it is considered a defect.

- *Fulfilment:* It is the ability to deliver service meeting the customers' quality expectations. In the case of streaming media, this would be the quality of the audiovisual experience of the user during the play session. It would be a function of encoding quality as well as rendering quality. A fulfilment score will depend upon recorded measurements and the algorithm used to compute the score.

11.4.3.3   Defects Per Million

For streaming media services, the three performance areas mentioned above – accessibility, continuity and fulfilment can be defined and monitored in terms of Defects Per Million (DPM) transactions as daily and monthly averages. In addition, the Unified DPM (UDPM) metric can also be defined as a single number to present the overall performance of a streaming media service based on the accessibility, continuity and fulfilment measures. DPM monitoring allows for identifying trends in service performance over time and detecting events when a DPM value exceeds the set threshold.

- *Accessibility DPM:* For the accessibility performance measure, 'active monitoring' can be employed, that is, tests are conducted to gather data. In active monitoring, a test platform initiates test transactions based on a predetermined sampling plan and the data is collected. The transactions are monitored for their success or failure to connect to the streaming media server. The DPM estimate is then calculated based on the number of failures and the number of attempted transactions on a daily basis.
- *Continuity DPM:* Continuity DPM can be calculated using 'passive monitoring', that is, no testing is done for this purpose. This is possible if data from the streaming servers is available indicating the number of transactions that were interrupted before reaching the end of the file and the total number of transactions (number of times the video was streamed). If the server data is not available, the data from 'active monitoring' can be used for the DPM estimate.
- *Fulfilment DPM:* Fulfilment DPM can be calculated using the data from 'active monitoring'. Using an appropriate algorithm and the data gathered for the test transaction, a fulfilment score can be computed to represent the audiovisual quality of that transaction. After collecting sufficient data to define the baseline, a performance threshold is established based on the transaction time for the 99th percentile after removing the outliers. The transactions with scores lower than the defined threshold are considered defects. The DPM numbers are then calculated and reported as their daily DPM value, and their average daily DPM value for each month. To calculate the DPM value, the fraction defective is converted to per million basis and then 10,000 (1%) is subtracted to compensate for the fact that while the 99th percentile was chosen to define the threshold, all transactions belonging to the population are considered a success, except for the outliers. If the subtraction results in a negative number, the DPM is considered to be zero.

11.4.4   *Factors affecting streaming media quality*

The Internet is basically a data network, that is, it was created to transmit data. Data networks are bursty in nature. This is because when a file or a message is sent, the computer/machine tries to send as much information as it possibly can, limited mainly by the network connectivity and capacity. As a result, network utilisation can change rapidly, particularly for smaller networks, and can lead to network congestion. Network congestion can then lead to impairments such as loss of packets, delay in packet transmission, and variation in delay (jitter). In the case of 'static' applications such as e-mail and web sites that involve transfer of text and still images, the performance degradation caused by the network impairments can be corrected by retransmitting the packets with the help of Transmission Control Protocol (TCP). The extra time added for the retransmission does not affect the user experience significantly for these static applications.

In the case of streaming media, however, a large, and more importantly, consistent bandwidth connection is needed for good performance. This is because the streamed packets are being played continuously, and the delay involved in retransmission of any lost packets using TCP could affect the performance just as much, or possibly more. As a result, UDP which does not use retransmission is used for streaming applications. The streaming media players use one of several algorithms to compensate for the lost packets and minimise performance degradation. Even so, depending upon the number of packets lost, performance is affected to some extent.

When we consider the above two facts together in terms of what the Internet has to offer and what the streaming media ideally needs for good performance, it becomes clear that the match is not perfect. Unlike the cable or satellite television broadcasts which have dedicated bandwidth channels, streaming media applications try to work with variable and sometimes inadequate bandwidth. This is a challenging proposition. An overall increase in the Internet backbone bandwidth has helped the cause of streaming media by reducing network congestion and, therefore, the associated packet loss. Increasing the 'last mile' bandwidth to the users' computer should also help the streamed media performance.

Creating the video file ready for distribution and setting up the distribution channel are the two primary areas that the streaming media producer has under control that can affect the service quality. Because creating a video file involves working with and optimising many parameters, it is also an art.

For creating a video file, there are four steps. The first step of a video creation for streaming involves lighting, camera movement, type of camera, types of lenses etc. and is beyond the scope of this section. In general, a steady source is preferable rather than rapid camera movement and fast transitions. The latter can result in jerky images due to the necessary encoding done later. Also, it is preferable to have shots that are not too wide, because viewers with modem connections have smaller viewing windows [16].

The next two steps relate to video capture and video editing. There are several products available on the market with different features, some more popular than others.

The next crucial step in creating a streaming video file is video encoding. The need for video encoding arises because of the huge size of uncompressed digital video files. A typical five minute uncompressed NTSC (the standard for television in the USA) video can take more than one gigabyte of space. Size makes the uncompressed video files unmanageable for storage as well as for transmission. Compression (encoding) reduces the file size so that the video can be stored and transmitted at the cost of lost resolution.

The goal in video encoding is to produce a high quality video that streams with minimal interruptions and has a quick download time. Creating such encoding involves trade-offs. On the one hand, if the video has a high frame rate, a large viewing window, and a high data rate, the video will be good quality, but will only be viewable by people with high bandwidth connections. On the other hand, if it is encoded with speed and delivery in mind, the perceived video quality will be lower. Finding a balance between video quality and connection speed that best serves the target audience becomes the key.

The following parameters need to be considered and decided upon during video encoding to achieve the best possible user experience at the receiving end.

- *Encoder:* The choice(s) would depend upon which viewer is most often used by the target audience. In addition, for narrowband (modem) connections, some encoders may work better than others. The video and audio codecs that are part of the encoder affect the encoded stream quality. While RealVideo, Microsoft Windows Media, and Quicktime have encoders that are proprietary, MPEG-4 with an open format is also emerging as an alternative.
- *Data speed:* Most 56 K modems connect at a lower speed in practice. The choice of data speed has to balance the need for frequent re-buffering against using most of the available bandwidth for a better user experience. Very often data speed around 40 K is used as a balance.
- *Window size:* This depends on the bandwidth. For 56 K connections, a 160×120 pixel window is often used. A larger window would necessitate the lowering of the frame rate to achieve similar bandwidth requirements.
- *Frame rate:* The usual 30 frames per second NTSC frame rate may be compressed to 5 or 6 frames per second for a 56 K modem connection. For broadband, 10–15 fps may be used [17].

When the streaming server starts sending the packet streams, some packets are likely to be lost due to network congestion – a fact beyond the video producer's control. What is possible to control at the server end is the connection between the server and the Internet. This connection should be a high bandwidth, dedicated connection to minimise any packet loss at the distribution end.

Distributing the streaming media using a good hosting service is another way to assure better quality. Hosting or content distribution services typically employ server farms to provide more streams and better reliability. They can also provide the servers at geographically diverse locations and closer to the viewers. Servers being geographically closer can reduce the packet loss to provide better quality.

11.4.5 Summary

Starting with e-mail and web sites, the progression of Internet applications has continued to other areas such as streaming audio and video, which have larger and more consistent bandwidth requirements, and are more sensitive to network impairments.

Even though the quality of video viewed on the Internet today is considerably inferior to that on television, the growth of streaming media has been phenomenal because of the ease of access. The applications include news clips, commercials and live events broadcasting. In addition, businesses often use streaming videos to enhance their enterprise communications with employees around the world.

The streaming media application works by creating an encoded file and making it available on a streaming media server. Viewers can access the file over the Internet and play it on the client machine.

The quality of streamed video is determined mainly by how the video was created (use of the camera), how it was encoded (choice of encoding parameters), and what the network impairments were at the time of playing the streamed video.

Streaming media players help capture performance data that can be used to assess the service performance after processing the captured data. Streaming media performance can be viewed in the QoS–Technical Quality matrix covering the accessibility, continuity and fulfilment measures. The service performance can then be expressed as DPM charts to monitor performance trends over time, as well as daily exceptions.

11.5 References

1 CHRISTENSEN, C. M.: 'The Innovator's Dilemma' (Harvard Business School Press, Boston, 1997)
2 'IP Telephony Market Forecast and Analysis, 2002–2007', Elizabeth Farrand, IDC, 2002
3 ITU-T Recommendation P.800: 'Methods for subjective determination of transmission quality' (1996)
4 ITU-T Recommendation G.107: 'The E-Model, A Computational Model for use in Transmission Planning' (2000)
5 Controlling Echo in the Worldwide Telecommunications Network: A Primer, Tellabs Operations, Inc., 1995
6 ITU-T Recommendation G.114: 'One-way Transmission Time' (1996)
7 ITU-T, Study Group 16, Contribution D.249: 'A High Quality Low-Complexity Algorithm for Frame Erasure Concealment (FEC) with G.711' (1999)
8 100 Million Potential On-Ramps: 'The Worldwide Fax Machine Installed Base and Market Forecast – 1999–2000', DAVIDSON, P. J.: International Data Corp., 1998.
9 CULLINANE, L. and MARWAHA, J.: 'Streaming Media: When Will It Take Off?', TelephonyOnline.com, 26 November 2001
10 Cisco Systems, Inc.: (1999) NetAid Web site sets world record for largest Internet broadcast; More than one thousand organizations join initiative to fight extreme poverty [Online].: http://www.cisco.com/netaid/pr_101299.html

11   Vision Consulting: (1999) Opportunities in streaming media [Online].: http://www.visionconsult.com/streamin.htm
12   Video Software Laboratory, http://www.video-software.com/
13   RealNetworks, Inc. http://www.realnetworks.com/company/index.html
14   Microsoft Corporation http://www.microsoft.com/windows/windowsmedia/default.asp
15   Keynote Systems, Inc. http://www.keynote.com/
16   'Live Webcast Case Study: Surf's Up', COURTNEY, C.: 12 April 2001, http://hotwired.lycos.com/webmonkey/01/15/index3a.html
17   'Streaming Video for the Masses', MARIONI, R.: 19 January 2001, http://hotwired.lycos.com/webmonkey/01/03/index4a.html

Further reading

1   ITU-T, Study Group 12, Contribution D.114: 'Results of a Subjective Listening Test for G.711 with Frame Erasure Concealment' (Source: AT&T) (1999)
2   PERKINS, M. E., DVORAK, C. A., LERICH, B. H. and ZEBARTH, J. A.: 'Speech transmission performance planning in hybrid IP/SCN networks', *IEEE Communications Magazine,* July 1999, pp. 126–31
3   BAKER, M. R.: 'Speech transport for packet telephony and voice over IP', in R. O. Onvural, S. Civanlar, J. V. Luciani (Eds): 'Internet II: Quality of Service and Future Directions', 11/1999. Proc. SPIE **3842**, pp. 242–51
4   COLLINS, D.: 'Carrier Grade Voice over IP' (McGraw-Hill ISBN 0-07-136326-2)
5   ITU-T, Study Group 12, Contribution D.110: 'Subjective Results on Impairment Effects of IP Packet Loss' (Source: Nortel Networks) (1999)
6   Rec. T.30, 'Procedures for Document Facsimile Transmission in the General Switched Telephone Network', 3/93.
7   Rec. E.450, 'Facsimile quality of service on public networks – General aspects', 3/98.
8   Rec. E.451, 'Facsimile call cut-off performance', 3/93.
9   Rec. E.452, 'Facsimile modem speed reductions and transaction time', 3/93.
10   Rec. E. 453, 'Facsimile image quality as corrupted by transmission-induced scan line errors', 8/94.
11   Rec. E.457, 'Facsimile measurement methodologies', 2/96.
12   Rec. E.458, 'Figure of merit for facsimile transmission performance', 2/96.
13   Rec. T.38, 'Procedures for real-time Group 3 facsimile communication over IP networks', 6/98.
14   Akamai Technologies, Inc. (1999) Our network [Online].: http://www.akamai.com/service/network.html
15   SAVOLAINE, C. G.: 'Quality of service issues for voice in an IP environment', presented at the IEEE GLOBECOM 2000 conference, November 2000.

Chapter 12

Quality of Service for non-real-time Internet applications

12.1 Introduction

This chapter describes the two major non-real-time (elastic) Internet services of 'electronic mail' (e-mail) and 'Web applications' for electronic commerce (EC) electronic business (EB) services. It first provides a high-level architecture of each of these services and then identifies, defines and classifies key Quality of Service (QoS) parameters of the corresponding service. It focuses on the end-users' viewpoint of QoS for each specific service. Further, it will put emphasis on service-specific QoS. Section 12.2 applies the methodology to generic e-mail, and in particular to the AT&T WorldNet(R) e-mail services offering. Section 12.3 applies the methodology to Web hosting as well as the Content Distribution Service (CDS), the two emerging services widely used by businesses for their EC and EB services.

12.2 E-mail application

E-mail is a major Internet application. It is the most popular tool of communication for businesses and consumers. Users have come to expect a high quality of service from the Internet mail services. Most of the intra-company e-mail traffic is on the Local Area Network (LAN) or Wide Area Network (WAN), and most of the inter-company e-mail traffic as well as most of the consumer/residential e-mail traffic flows over the Internet. The growth and use of Internet mail has prompted an interest in the mail services provided by major Internet Service Providers (ISPs).

This section provides a generic architecture of Internet mail and how it works, describes an implementation example – the AT&T WorldNet(R) mail service and formulates the measurement of QoS of e-mail services.

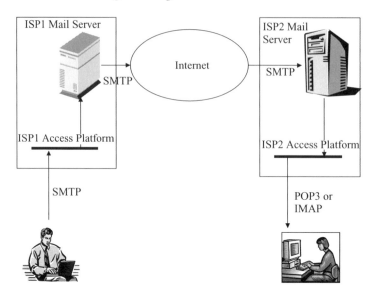

Figure 12.1   A high-level e-mail architecture

12.2.1   E-mail architecture

This section describes a generic e-mail pathway, that is, how a mail message would travel from one user to another user. To make the case more general, assume that the two users are being served by different ISPs. The end-to-end communication between the users can be divided into three parts: (1) mail travelling from the sender's computer to the ISP computer, (2) mail travelling from one ISP computer to the other ISP computer, and (3) the receiver accessing the mail using their computer, from their ISP computer. Figure 12.1 shows a high-level architecture of e-mail [1].

Each of the computers runs software applications to perform the tasks of establishing a connection, authentication and mail transfer as needed. The user computers run the software that allows them access to their respective ISP networks. The access is controlled by the ISPs for billing and security reasons. The user computer thus has to first establish a connection to the ISP access network from where the user can then access the mail service. The ISP access network connection is made either over a normal phone line and modem dial-in, or over a broadband connection – either via cable or DSL connection, described in Chapter 10. In the corporate environment, the network access is typically setup over a LAN.

A mail client is a software application that runs on the end-user machine. The client allows the user to create and send messages, as well as receive messages. A mail client is also known as Mail User Agent (MUA).

Each ISP has a set of computers running mail applications within its network. An Internet mail server is a software application that provides the mail services of transmitting and receiving mail across the Internet. It is also known as Mail Transfer Agent (MTA). The mail server works with client software to send and receive messages from the users.

The machines transfer mail messages using a common protocol. For the first two parts – mail travelling from the sender's computer to the ISP computer and from one ISP computer to another – Simple Mail Transfer Protocol (SMTP) is used. For the third part, the end-user computer accessing mail from the mail server, a few different protocols exist. The Post Office Protocol Version 3 (POP3) has been used most commonly, while Internet Mail Access Protocol (IMAP) is now growing in usage [2,3].

With POP, a personal computer user periodically connects to the 'server' machine and downloads all of the pending mail to the 'client' machine. Thereafter, all mail processing is performed local to the client machine. POP provides a store-and-forward service, intended to move mail (on demand) from a server to a user machine, usually a PC. Once delivered to the PC, the messages are typically deleted from the POP server. This operation may be referred to as offline mode because once the mail is downloaded the user can process the mail without being connected to the network. In this case, the mail is typically available from just one designated computer.

The IMAP protocol is designed to permit manipulation of remote mailboxes as if they were local. With IMAP, the mail client machine does not normally copy it all at once and then delete it from the server. It is more of a client–server model where the IMAP client can ask the server for headers or the bodies of specified messages, or to search for messages meeting certain criteria. With IMAP, all e-mail, including the inbox and all filed mail, remains on the server at all times. The client computer must be connected to the server for the duration of the e-mail session. A temporary copy of the mail or part of it is downloaded to the client computer while the user reads it or sends it, but once the connection is finished, the copy is erased from the client computer. The original remains on the server. IMAP thus works in an online mode, and the mail is accessible from multiple computers.

12.2.2   *AT&T WorldNet(R) e-mail service*

The generic description of mail service provided above translates into specific implementations by the ISPs. An overview of AT&T's WorldNet(R) mail service is covered here as an example. The WorldNet(R) service is a dial-up service that uses modems and phone lines. Broadband and dial-up services function similarly beyond the ISP network access. The main difference between the two is the ISP network access. In the case of broadband, the high-speed ISP network connectivity through a cable or a DSL is available on the user premises.

The WorldNet(R) mail system performs the following functions:

- receives e-mail from the Internet or from WorldNet users via AT&T Common Backbone (CBB);
- temporarily stores mail until retrieved by the user;
- downloads mail to user's PC on demand;
- routes outgoing e-mail from WorldNet(R) users to the Internet or other WorldNet users;
- allows multiple mailboxes for a single user account;
- stores and updates user account information.

12.2.3   AT&T WorldNet(R) architecture

A typical user accesses the AT&T WorldNet(R) service from the computer via a dial-up modem by dialing one of the modem pool numbers to first reach AT&T's Dial Platform. Once the modem pool is reached and the modem communication is established, an IP/PPP (Point-to-Point Protocol) connection gets set up to allow the user computer to communicate with the Network Authentication and Authorization Services (NAAS) server. The user computer then provides the login name and password to gain access to the Dial Platform. The access to the mail services from the Dial Platform is depicted in Figure 12.2 [4]. Once the user has successfully logged on to the Dial Platform, access to mail services can proceed as described below.

The WorldNet(R) mail system is comprised of various interconnected server machines running their associated software processes. The category names and basic functions of the various machines with the respective server applications are as follows.

1 *Mail Gateway*: Routes incoming mail from the Internet to the PostOffices, bounces errored mail, and prevents Spam (unsolicited e-mail).
2 *Mailhost*: Sends outgoing mail to the Internet and sends mail for WorldNet customers to the PostOffices.
3 *PostOffice*: Holds the incoming mail for WorldNet users in their respective mailboxes.
4 *POPServer*: Authenticates mail access and delivers user mail from the PostOffice mailbox to the user.
5 *WebMail Server*: Provides mail access via browser.
6 *ISD (Integrated Services Directory)*: ISD provides detailed user account information such as mailbox names and passwords and is used to update all the directory cache databases.

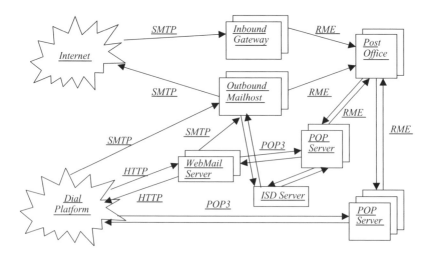

Figure 12.2   WorldNet(R) mail system architecture

12.2.4 E-mail transactions

There are two primary transactions that a user performs with reference to an e-mail application:

- *Sendmail*: Sending one or more mail messages to other users on the Internet or on WorldNet(R) service.
- *Getmail*: Accessing and/or getting mail, downloading messages from the user's mailbox on the PostOffice.

Mail services are accessed via mail client software such as Outlook Express, Eudora, or a browser residing on the user's computer. The mail client software runs the necessary scripts to communicate with the mail server. The details of this interaction are transparent to the user.

Sendmail

Using the SMTP protocol, the mail client sends the mail message(s) to the Mailhost. The Mailhost forwards the mail messages to the appropriate PostOffice servers if they are addressed to WorldNet users. If not, they are forwarded to the Internet. The client sends a sequence of commands to the server in the process of sending the mail. If any of these commands fail, the server log file creates an error entry.

Getmail

Mail from the Internet arrives at the Mail Gateway machines. Mail Gateways screen the mail for errors and Spam (unwanted and unsolicited e-mail) and then forward it to the appropriate PostOffice servers where the user mailboxes are located.

The primary mode of accessing the mail is the POP3 protocol. Using the POP3 protocol, the mail client establishes a connection with a POPserver, from which the mail is accessed. The user id and password are provided to the POPserver. Based on information from the ISD server database, the user is authenticated and the POPserver knows the user mailbox (Message Store ID). The mail client then issues a sequence of commands to download the mail. The POPserver then forwards the mail residing on the PostOffice mailbox to the client. Any failures occurring during the process of retrieving mail are written to a log file residing on the server.

12.2.5 E-mail QoS

Because Internet mail service is so important for the users, it has become important for the ISPs to improve their services. Service improvement starts with monitoring the existing service. In the United States, independent performance monitoring companies such as eTesting Labs have started monitoring and reporting ISP performance for e-mail services.[1] The testing includes a sample of dial-up platform POPs in various metropolitan areas across the country for the different ISPs. The reports are issued monthly. The measures reported include the failure rates to access the dial-up

---

[1] eTesting Labs E-Mail Benchmark Report, US, Dial Edition, July 2001.

platform and mail throughput rates. However, the industry is still in the process of defining and standardising a universally acceptable set of parameters related to the user's perspective of the e-mail QoS. Additionally, while these external QoS measurement benchmarks are helpful in knowing how one ISP performs in comparison to others, extensive internal testing within the ISP is done to identify areas of concern for the purpose of performance improvement.

This section identifies, defines and formulates parameters of dial-up access platform for e-mail services. As far as the e-mail QoS parameters are concerned, the dial-up access encompasses all other access technologies. As indicated in Chapter 10, user's perspective of QoS of Internet-based applications is characterised within the accessibility, continuity, and fulfilment framework.

1   *Accessibility*: It is the ability of a user to access the application and initiate a transaction (Getmail or Sendmail). For both Getmail (receiving mail) and Sendmail (sending mail), the access consists of two parts – first an access to the e-mail service Dial-Up Platform, and then the access to the appropriate mail server. A user will experience e-mail failure if either a Getmail or Sendmail transaction attempt fails in either one of the above two access stages – access to the Platform or access to the mail server, given access to the Platform has been successful. This has led to the identification and definition of six dichotomous QoS parameters specified in Section 12.2.6.
2   *Continuity*: It is the ability of delivering the service (e-mail) with no interruption, given that accessibility has been successful and a transaction (sending or receiving e-mail) has been initiated. If the message gets sent or received completely with a proper connection closure, it is considered a successful transaction; otherwise it is classified as a failure. There are two dichotomous continuity parameters defined in Section 12.2.7.
3   *Fulfilment*: It is the ability of delivering the service meeting the customers' quality expectations. In the case of e-mail, the quality of a transaction can be expressed in terms of throughput or speed. Once an access is established for Getmail or Sendmail, the time required to complete the transaction (send or receive e-mail) determines fulfilment. E-mail transactions experiencing a longer time are of lower QoS. There are two continuous parameters defined in Section 12.2.8.

12.2.6   Accessibility QoS parameters

For accessibility, 'active' monitoring is employed, that is, e-mail tests are conducted to gather performance data. The measurement system actively initiates e-mail (Getmail or Sendmail) test transactions and monitors and records success or failure of the transaction at each of the access stages. The test transaction to access mail is monitored at each step for its success or failure, and the data is used to calculate accessibility completion (or failure) rates. The following six dichotomous accessibility parameters are measured, estimated and monitored over each specified time horizon such as hourly, daily, weekly or monthly.

1 *Line failure rate – Line FR*: This is failure to connect to the modem because of line problems (busy, no carrier, ring-no-answer). Let A be the number of attempts to access an e-mail service Dial-Up Platform (accessing the Dial-Up Platform includes available local telephone line, modem connect, modem synchronisation, and IP/PPP setup and authentication). Let $F1$ be the number of attempts failed to connect to a modem. Then,

$$\text{Line FR} = F1/A$$

2 *Modem synchronisation failure rate – Modem Sync FR*: This is failure of the modems (user's and dial-up Platform) to communicate properly even though connected. Let A be the number of attempts to access the Dial-Up Platform, and $F2$ be the number of modem synchronisation failures. Then,

$$\text{Modem Sync FR} = F2/(A - F1)$$

3 *IP/PPP setup and authentication failure rate – IP/PPP Setup/Auth FR*: This is the IP or PPP negotiation failure or authentication to the e-mail service (even though the modem connected and synchronised). Let A be the number of attempts to access the Dial-Up Platform, and $F3$ be the number of IP/PPP setup or authentication failures. Then,

$$\text{IP/PPP Setup/Auth} = F3/(A - F1 - F2)$$

4 *POPserver connect failure rate – POPserv Connect FR*: This is failure to connect to the POPserver after Dial Platform connection. Let B be the number of attempts to access the mail (i.e. connect to the POPserver and authenticate) and $F4$ be the number of failures to connect. Then,

$$\text{POPserv Connect FR} = F4/B$$

5 *POPserver authentication failure rate – POPserv Auth FR*: This is failure to authenticate to the POPserver even though connected. Let B be the number of attempts to access the mail and $F5$ be the number of failures to authenticate to the POPserver. Then,

$$\text{POPserv Auth FR} = F5/B$$

6 *Mailhost connect failure rate – Mailhost Connect FR*: This is failure to connect to the Mailhost after the Dial Platform connection and authentication has been completed. Let B be the number of attempts to send the mail (i.e. connect to the Mailhost) and $F6$ be the number of failures to connect. Then,

$$\text{Mailhost Connect FR} = F6/B$$

12.2.7   Continuity QoS parameters

For continuity, the 'passive' monitoring approach is utilised, that is, the performance data is collected from network elements (mail server log files). The mail servers keep log files that provide information on each mail transaction as to whether or not the transaction was successful. Mail server log files also indicate failure causes. Root cause analysis is often conducted using the passive data. The information from all mail servers is gathered on the Log server. The Log server has a complete record of the total number of attempts and failures on each server. The following two dichotomous continuity parameters are measured and monitored over each specified time horizon such as hourly, daily, weekly or monthly.

1   *Sendmail continuity failure rate*: This is the failure to complete the transaction of sending mail from the client to the mail server, once the transaction was initiated. Let $C1$ be the total number of Sendmail requests initiated on all Mailhost servers. Let $F7$ be the total number of failures while sending the mail. Then,

$$\text{Sendmail Continuity FR} = F7/C1$$

2   *Getmail continuity failure rate*: This is failure to complete the transaction of receiving mail from the mail server to the client, once the transaction was initiated. Let $C2$ be the number of initiations to receive mail on all POPservers and $F8$ be the number of failures while receiving the mail. Then,

$$\text{Getmail Continuity FR} = F8/C2$$

12.2.8   Fulfilment QoS parameters

For the fulfilment, the 'active' approach is used to measure the QoS parameters. E-mail test transactions are sent and received – just the way a typical end user would. The time taken to send and receive the test e-mail is recorded. There are two continuous fulfilment QoS parameters: Sendmail response time and Getmail response time.

1   *Sendmail response time*: This is the time it takes to complete the transaction of sending mail from the client (user's computer) to the mail server. In practice, mean, median and certain percentiles are used to monitor the Sendmail response time. Sometimes a certain threshold, say X, is specified and Sendmail transaction times exceeding the threshold X are classified as failed transactions. Sendmail failure rates are then calculated as follows. Let $G1$ be the number of 'Sendmail' test transactions completed, and $F9$ be the number of transactions for which the Sendmail time exceeded the threshold X. Then,

$$\text{Sendmail FR} = F9/G1$$

2   *Getmail Response Time*: This is the time it takes to complete the transaction of the client receiving e-mail from the mail server. Similarly, if Y is the threshold, $G2$ is the number of 'Getmail' test transactions completed, and $F10$ is the number of transactions for which the Getmail time exceeded the threshold Y. Then,

$$\text{Getmail FR} = F10/G2$$

Table 12.1   QoS parameters of Dial-Up e-mail

Function	Dichotomous parameters	Continuous parameters
Accessibility	• Line Failure • Modem Synchronisation Failure • IP/PPP Setup and Auth. Failure • POPserver Connect Failure • POPserver Auth. Failure • Mailhost Connect Failure	
Continuity	• Sendmail Continuity Failure • Getmail Continuity Failure	
Fulfilment		• Sendmail Response Time • Getmail Response Time

Table 12.1 summarises the QoS parameters of a Dial-Up e-mail service within the accessibility, continuity, and fulfilment framework:

12.3   Web applications

E-commerce is an emerging business concept for buying, selling or exchanging goods, services, information or knowledge via the Internet. Individual users reach the Web sites of e-commerce businesses through the Internet infrastructure introduced earlier. The User (client) is a service-requesting computer that generates a request and receives and reads information from a service-providing computer. The service-providing computer (Web server) receives, processes and responds to the user's (client's) requests. Depending on their function, servers may be classified into 'Web server', 'application server' or 'database server'. The application server works in conjunction with the Web server. It may process the user's orders, and check the availability of desired goods. The database server provides access to a collection of data stored in a database. A database server holds information and provides this information to the application server when requested.

E-commerce businesses need to build, operate and maintain servers within the Web. They need to acquire the hardware and software necessary to operate web servers. Web sites often require data storage equipment, database servers, security firewalls and other related software and hardware. Building and operating Web sites in-house may be very costly and require technical expertise. For this reason many organisations choose to outsource their Web sites to Web hosting service providers. There are many Web hosting companies that host Web sites for their customers and provide space on a Web server and frequently offer an entire suite of services to remove the technical burdens and allow e-commerce businesses to focus on their core competency. Section 12.3.1 provides a short description of a suite of Web hosting services currently offered by many service providers. The description focuses on segments of Web hosting that are related to identification and formulation of the QoS of a Web application from the viewpoint of the end users' experience.

12.3.1 Web hosting services

Web hosting service provides the means for businesses to, totally or partially, to outsource support for, and management of their Web sites. It stores customers' content and provides users with a reliable and fast access to the customer's site and the stored content. The content could be data, audio, video or multimedia. Here, the 'customer' is defined as the e-commerce business or organisation that outsources the Web site to a hosting service provider and user is defined as an individual who accesses the customer's Web site through the Internet. End 'user' is the person or system, who attempts to access the Web site, 'customer' is the e-commerce business and/or content provider, and 'service provider' is the telecommunication company that maintains the Web site and provides access to customer's users.

The hosting service provider may offer *shared, dedicated* and/or *co-location* support which are differentiated as follows [5]:

- *Shared hosting service* – Multiple customers' Web sites are hosted on a single Web server.
- *Dedicated hosting service* – A single customer's Web sites are hosted on one or more Web servers exclusively assigned to that customer.
- *Co-location hosting service* – Customer uses service provider's rack space within the Internet Data Centres (IDCs) and service provider's bandwidth for Internet connectivity. Customer supplies own Web servers and manages server content updates.

12.3.1.1 Shared hosting service

Shared service is targeted towards small and medium size businesses. The shared service offering may include (1) internet connectivity, (2) domain name registration, (3) basic Web design templates and support, (4) twenty-four by seven (24 × 7) site monitoring, (5) basic reports and (6) credit card transaction processing [6].

12.3.1.2 Dedicated hosting service

Dedicated service is targeted towards businesses with high volume Web site traffic; businesses whose Internet presence plays a mission critical role and businesses that cannot afford delays associated with shared servers. Dedicated service may be further differentiated as follows [5]:

- *Simple dedicated service* – A single customer's Web sites are hosted on a single dedicated Web server.
- *Complex dedicated service* – A single customer's Web sites are hosted on multiple dedicated Web servers using hardware and software supported by the service provider's standard offerings.
- *Custom dedicated service* – A single customer's Web sites are hosted on dedicated Web servers using some hardware and/or software customised to specific needs.

The dedicated service may include (1) internet connectivity, (2) domain name registration, (3) customised Web site design, (4) server capacity load balance, (5) firewalls, (6) 24×7 monitoring, (7) systems administration services and (8) project management services [6].

12.3.1.3 Co-location hosting service

Co-location service is targeted towards businesses that desire high-speed site delivery and performance, but want to control server applications and content updates. The co-location customer basically uses the service provider's rack space within the IDC and the service provider's bandwidth for Internet connectivity. The customer may use some minimal service provider monitoring services, but generally manages its servers and services.

Web hosting providers capable of complementing service through extending their offerings to meet these more sophisticated customer needs or able to establish part-nerships to acquire support to meet these customer needs view these rapidly emerging services as synergistic to their existing businesses. Other Web hosting providers may view these trends as competitive to their existing businesses.

12.3.1.4 Internet data centres

Internet Data Centres are secure facilities in which hosting services are provided and managed. These data centres are protected with multiple power back-ups and servers are secured against external intrusion with firewalls. The content on the servers is backed up on multiple sites for disaster recovery. All IDCs have high-speed connectivity to the common backbone.

As indicated in Chapter 10, users connect to the Internet in many different ways including dial-up, Cable, DSL or dedicated access. Users' requests are routed from their ISPs through the peering links and backbones to the hosting service provider's IDC. The server that has the requested content retrieves it and forwards it to the user. Figure 12.3 shows a high-level architecture of a typical web application:

12.3.2 Content distribution service

Web hosting service enables customers to serve their users with a reliable and fast web site and content access. However, for some customers with a large global user base, Web hosting in a few IDC servers is not very efficient. Although Web hosting is more easily scalable and will provide faster and more reliable access to a customer's Web site, it can become a victim of its own success. It still does not provide the fastest and the most efficient access if Web requests were viewed from a geographic perspective. Web acceleration service – also known as Content distribution service (CDS) – is a clever solution for these and other problems.

The CDS moves the delivery of content close to the end user in order to reduce network cost while improving the user experience. Figure 12.4 shows a high level diagram of how the CDS works. It provides virtual server capacity and virtual band-width to the CDS customer, thus relieving the origin server capacity and connectivity constraints. It is based on the notion of caching 'cacheable' content at various nodes.

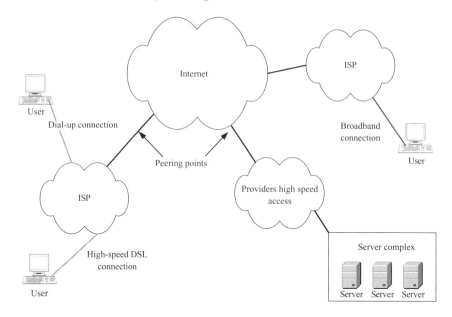

Figure 12.3   A high-level architecture of a typical web application

In a regular hosting scenario, a user's request to a particular 'page' (URL – Uniform Resource Locator) is directed to the origin server where the content resides, by a series of Domain Name System (DNS) resolutions. Thus a client and server set-up a hypertext transfer protocol (http) session to transfer all contents of the page to the client. In the case of CDS, the DNS resolution process will direct the user's request to an CDS node and the http session will be established between this CDS node and the user. The CDS node is chosen dynamically based on some metrics such as server health, network health and proximity. CDS is a powerful tool in this regard and could provide marked improvements to a customer origin server, especially under heavy loads. CDS directs Web requests away from heavily loaded Web servers to better manage the load on servers and Web traffic on the network. In Figure 12.4, a user Web request from ISP A may go to CDS Server Site X, whereas the same request from a user on ISP B may go to CDS Server Site Y.

The main components of the CDS node are the cache servers (which contain faithful, reliable, and up-to-date replica of the content) and the L4 switch, which provides both intra- and inter-nodal (global) server load-balancing. Depending on the particular implementation of the global server load balancing, any one of the L4 switches can act as the authoritative DNS for CDS customers' web pages. The customer's page(s) is (are) provisioned in all the nodes.

12.3.3   Web QoS parameters

An end user's request to a Web site initiates a sequence of sub-transactions that take place between the user's computer and all computers and network elements within

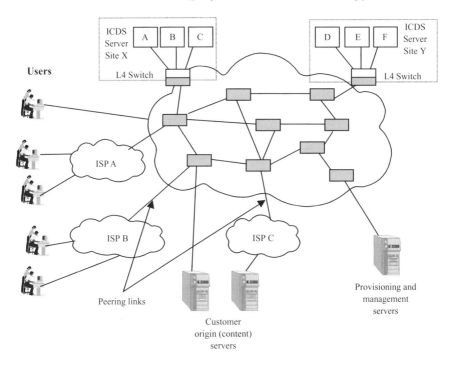

Figure 12.4   ICDS high-level architecture

the end-to-end path that links the user to the server. There are four main phases
associated with each user transaction. These include DNS Resolution Phase, TCP
Connect Phase, HTTP Base-Content Phase, and HTTP Embedded-Objects Download
Phase. The four phases are seamless to the users. Users simply want reliable, fast and
accurate responses to their request. As far as users are concerned they want to be able
to access the site, and fulfil their request with no interruption. This again falls within
the Accessibility, Continuity, and Fulfilment Framework introduced in Section 5,
Chapter 10.

This section uses the framework of Section 10.5, Chapter 10 and the fundamentals
of Section 10.6, Chapter 10 to identify, define and formulate QoS parameters of Web
applications from the perspective of the users' experience.

12.3.3.1   Accessibility QoS parameters

Accessibility is the ability of a user to access the service. For Web applications there
are two phases to the services accessibility: server/network accessibility and page
accessibility.

Server/Network accessibility parameters

Server/Network accessibility determines whether or not a user's request is properly
resolved by the server in a timely manner. It involves resolving the URL via various

DNSs, as well as reaching the server that is hosting the content. Occurrence of any of the following errors will result in the failure of the service in terms of server/network accessibility. There are two types of server/network accessibility failures:

- *DNS lookup failure*: This is the failure of the DNS to resolve the request in a timely manner or to supply the correct resolution of the request. It could be caused by user's ISP DNS complex, or customer's DNS server or hosting DNS server being configured improperly. Failure will be determined either by the user receiving an error message or by exceeding a maximum lookup time. The DNS lookup failure rate can be calculated as

$$\text{DNS Lookup FR} = \frac{\text{Number of DNS Lookup Failures}}{\text{Total Number of Attempted Requests}}$$

- *TCP connection timeout*: Once the DNS request is fulfilled, the browser tries to make a connection to that web site. If the connection acknowledgement is not received within the timeout window set by the browser, a timeout occurs and the web request is aborted.

$$\text{TCP Timeout FR} = \frac{\text{Number of TCP Timeouts}}{\text{Total Number of TCP Connect Attempts}}$$

Page accessibility parameters

After the TCP connection is made, the browser tries to download the page from the server. Even though the connection between client and server is established, this does not guarantee that the base page will be accessible. Page accessibility is measured by the percentage of requests that complete successfully (get access to requested base page), out of the total requests that successfully access the server. There are two types of base page accessibility failures:

- *Page not found*: The server returns a page not found error when the user's browser request for the base page of a Web site is not successful. This error occurs if the requested page does not exist or could not be found on the origin server.

$$\text{PNF FR} = \frac{\text{Number of Page Not Found 'Errors'}}{\text{Number of Page Download Attempts}}$$

- *Page access timeout*: The browser waits for a certain amount of time after issuing a request for a Web page. If that request is not fulfilled within this time, a page timeout occurs.

$$\text{PATO FR} = \frac{\text{Number of Page Access Timeouts}}{\text{Number of Successful Page Download Attempts}}$$

12.3.3.2   Continuity QoS parameters

Service continuity is defined as the user's ability to successfully receive the base page and all the content on that base page. It is measured in terms of success or failure of the customer's requested page content to completely fill the user's screen. There are three parameters to track service continuity:

- *Content Error*: This error occurs when the user receives a partial Web page and gets errors downloading objects or images embedded in the base page.

$$\text{ContErr FR} = \frac{\text{Number of Content Errors}}{\text{Number of Content Download Attempts}}$$

- *Connection Reset*: During the base page or content download, the browser and the server exchange protocol information. If the server does not receive appropriate response, it resets and closes the TCP connection. The Web request is aborted.

$$\text{Conn Reset FR} = \frac{\text{Number of Connection Resets}}{\text{Number of Page or Content Download Attempts}}$$

- *Content Timeout*: Content timeout is encountered when the user's computer displays an error message indicating that the download time has exceeded the user's download threshold.

$$\text{ContTO FR} = \frac{\text{Number of Content Timeouts}}{\text{Number of Content Download Attempts}}$$

12.3.3.3   Fulfilment QoS parameters

Fulfilment is the ability of delivering the service meeting the user's quality expectations. A Web request may be successful based on the parameters listed in the previous sections. A user's experience of a Web request may be classified as unsatisfactory if various response times exceed the user's expectations. Below are the six response time parameters for fulfilment:

- *DNS Lookup Time T1*: The time it takes for a DNS server to respond to the request from a user's browser for an IP address for a particular Web site.
- *TCP Connect Time T2*: The time it takes for the user to receive TCP acknowledgement from the server. Once the IP address is obtained, the browser makes a TCP connection to the Web server. There is handshaking involved before this connection takes place.
- *Time to First Byte T3*: After the connection is established, the browser sends a request for information. The time from when the request is sent to the time the first byte is received is the time to first byte. This parameter gives information about a server's latency.
- *Base Page Download Time T4*: As the name implies, this is the time to download the base page of a Web site. The size of the page affects this parameter. The larger the content, the longer it takes to download.
- *Content Download Time T5*: If there are images or other components embedded in the base page, their download time is measured here. Many times these images are located on servers other than the main server so their performance is also important.

Table 12.2   Web QoS parameters

Function	Dichotomous parameters	Continuous parameters

Accessibility	• Server/network accessibility:
	  • DNS lookup failure
	  • TCP connection timeout
	• Page accessibility:
	  • Page not found
	  • Page access timeout	

Continuity	• Content error • Connection reset • Content timeout	
Fulfilment		• DNS lookup time, $T1$ • TCP connect time, $T2$ • Time to first byte, $T3$ • Base page download time, $T4$ • Content download time, $T5$ • Total response time, T

- *Total Response Time T*: This time is the total time it takes to complete the desired Web transaction. Users are very sensitive to the total response time. It is measured as

$$T = T1 + T2 + T3 + T4 + T5$$

A Web application QoS program would require measurement and monitoring of parameters. Table 12.2 provides a short list as well as a classification of these parameters.

12.4   Summary

This chapter introduced a systematic approach for monitoring, quantification and improvement of the QoS of emerging Internet-based services. It focused on the service specific and end-users' perspective of the QoS – the way Internet service users view the quality. It first described e-mail and Web applications architecture and then identified and defined user's perceivable service quality for each of these applications. Finally, it classified parameters within the QoS framework of accessibility, continuity, and fulfilment.

12.5   References

1   "Internet Messaging Primer" QUALCOMM Inc., Eudora Division. http://www.eudora.com/pdf_docs/primer.pdf

2 'Internet EMail Protocols: A Developer's Guide', Kevin Johnson, Addison-Wesley.

3 'Internet e-mail: Protocols, Standards, and Implementation', Lawrence Hughes, Artech House.

4 BURNS, H., OPPENHEIM, R., RAMASWAMI, V. and WIRTH, P.: 'End-to-end IP service quality: are DPM service transactions the right measure?', *Journal of the IBTE*, 2001, **2**(2).

5 International Data Corporation. *Web-Hosting Services: Market Review and Forecast, 1998–2003*, Analyst: Stephen Murray.

6 The Yankee Group, 'What's Happening in the US Web Hosting Market?', Presenters: Joanna Makris and Michele Pelino, AT&T Labs IRC program, May 3, 2000.

7 VERMA, DINESH, C.: 'Content Distribution Networks: An Engineering Approach' (John Wiley & Sons, Inc. ISBN 0-471-44341-7).

8 The Yankee Group, 'A Framework to Manage Pervasive Content at the Edge of the Network'. Internet Computing Strategies Report Vol. 4, No. 15, December 1999.

9 PSINet, 'Web Hosting Tutorial', The International Engineering Consortium, August 3, 2000.

10 Global Sales Consulting, Web Hosting FAQ's, http://www.globalsc.com

11 The Yankee Group, 'Content Delivery and Web Hosting: Friends or Foe?', E-Networks & Broadband Access Report Vol. 1, No. 5, July 2000.

12 Communications News, 'New business model for ASPs', Timothy Ruberti and Ted Ruberti, pp. 21–23, April 2000.

13 The Yankee Group, 'An Introduction to NetSourcing', Internet Computing Strategies Report Vol. 5, No. 1, March 2000.

14 The Yankee Group, 'Content Delivery Services: Coming to a Server Near You', Data Communications Report Vol. 14, No. 22, December 1999.

15 Telecom Business, 'The Bandwidth Plays On: The Fast Connection Between ASPs and Resellers', Ilene Kaminsky, March 2000.

16 International Data Corporation, 'Web Hosting Trends: ISPs Jump on the ASP Bandwagon', Analyst: Stephen Murray, Document #19324, June 1999.

Exercises

1 Identify and define e-mail QoS parameters for Cable Modem users. Specify accessibility, continuity, and fulfilment parameters and build Table 12.1 for Cable Modem users.

2 Identify and define web application QoS parameters for Dial-Up users. Specify accessibility, continuity, and fulfilment parameters and build Table 12.2 for Dial-Up users.

3 Discuss how to determine threshold(s) for the response time of each of the access technologies of Dial-Up and Cable Modem so that user's response times exceeding a threshold(s) may be classified as a failure. Show how to calculate total response time failure rate.

Chapter 13

Quality of Service in
mobile communication systems

13.1 Introduction

The last few years or so have seen two great revolutions in communications – mobile and the Internet. The Internet phenomenon is actually much wider than just being able to browse the World Wide Web and send e-mail from a home PC – it is really a complete data technology revolution. There is the prospect that IP technologies will be able to integrate the networking of all future applications and services when a QoS enabled IP infrastructure is developed and deployed. Current mobile systems, however, overwhelmingly offer a single service – voice. They are more usefully compared to the Public Switched Telephony Network (PSTN) than IP networks. However, with the launch of so-called third generation (3G) mobile over the next year or so users will be able to access the Internet, including corporate e-mail and so forth, from high bandwidth, quality-enabled, mobile links. The combination of mobile and Internet technologies/applications is considered by many industry commentators to be the next great communication revolution.

Originally, the first public voice systems were based on analogue technology. As such they were not secure, did not efficiently use bandwidth and most countries developed different, incompatible systems, which prevented roaming and increased prices. An example of these, so-called first generation (1G), systems was the UK TACS (Total Access Communications System) platform. The first major advance in mobile technology was the move to digital systems – these offered better performance, both in the number of users able to share a particular block of spectrum and also to the transmitted voice quality, and better security. Moreover one of these second generation systems – GSM – has come to dominate the market for cellular voice systems with 60% of the world market and systems in over 150 countries. The ability of GSM users to roam onto foreign networks, coupled with the reduced costs produced by tight standardisation and high volumes has meant that operators have been able to offer pre-pay mobiles for as little as €40. It is factors such as these that have driven up second generation (2G) mobile phone penetration rates to 80% in the UK and over 100% (more than one per person on average) in Scandinavia. In some countries GSM has been enhanced to provide higher speed access for data traffic: GPRS (GSM Packet Radio Service) also offers per packet charging and an 'always-on' service for data services. GPRS is an example of a 2.5G system.

The next major development for public mobile cellular systems is the introduction of 3G systems offering multimedia, with transfer speeds up to 384 kbit/s or so, as well as traditional voice services. There are two major 3G systems, the European/Japanese Universal Mobile Telecommunication System (UMTS) development and the North American cdma2000 system: both systems are similar and there are development moves to offer roaming between them in the future. As far as QoS is concerned, 3G systems are important in that they offer QoS support for real-time multimedia services – something that will not be available on the Internet for many years.

Looking further ahead it is clear that, as well as higher speeds, fourth generation mobile (4G) will not distinguish between circuit and packet traffic (as 3G does). 4G will utilise a single IP network that provides appropriate QoS for all traffic; in addition, it will no longer have a separate network dedicated to a single mobile access technology but will be flexible enough to allow multiple access technologies.

Mobile systems are not limited to just the public cellular systems that we have been discussing above. BT has just announced that it is rolling out wireless LAN access, offering corporate and fast Internet access, in 20 trial hot spots in 2002, rising to 400 in 2003 and 4000 in 2004. These will be installed in cafes, airports and railway stations. The combination of UMTS and current WLAN technologies is often called a 'system beyond 3G'. Table 13.1 provides a summary of these various mobile generations.

13.2   Difficulties of providing QoS in mobile systems

13.2.1   Introduction

It is convenient to use the OSI 7 layer model or, for IP data, the Internet stack (Figure 13.1) when discussing mobile systems – basically because the mobile-specific challenges are confined to only two layers. This does not make the problem any simpler, often the consequences are felt across all seven layers and custom solutions (vertically integrated) have been the norm mobile systems up to the very latest 3G developments.

The first major challenge is the physical layer – for all practical mobile systems this is a radio channel (typically in the range 500 MHz to 3 GHz). We will look at the vagaries of the radio channel in detail in the next section but, in summary, the radio channel typically has a very high and variable error rate compared to fixed connections. This is in fact becoming more of a problem as fixed networks now typically run over almost error free networks (fibre systems provide error rates of 10^{-9} compared to 10^{-3} for a typical wireless link) and most common applications/protocols typically assume that errors are negligible.

Mobile systems that are providing wide area coverage, such as GSM, divide coverage areas into cells (Figure 13.2) and use, in the case of GSM, a different

Table 13.1   Summary of different generation mobile standards

	1G	2G	2.5G	3G	Beyond 3G	4G
Date of UK launch	1980	1990	2001	2003	2004	2010
Services	Analogue voice	Digital voice Short Message Service (SMS)	Fast access to WAP Internet/Intranet Access	Digital voice, video and multimedia	UMTS services and very fast access to Internet/Intranets in hot spots Integrated services 802.11 WLAN and UMTS	Ad hoc network extensions IP multicast services
Example system	TACS	GSM	GPRS	UMTS		BRAIN [1]
Structure	Cellular	Cellular	Cellular	Macro and micro cells	UMTS and WLAN picocells	Enhanced UMTS with Hiperlan and enhanced Hiperlan Micro and pico cells
Connection speed (data)	8 kbit/s	9.6 kbit/s	14.4–64 kbit/s	384 kbit/s	1 Mbit/s	10 Mbit/s+

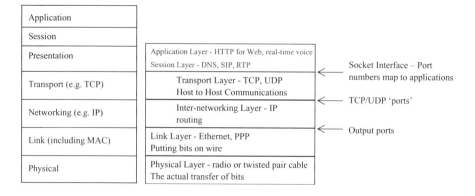

Application
Session
Presentation
Transport (e.g. TCP)
Networking (e.g. IP)
Link (including MAC)
Physical

Application Layer - HTTP for Web, real-time voice
Session Layer - DNS, SIP, RTP

Transport Layer - TCP, UDP
Host to Host Communications

Inter-networking Layer - IP
routing

Link Layer - Ethernet, PPP
Putting bits on wire

Physical Layer - radio or twisted pair cable
The actual transfer of bits

Socket Interface – Port
numbers map to applications

TCP/UDP 'ports'

Output ports

Figure 13.1   The OSI seven layer stack and the Internet equivalent

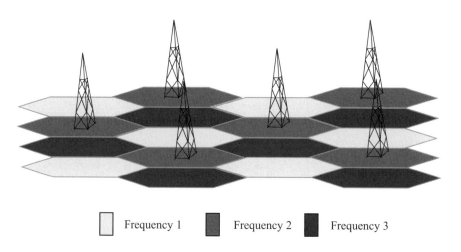

Frequency 1   Frequency 2   Frequency 3

Figure 13.2   Typical cellular mobile system

radio frequency for each cell. Frequencies are reused about every 12–20 cells and this results in interference being created in distant cells using the same frequency. Base stations also serve a number of mobiles and need a Multiple Access Control (MAC) scheme to assign resources to the mobiles for voice/data transmission. The link layer then needs a complex MAC component and also heavyweight error correction protocols that protect against the high error rate on the underlying physical radio layer. Readers who have followed the European 3G spectrum auctions [2] will know that spectrum is usually scarce and expensive. Cellular systems need to operate in the frequency band 500 MHz to 3 GHz to take advantage of low cost silicon technology and reasonable propagation characteristics. However, these frequencies are crowded with many different users, such as satellite links, maritime communications and military applications, and agreeing, globally or even regionally, to re-farm spectrum is a long and complicated process [3]. The cost and shortage of spectrum, as compared to the burgeoning capacities available in fixed networks, means that the radio link is always the bottleneck in mobile systems and its resources must be carefully and efficiently used to provide an acceptable QoS for as many users as possible.

The network layer also needs greater functionality in mobile networks, when compared to the PSTN or fixed computer networks, say, since it must control the admission of new requests for network and radio resources and also provide assistance for handover of active voice/data sessions between base-stations when users move. A mobile that detects the radio channel deteriorating may identify a stronger signal from one or more nearby base-stations and request (or suggest) a handover; if resources are available (or the user priority sufficiently high) then the new base-station will take over the connection/session. The incoming part of the call/session (from the correspondent host/user) to the mobile must be re-routed to the new base-station and further network resources, such as buffers or temporary connections, might be needed to ensure there is no data/speech loss. In any case all base-stations have to reserve some resources purely for mobiles handing over from nearby base-stations and this requirement further reduces the overall utilisation of the spectrum.

The very fact that users are mobile also means that the traffic statistics, the probability of users making calls (voice), sessions (data), are subject to higher variability. For example, the base-stations around a football stadium will only be heavily used once a week, say, and the traffic must rise 10,000 fold for a couple of hours around the game and at half time.

Users of GSM systems might also have tried Wireless Application Protocol (WAP) the mobile equivalent of browsing the Internet. WAP is actually a whole new suite of protocols from application layer descriptions of WAP pages through to WAP Transport Protocol (WTP). The obvious question is why not use HTML over HTTP and TCP/IP (the normal Internet protocols) for displaying Web pages on mobile phones? The answer(s) in fact reveal mobile issues at the transport layer and above.

First, mobile devices are small, battery powered and, hence, tend to have small displays. Although this might change in the future the 17 inch pocket monitor will not be available anytime soon! So pages have to be adapted and WAP pages are specifically coded to fit in small displays. Currently WAP pages must be downloaded over a GSM speech circuit, which provides only a 9.6 kbit/s data rate, and so pages must be small. Finally, typical IP transport protocols, such as TCP are designed on the assumption that they are running over low loss fixed connections – any packet loss is assumed to be due to congestion and causes the TCP protocol to reduce the transmission rate. On wireless links, even the most heavyweight link layer protocol cannot protect data packets from all losses, and transport protocols, such as TCP, perform poorly when a wireless link is included in the path. Hence the need for WAP to replace TCP with a more friendly wireless protocol. The subject of TCP performance over wireless link is taken up in greater detail in Chapter 14.

All in all, the extra variability of mobile traffic, the need for handover support and the fact that the radio link is a bandwidth bottleneck means that mobile systems have to manage resources very carefully and it is very worthwhile to develop complicated resource admission, control and handover procedures. It has also been the case with mobile systems to date that the whole stack – including the applications (speech, messaging, WAP browsing) – have been completely integrated and tailored especially for optimum performance in the mobile environment. In 3G we will see a move to more generic support for data – in particular for QoS-supported IP packet transport that is not tied to a particular service. In 4G the process will probably be completed with no mobile specific applications – rather, IP session negotiation protocols (such as Session Initiation Protocol (SIP) [4], higher bandwidths and more mobile friendly transport protocols will mean that applications will not need to be specifically written for mobile environments and mobile systems will become less tightly integrated.

At the end of the day it is only the QoS at the application layer – be it for voice, video or web browsing – that the user actually experiences. For voice, users want good coverage, good availability (so that the network is never busty), low drop rate (on handover) and intelligible speech. When analogue mobile phones were launched many commentators thought they would be used for emergencies use only, with users switching to a fixed line as soon as possible. In this chapter, we will try to make the connection between the user perceived QoS and how the underlying network provides this.

Inevitably QoS must be traded against cost – base-stations can be erected on every corner to give fantastic capacity, never a dropped call and so forth but it is doubtful that anybody would be prepared to pay the cost incurred. Mobile systems have to set the QoS/cost compromise point very carefully – the radio spectrum is always a scarce resource and this puts resources at a premium.

13.2.2   The radio channel

The simplest possible radio transmission path would be the line of sight between a transmitter (on a base-station say) and a receiver (on a mobile terminal). If the transmit and receive antennae both have a very narrow field of view (or conversely a large gain) then no radiation will be reflected off nearby buildings or trees and so on and, even if it is, then it will not be picked up by the receiver. The signal loss will then be inversely proportional to the distance between the transmitter and receiver squared and will not vary with the wavelength used.

Such a system is, obviously, totally impractical for the sort of mobile systems we are considering – there is hardly ever a direct line of sight from your mobile to the base-station, and to allow full mobility the receive and transmit antennae must cover large angles. The result is that for typical cellular systems multiple, reflected and diffracted, signal paths exists. In a city the signals from the base-station will, typically, have bounced off several buildings and cars/lorries before reaching a mobile. Since these multi-paths have varying lengths, the signals they carry will arrive with different amplitude and phase at the receiver (Figure 13.3) and may combine either constructively or destructively. This is known as multipath fading. Imagine the receiver is moving – perhaps the user is in a car or on a bicycle – then, the received signal will fluctuate in time. It will actually fluctuate quite rapidly – the typical coherence length of the received signal being a wavelength (40 cm at typical cellular frequencies) – so that 10 m/s covers 25 different signal levels or so.

Figure 13.3   Multipath fading

The traditional solution to multipath fading is to use diversity – two or more different signals from the transmitter to the receiver that are uncorrelated – meaning the probability of deep fade for both is very low indeed. Diversity can be achieved by

- space – two receivers or transmitters separated by a wavelength or more;
- polarisation – two orthogonal polarisations;
- time – transmitting the same data at a time when the mobile has moved at least a wavelength;
- frequency – transmitting the data on a different wavelength that has a different fading pattern.

This latter point is very important – if we observe the received multipath signal at a given (static) point, by how much do we have to shift the wavelength to move from constructive to destructive interference? The coherence of the radio channel is actually the reciprocal of the RMS delay spread (D) of the various multipath components.

One solution to multipath fading is therefore frequency diversity – simply transmit data fast enough so that the bandwidth exceeds $1/D$ and some part of the signal will always be in a constructive phase. Unfortunately, the different multipath components are now separated in time by more than a symbol period and cause signal distortion or Inter-symbol Interference (ISI) when they arrive out of synchronisation at the receiver. Viewed in the frequency domain this is dispersion.

Dispersion and multipath fading are, therefore, two sides of the same coin. A typical solution to multipath dispersion is to employ an equaliser. This is a device that essentially estimates the frequency response of the channel and applies a filter to make the channel response appear flat with wavelength. Obviously, it takes time to estimate the channel and if the mobile is moving at high speed then, by the time a good estimate of the channel has been obtained, the channel will have changed. Nevertheless, many cellular mobile systems are designed to be wide-band and employ some form of equaliser to negate dispersion.

Interference is the other major issue in considering the radio channel performance. The performance of an isolated transmitter and receiver might only be limited by the thermal noise of the detector and noise in the amplifier, say. In a cellular system there will be a base-station and a number of mobiles. The MAC scheme will use frequency, time, code or some other means to separate the different transmissions but, inevitably timing/filtering etc. will not be perfect and interference will be generated for neighbouring mobiles. In addition, there will be a scheme to separate transmissions in one cell from those in another. As a concrete example we will see that GSM uses a particular frequency in a given cell and then reuses it every 12–20 cells. Mobiles in distant cells using the same frequency generate interference directly. The mobile system designer needs to reuse frequencies as often as possible to increase capacity but must ensure a sufficient signal to interference ratio even at the edge of cells.

13.3   QoS and 2G mobile – GSM

13.3.1   Introduction to GSM

Most people are familiar with second generation mobile phones – the global GSM system, the North American D-AMPS system and the Japanese PDC are all examples of 2G digital systems. They primarily offer voice services and their architectures/protocols are optimised for efficiently delivering voice in a mobile environment.

The major 2G systems listed above are all quite similar – Figure 13.4 shows the GSM architecture. Basically mobiles connect by radio link with Base Transceiver Stations (BTS) that are located every kilometre or so. The BTSs are managed by a Base-Station Controller (BSC), BSCs manage the radio resources for several BTSs – handling radio channel set-up and handovers.

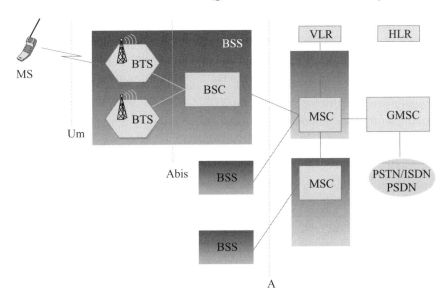

Figure 13.4   GSM system architecture

The Mobile Switching Centre (MSC) is effectively a mobile-enhanced telephone exchange. The MSC handles authentication and registration of users, generates call records for billing, location updating and incoming call routing to mobile users. In addition the transcoder that converts the low rate coded speech to normal 64 kbit/s digital circuits is often located at the MSC. A Gateway MSC (G-MSC) interfaces with the fixed PSTN and also with other mobile networks, acting as a speech and signalling gateway – talking the normal PSTN signalling language. When an incoming call is received, the GMSC consults the Home Location Register (HLR) (see below) – a database that keeps a record of the current MSC with which the user is associated – and routes the call signalling to the appropriate MSC.

Each GSM network has a HLR which is basically a large database that contains information about users – when users try to connect to the network they are authenticated by the MSC with the HLR – the MSC also downloads (to a local database – the Visitor Location Register or VLR) data about the services the user is entitled to (call barring, remaining credit and so forth).

13.3.2   User level QoS in GSM

In GSM the primary service is voice and the user level QoS can be broken down into several parts:

- Coverage – is there sufficient radio signal strength to make and receive calls?
- Availability of service – if there is a good signal strength, then, can calls be made or received?

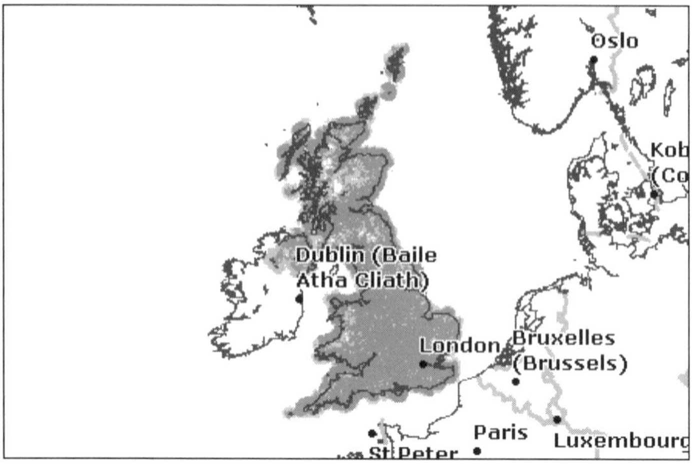

Figure 13.5   BT Cellnet GSM network coverage

• Call dropping – if a call is started does it continue to completion?
• Speech quality – is the speech quality sufficient?

In the next four sections we will explore the parts of GSM network operation and planning that directly support these four aspects of user QoS.

13.3.3   Network planning

The coverage of a cellular radio network is determined by a number of factors not least of which are the number and location of the base stations. Radio planners use sophisticated software tools, in conjunction with cartographic data on the terrain, to predict the signal strength within a proposed cell. If potential coverage is poor then alternative base-station sites may need to be found – although these can be hard to find, requiring a high aspect, power and acceptable environmental impact.

Cells must be planned over a wide area to ensure that interference from distant cells reusing the same frequency does not degrade performance at the cell edges. Costs are also a factor, particularly in rural areas, where traffic levels may not be sufficient to justify 99% coverage (Figure 13.5). Even in urban environments, there will always be radio black spots, such as railway tunnels, where it is impossible to guarantee coverage.

13.3.4   Radio resource management

Assuming a good signal level is present, a mobile can make a request to place a call. This request is handled by a distributed function called the Radio Resource Management (RRM) function. The allocation of radio resources is an all or nothing thing – since basic GSM is a circuit orientated technology the resources will then be tied up until the call either fails (e.g. the called party is not in) or either user hangs up.

The RRM must make sure that in allocating the traffic channel there are still sufficient spare channels for any users handing over (mid-call) from a neighbouring cell.

If the RRM has a spare radio traffic channel then this is allocated to the mobile and it uses this channel to send signalling messages to the MSC which carry the dialled number. The MSC checks that access to the number is allowed – premium rate or international numbers might be blocked – and decides where to route the signalling for the call. If the call can be completed, then the MSC must set up a bearer path from the BTS, through the MSC to the GMSC. As noted earlier the transcoder unit will most likely be at the MSC and so a 13 kbit/s circuit is required from the BTS to the MSC and a 64 kbit/s path from the MSC to the GMSC. This bearer set-up must take place before the called phone is made to ring and before the MSC signals the mobile to produce a ringing tone (otherwise this step may fail and the phone will ring and then stop or be answered and go dead). Normally, the bottleneck of resources in a mobile network is designed to be the radio link and sufficient back-haul capacity is provisioned to make a failure at this stage a rare occurrence.

13.3.5 Handover

Whilst a user is moving around a cellular network, the mobile and the base-station are monitoring the link quality (signal strength, interference level, error rates and so forth). The mobile is also monitoring neighbouring base-stations and reporting their identity to the serving base-station. GSM uses a mobile-assisted handover process meaning that the mobile makes reports to the base-station but it is the base-station that controls the handover procedure. For each handover the base-station runs a handover algorithm the details of which are left to the specific manufacturers implementation. However, a typical example is given in GSM 05.08 Annex A. The algorithm uses a combination of the following criteria to decide that a handover is required:

- signal strength;
- signal quality;
- mobile to base-station distance (as measured by a timing advance);
- power budget (if a neighbouring cell offers a lower loss path);
- interference (as detected in empty time slots);
- congestion.

In the case of congestion the base station is heavily used and handover mobiles with a large timing advance (i.e. those at the edge of the cell) are handed over, even if they have acceptable quality links.

The algorithm will also create a candidate list of neighbouring cells, which might take the call. The path loss to each neighbour is estimated by the mobile which listens to the broadcast channels of neighbouring cells in certain timeslots and reports back their strength and cell id to the base-station. The candidate list is ordered by path loss and sent to the MSC which considers the general traffic congestion situation before selecting the target handover cell. Sometimes, however, it may not be possible to handover if all neighbouring cells are suffering congestion. To some extent this is another aspect of the cost versus quality trade-off inherent in any mobile network.

13.3.6   Speech quality

In this section we look at what GSM does to provide quality, in the face of radio challenges over the wireless link. First, the speech is coded at a lower rate than in the PSTN – 13 kbit/s compared to 64 kbit/s. This is really just an attempt to pack in more users on a given bandwidth and results in lower quality speech than is typical on PSTN–PSTN calls. However, the algorithm used is an advanced linear predictive coding algorithm that is able to predict future values and sends only correction signals, reducing the bandwidth considerably with only a small loss of perceived speech quality.

The digitised speech is then protected from errors by a Forward Error Correction scheme using a convolutional code (essentially redundant bits are added such that error recovery is possible from some lost bits). The speech coder produces 260 bits for each 20 ms of speech digitised – however, because of the characteristics of the speech coder it has been found that not all these bits are equally important to speech quality and so some are preferentially encoded:

- Class 1a: most sensitive (50 bits), 3 bit cyclic redundancy code and $\frac{1}{2}$ rate coding;
- Class 1b: less sensitive (132 bits), 4 bit tail sequence added and $\frac{1}{2}$ rate coding;
- Class 2: least sensitive (50 bits), no protection.

Each 260 bit speech sample is therefore transformed into 456 bits – equivalent to a raw bit rate of 22.8 kbit/s. Notice no backward error correction is used – by the time the speech samples had been retransmitted due to errors they would have missed their playback slot.

Even with equalisation it is possible that mobiles move into deep fades and quality is very poor. To provide further diversity the GSM system uses slow frequency hopping to change frequency every frame – meaning only a small loss of quality even if one frequency is in deep fade at a particular location. Typically operators have 25 MHz of spectrum and so have 125 frequency blocks of 200 KHz available – even with a reuse pattern of 20 this allows 6 frequencies per cell for hopping.

Another technique to improve quality on the radio link is discontinuous transmission – basically only transmitting when a person is talking. This has the advantage, as well as saving battery power, of reducing interference to other mobiles using the same frequency. It is achieved by the use of a voice activity detector and, typically, reduces transmissions by 50%. Noise has to be added at the receiving mobile, however, as total silence is found to be disconcerting.

13.4   3G networks – UMTS

13.4.1   Introduction to 3G

Third generation mobile is about multimedia at bandwidths up to 384 kbit/s. There are two major 3G systems – UMTS from Europe and Japan and cdma2000 from North America. Each system offers conventional voice circuits; higher bandwidth circuits and data transfer with volume-based billing and an 'always on' capability.

Bandwidths for data and circuits will be initially 64–144 kbit/s, rising to 384 kbit/s or more as more capable equipment and base-stations are rolled out over the next few years. Operators will offer some multimedia services, such as video telephony, video post-cards and also access to the Internet and corporate Intranets.

In terms of QoS 3G is interesting in that it will offer data users a variety of QoS classes and parameters, such as bandwidth and delay, that the network will support. The air interface, access network and core network all have adaptations to support these QoS classes, as well as dealing with the consequences of mobility. 3G QoS is also interesting since there are, currently, no large-scale QoS-enabled multimedia networks in operation.

In this section we will look at UMTS as a representative 3G system, and describe these adaptations for QoS support. We will do this by first looking at the QoS that is supported, seeing how it is signalled between the mobile and the networks and then looking at the core and access and, finally, the radio link (since the problems and countermeasures are quite similar to those of GSM).

13.4.2   UMTS system architecture

The system architecture of UMTS is similar to GSM (Figure 13.6) – there are base-stations (called node Bs) base-station controllers (now Radio Network Controllers (RNCs)) and MSCs/GMSCs (upgraded from their GSM equivalents). This is the so-called circuit-switched domain. In addition there is a new domain, the packet-domain, within the core network to handle data – specifically the Serving GPRS Support Node (SGSN) and the Gateway GPRS Support Node (GGSN) have been added. These nodes act in a very similar way to the MSC/GSMSC for packet data: the SGSN handles and generates billing records, sets up data bearers offering the correct QoS – after querying the HLR – and pages idle mobiles if an incoming data session is detected; the GGSN acts as a gateway to external IP networks offering security, firewall and DHCP services as well as handling incoming data sessions. The term GPRS refers to a GSM enhancement that offers an always-on packet data service with volume billing and data rates up to 56 kbit/s; the UMTS SGNs are enhancements of their GPRS counterparts. GPRS is often called a 2.5G system but it only offers best effort QoS and its long delays mean it is not suitable for real-time traffic.

One new feature of UMTS is that the Radio Access Network (RAN) now handles (almost) all aspects of mobility, freeing the core network from the task. Underlying the RAN is an ATM network – using a specially developed adaptation layer and signalling layer to set up connections within the RAN – that provides QoS to multiplexed circuit and packet data. Another new feature, compared to GSM, is the Code Division Multiple Access (CDMA) air interface. Instead of time slots being used to multiplex many users within a single block of spectrum, UMTS 'spreads' the signal from each user across the entire spectrum but uses orthogonal codes to recover the signal at the receiver. As we shall see at the end of this section, CDMA can provide improved capacity compared to TDMA systems as well as offering an easy way to vary QoS on the radio links.

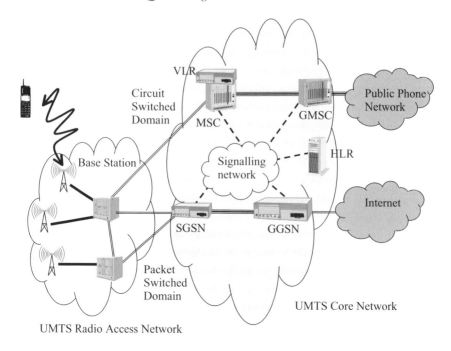

Figure 13.6   UMTS network architecture

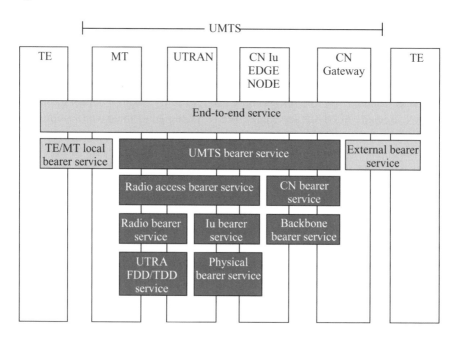

Figure 13.7   UMTS QoS

13.4.3 User QoS specification UMTS

UMTS provides QoS through a layered approach (Figure 13.7). To quickly illustrate what this layered QoS really means imagine a user making a voice call on a 3G mobile. They may have a blue tooth link from their tiny headset to the mobile and this would be a local bearer service. From the GGSN to a PSTN phone, likewise, would be an external bearer service. The UMTS bearer service proper is three basic bearers: one over the radio link; one over the RAN and one over the core network. Each of these actually runs over a physical bearer. The RAN QoS is actually provided by ATM Adaptation Layer 2 (AAL2) over ATM switched circuits – although in the standard (and in Figure 13.7) it is shown only as defined at the interface between the core network and the RAN (the Iu interface). This is because UMTS standards are all about making equipment from different manufacturers inter-work and providing end functionality. How it works in the middle is an implementation issue. Thus, it is, in principle, possible to have an IP-based RAN. Indeed the UMTS standards specify that the backbone bearer may be either based on ATM virtual circuits or IP DiffServ.

When QoS is requested by users it describes directly the QoS at the UMTS bearer level and classifies the QoS requested into four broad classes depending on the maximum delay that the data/connection can tolerate (Table 13.2).

Users employ seven parameters, in addition to the basic class, to request QoS in UMTS:

- maximum bit rate (the peak rate of traffic that is only supported if resources allow);
- guaranteed bit rate (the rate the network guarantees – conversational and streaming classes only);
- traffic handling priority (used to create sub-classes within the interactive class);
- delay;
- reliability – are errored packets delivered?
- maximum packet size;
- packet ordering – are packets delivered in order?

Only certain values of these parameters are allowed – details can be found in the section on further reading at the end of the chapter.

13.4.4 UMTS radio link

We have looked at how UMTS QoS is specified. In this section we will look briefly at how UMTS actually transports packets/connections to deliver QoS. Obviously the most important part of this, as in all mobile networks, is the radio link.

CDMA systems work differently compared to time or frequency division multiplex systems – they allow each user to transmit across the full bandwidth allocated to the cell (Figure 13.8). The user multiplies his data stream of 1 s and 0 s with a (much faster) repeating pattern – the spreading code and the resultant signal is transmitted. The receiver hears the signal from every other mobile but, when multiplied by the original spreading code, the required signal is recovered and the remaining transmissions converted to noise (providing the spreading codes are carefully chosen and are orthogonal – Figure 13.9).

Table 13.2   UMTS traffic classes

Traffic class	Conversational class: Conversational RT	Streaming class: Streaming RT	Interactive class: Interactive best effort	Background: Background best effort
Delay	≪1 s	<10 s	Approx. 1 s	>10 s
Example: Error tolerant	Conversational voice & video	Streaming audio & video	Voice messaging	Fax
Example: Error intolerant	telnet, interactive games	FTP, still image, paging	E-commerce, WWW browsing	E-mail arrival notification
Fundamental characteristics	Preserve time relation (variation) between information entities of the stream Conversational pattern (stringent and low delay)	Preserve time relation (variation) between information entities of the stream	Request response pattern Preserve payload content	Destination is not expecting the data within a certain name Preserve payload content

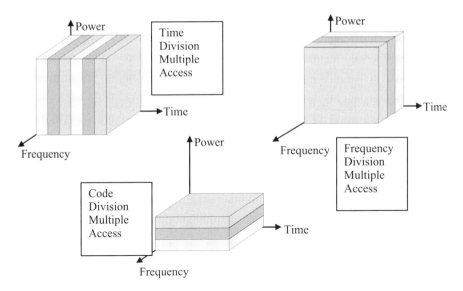

Figure 13.8   Multiplexing schemes

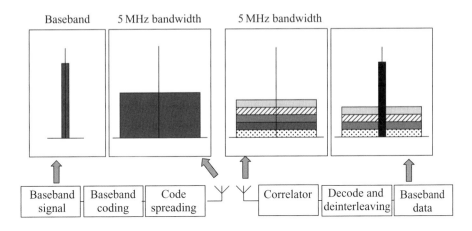

Figure 13.9   CDMA spreading and de-spreading

3G selected CDMA over GSM enhancements because

- it gives more efficient use of the spectrum;
- receiver performance is enhanced by the wide-band transmission (5 MHz);
- the GSM TDMA timeslot structure restricted flexibility to allocate capacity to users;
- the higher bandwidth (5 MHz) of the UMTS spectrum blocks allows higher bandwidth transmission than the GSM block (200 KHz);
- it allowed harmonisation towards a world standard based on CDMA.

In CDMA all users transmit on the same frequency – even those in neighbouring cells – and appear as interference to one another. Cells are separated by scrambling codes – these scramble transmission from different cells allowing reuse of channellisation codes (so the final spreading code is: scrambling code ∗ channellisation code) – and frequency planning is not required.

The resource that each mobile consumes in transmitting is interference – the greater the power used the higher the interference and the more the overall system capacity is used up. However, in CDMA the higher the bit rate of the transmission the more the power required for a given bit error rate (lower rate transmissions have a larger so-called processing gain – the rate of the bit rate of the transmission compared to the chip rate – 3.84 Mchip/s in UMTS). Thus CDMA allows capacity to be allocated to users from one big pipe, and even allows movement of capacity between cells to a certain extent. As long as the system stability is maintained (i.e. it is not run too close to the capacity limit) there is great flexibility in allocating capacity to users.

In the RAN the Radio Resource Management (RRM) functionality is distributed between the terminal, base station and RNC. It basically consists of algorithms and procedures to control the following:

- admission control
- power control
- code management
- packet scheduling
- handover

The RRM system is also responsible for mapping the packet/connection flow to the correct error correction mechanism on the link layer. Both forward and backward error correction are offered in UMTS. Backward error correction allows radio frames (as they are called at the link layer) to be retransmitted if they are corrupted or lost. Obviously this adds to the delay and is not suitable for real-time services (such as voice) where it is often better to lose the data than wait for it to be retransmitted.

13.5　4G systems

13.5.1　*What is meant by 4G*

There is no current industry consensus as to what constitutes 4G mobile. A number of different views are currently being debated: First, that it is a complete new air interface with new spectrum offering connection speeds of 2–20 Mbit/s [4 and 5]. Here, the emphasis is on more efficient use of the spectrum with advanced multiplexing techniques, such as Orthogonal Frequency Division Multiplexing (OFDM), and smart antennas being employed. An all-IP access network is a feature of this view.

Second, is the view that a 4G is a network of networks. In this view, a great number of access technologies – such as WLANs, cellular and satellite – all provide QoS-enabled IP connectivity. User services are tailored to the particular link they are connecting from by edge of network intelligence. One example of this view is provided

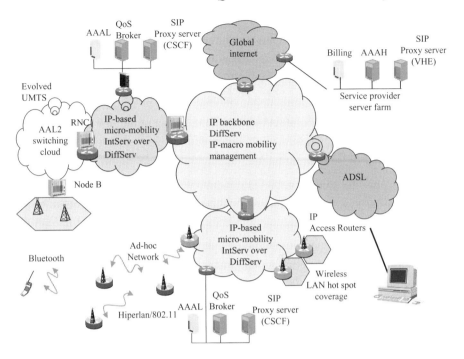

Figure 13.10   BRAIN architecture

by the EU IST Broadband Radio Access over IP Networks (BRAIN) project [1] and a general architecture is shown in Figure 13.10.

Finally, there is the view that 4G will be user-driven [6]. Users will have access to a range of competing public and public wireless access technologies and will be able, for example, to route their traffic though overlapping private WLANs before entering the fixed network. They will frequently form ad hoc networks and share network connectivity.

13.5.2   IP QoS solutions

As an example 4G system the IST BRAIN project has looked at the general problem of designing an IP access network that has QoS, mobility and security functionality and can connect to a number of wireless or wired access technologies. In the BRAIN network all services (including voice and video) are carried in IP packets. Figure 13.10 shows the overall BRAIN architecture.

In today's Internet there is no QoS offered – all packets are accepted on a best efforts basis and may be delayed, re-ordered or lost. QoS for IP packets can, of course, be provided by either over-provisioning or by using an underlying layer 2 switching technology with inbuilt QoS such as ATM or Multi-Protocol Label Switching (MPLS).

However, these approaches will never be the most flexible or efficient for future IP networks – how will multicast work? How will end terminals achieve end-to-end QoS? What is needed is a QoS solution at the IP layer that enhances the routers to allow them to treat some packets different from to others – Internet QoS is poor because of delays in router queues and packet loss due to dropping from queues when buffers are exhausted. Elsewhere in this book you can find descriptions of the basic approaches – epitomised by IntServ and DiffServ – being proposed as solutions for IP QoS in fixed networks. Here, we will look at the BRAIN access network and the solution adopted to cope specifically with the mobility of terminals. This solution is capable of inter-working with a core network QoS solution.

BRAIN provides a base-line QoS architecture – something that is simple, resilient and easy to deploy – providing a basic QoS with limited guarantees. Further extensions are described to allow extra functionality and harder guarantees, as we will describe below. The baseline architecture is based on the IETF Integrated Services over Specific Link Layers (ISSLL) working group framework for sending RSVP-controlled traffic through a differentiated services network (Figure 13.11). Much of the functionality required – such as RSVP in the terminals and DiffServ capability within the routers is available today.

The baseline QoS architecture provides per-application resource allocation with RSVP and admission control, and traffic shaping takes place on a per-flow basis at the edge of the network. The edge nodes are responsible for allocating appropriate DiffServ codepoints; inside the access network the flows are highly aggregated (making the network scalable) and treated according to these code-points. The use of RSVP means that end-to-end QoS provision is possible, provided the other domains traversed can support QoS. The baseline architecture also supports mobility because DiffServ uses per-hop behaviour that is not affected by routing changes.

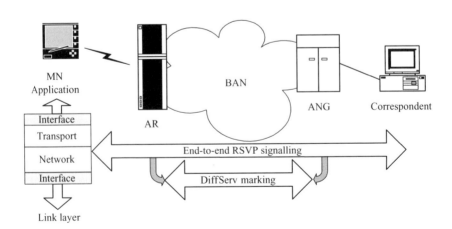

Figure 13.11   BRAIN baseline architecture

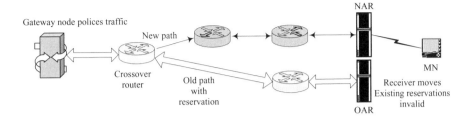

Figure 13.12   DiffServ handover markings

The baseline architecture does not, however, support hard guarantees of maximum packet delay – to do this requires three extensions:

- Introduction of the bounded delay DiffServ class. The bounded delay DiffServ class guarantees an absolute maximum delay at each router. Hence the maximum delay across the network can be calculated (and is made sufficiently low for all practical real-time applications). Jitter is not specifically controlled and end systems use the assumption: maximum jitter = maximum delay.
- Providing admission control within the access network. Edge admission is not sufficient for hard QoS guarantees – particularly when new, high bandwidth air interfaces are considered – the bottleneck might well move into the access network. If hop-by-hop admission and reservation of resources is made then the guarantees are greatly strengthened.
- Improving QoS support for handover. When a handover occurs packets are forwarded from the old to the new access routers and, additionally, a new path is required from the cross-over router to the new access router (NAR) (Figure 13.12). Packets can be delayed or lost in the time it takes to set up the new path leg from the cross-over router to the new access router) and in the forwarding. By introducing a new DiffServ class – high priority handover – this problem can be overcome.

The whole area of IP QoS and mobility management is a very active research area in the IETF and interested readers should look at some of the latest proposals at www.ietf.org.

13.6   Summary

Providing QoS on mobile systems is difficult for the following reasons:

- The radio link has totally different characteristics to typical wired links. Error rates are much higher and the statistics of errors are very different.
- Handover is needed to cope with the mobility of terminals. Resources may or may not be available to accept the handover.
- Mobile terminals typically have smaller displays, are battery powered and have limited processing capabilities.

- Spectrum is a scarce resource and there is always a tension between quality and efficiency.
- Mobile systems have to introduce a number of systems to overcome these fundamental problems:
  - multi-path fading countermeasures;
  - powerful forward or backward error correction mechanisms;
  - RRM algorithms for admission/handover and resource control;
  - a high level of system integration – including mobile specific applications.

Second generation cellular mobile systems – epitomised by GSM – offer only voice and text messaging and are highly optimised to do this. 3G systems, such as UMTS, will offer a much wider range of packet and connection-based transport services. A new air interface – CDMA – allows much more flexible allocation of radio resources. 3G will also increase bit rates up to 384 kbit/s and beyond as systems are rolled out.

Looking into the future, it is likely that, by the time 4G is rolled out, all applications will be carried in IP packets and 3G networks will transport these in a native way – using IP QoS mechanisms to provide quality. Session layer protocols, such as SIP, will be needed to set up and modify sessions on the PDA and smart-phones of the future.

13.7   References

1   EU IST BRAIN Project www.ist-brain.org
2   www.3gnews.org
3   HEWITT, T.: Radio Spectrum for Mobile Networks, *BT Technology Journal*, 1996, **14** (3), pp. 16–28
4   NAKAJIMA, N. and YAMAO, Y.: 'Development of 4th generation mobile communication', *Wireless Communications and Mobile Computing* 2001, **1** (1) pp. 3–12
5   MOHR, W. and BECHER, R.: 'Mobile communications beyond third generation', in Proceedings of VTC 2000
6   PEREIRA, J.: 'Fourth Generation: Now, it is Personal', Proceedings of 11th International Symposium on *Personal, Indoor and Mobile Radio Communication*, 18–21 September 2000, IEEE, pp. 1009–16, vol. 2

Further reading

General mobility

BT Technology Journal, 2001, **19** (1) – Special issue on Future Mobile Networks 'IP for 3G', WISELY D. *et al.* (Wiley, 2002), ISBN 0471 48697 3

GPRS

'GPRS Support Nodes' – EKEROTH, L. and HEDSTROM, PER-MARTIN: Ericsson Review, No. 3, 2000

'GPRS: Architecture, interfaces and deployment' LIN, Y., RAO, H. and CHLAMTAC, I.: *Wireless Comms and Mobile Computing*, 2001, **1**, pp. 77–92.

'GPRS General packet radio service' – GRANBOHM, H. and WIKLUND, J.: Ericsson Review, No. 2, 1999

UMTS

'UMTS network: architecture, mobility and services', KAARANEN, H. *et al.* (John Wiley and Sons Ltd)

'The UMTS network and radio access technology', CASTRO, J.: (John Wiley, 2001) ISBN 0 471 81375 3

'The complete solution for third-generation wireless communications: two modes on air, one winning strategy', HAARDT, M. and MOHR, W.: *IEEE Personal Comms*, December 2000, pp. 18–24

'Third generation mobile communications systems', PRASAD, R., MOHR, W. and KONHAUSER, W. (Eds), Artech House, 2000

UMTS QoS

'Supporting Ip QoS in the general packet radio service', PRIGGOURIS, G. *et al*, IEEE Network, September/October 2000, pp. 8–17

Cdma2000

'The cdma2000 packet core network', Ericsson Review no. 2, 2001, Tim Murphy

Web sites

Manufacturer information

1 **www.ericsson.com** – Ericsson Review available on line
2 **www.nokia.com** – white papers to download

Mobile standards bodies and organisations:

3 3GIP – **www.3gip.org** – IP pressure group for UMTS
4 MWIF – **www.mwif.org** – Mobile Wireless Internet Forum
5 3GPP2 – **www.3gpp2.org** – cdma2000 standards
6 3GPP – **www.3gpp.org** – UMTS standards and technical specifications
7 UMTS Forum **www.umts-forum.org** – Information on UMTS from suppliers and operators
8 ITU – **www.itu.int and www.itu.int/imt** for imt-2000
9 IETF – **www.ietf.org** – Internet drafts and RFCs

BRAIN project

10 **www.ist-brain.org**
11 **www.ist-mind.org**

Exercises

1   Using the OSI stack (Figure 13.1) think of at least two possible difficulties in providing QoS for an application running over a mobile network, as compared to a fixed network, for each layer of the stack. How does the answer change with the type of application (voice, video, web browsing etc.)?

2   What are the advantages and disadvantages in using voice over IP and an IP access network compared to a traditional, tightly integrated system such as GSM? How would you define end user quality for voice over IP? And what network features might be required to support it?

3   Imagine you have been asked to design a cellular mobile system, such as GSM, for a small country. How would you define user quality for this system? List the steps you would take to design a system that would fit the purpose. Which ones do you think are the most critical?

Chapter 14

Quality of Service of satellite communications

14.1 Introduction

Use of satellite links as a part of a telecommunications system introduces a set of new Quality of Service (QoS) issues. It can also emphasise (normally make worse or more noticeable) the effects of QoS issues that exist, in any case, with terrestrial systems. Essentially all satellite systems consist of a pair of ground installations that communicate with each other through a satellite-based relay system. The ground segment ranges between large capacity installations (ground stations) that are themselves connected through the terrestrial networks to individual users through to individual handsets that can communicate directly with satellites.

As will be seen in the following paragraphs the design of satellite (and associated ground segment) systems is a complicated and highly expert task requiring many levels of complex modelling. This feature in itself is not uncommon in the design of all networks. However, the major added complication with satellite-based systems is that once the constellation has been launched (or even partially launched) there is very little scope for improvement and optimisation.

Understanding the effect of longer delays on the overall performance of TCP/IP systems is critical to understanding the QoS offered by data networks using satellite links. A section expanding on the relationship between TCP/IP QoS and latency is, therefore, included in this chapter.

One of the other major influences on satellite link QoS is the existence of terrestrial radio communication systems, in the same bands as some satellite systems, using tropospheric scattering as the propagation mechanism. A section on tropospheric propagation is, therefore, also included in this chapter.

14.2 GEOs, MEOs and LEOs

The satellite systems themselves are commonly categorised into three different types depending on the distance of their orbits from the Earth. The three types are referred to as Geostationary or Geosynchronous Earth Orbit (GEOs), Mid Earth Orbit (MEOs) and Low Earth Orbit (LEOs). Although the QoS issues remain essentially the same the actual impact on overall QoS measures varies considerably by satellite system type.

- A GEO satellite follows a circular orbit in the plane of the equator revolving in the same direction and with the same period as the Earth's rotation at a height of 35,787 km above mean sea level. An object in a geostationary orbit appears to hover over a fixed point on the equator.
- MEOs are in orbits between 6000 and 20,000 km from the Earth's surface.
- LEOs are in orbits between 500 and 1600 km from the Earth's surface.

14.3   Visibility coverage

14.3.1   Line of sight

Communications satellite systems all operate in UHF and microwave radio frequency bands. A signal path between the ground station or handset and the satellite aerial only exists when there is a clear line of sight between the two. For handsets this automatically means that there is no in-building coverage. It is possible for in-building coverage to be engineered using Wireless LAN and Bluetooth technologies and window or roof aerials to repeat signals to and from satellites. For most satellite systems the ground station aerials are highly directional. Satellite-based aerials are also designed to illuminate a defined footprint on the Earth's surface. With GEO satellites this footprint stays in the same place on the Earth's surface. Three GEO satellites are enough to cover most of the globe, and mobile users rarely have to switch from one satellite to another. Some GEO satellites, such as those supporting Very Small Aperture Terminal (VSAT) services have spot beam aerials which can be directed at specific ground stations located within the footprint either permanently or for a specified duration. Each satellite aerial has its own characteristic footprint, the size and shape of which is defined by the level of Effective Isotropic Radiated Power (EIRP) measured on the Earth's surface at specific points. Maps of the footprints of most satellite systems are widely available on the Internet. The customer's location within the footprint for a specific GEO will determine the QoS at a particular site.

With MEOs and LEOs the footprint illuminated by each satellite in the system moves over the Earth's surface as the satellite moves. For a signal to be successfully received and/or transmitted the ground station must be within the ground footprint of the satellite and there must be no obstructions in the line of sight. From the user's point of view, the satellites move across the sky at a comparatively high speed, often requiring a switch by the handset and/or ground station from one satellite to another in mid-communication resulting in planned or unplanned interruptions to calls. In general, the number of satellites required to cover the Earth's surface depends on the height of the constellation orbit above the Earth. The lower the orbit the greater the number of satellites required to achieve complete and continuous coverage. MEO and LEO systems also use orbits inclined to the equator so as to cover the higher latitudes.

For example,

- the planned ICO (MEO) constellation has 10 operational satellites at 10,000 km above the Earth's surface in 45° orbits;
- the Globalstar (LEO) constellation has 48 satellites at 1414 km above the Earth's surface in 52° orbits;
- the Iridium (LEO) constellation has 66 satellites at 760 km above the Earth's surface in near polar (90°) orbits;

14.3.2   Latitude

From the customer's perspective the latitude of where he is located affects his QoS not only due to variable visibility but also due to the nature of the satellite constellation. For instance, access to the Inmarsat GEO systems gets significantly more difficult at high latitudes since the one satellite in view appears closer to the horizon the further the customer is from the equator. Thus, obstructions such as buildings, trees and hills become more difficult to avoid. The advantage that a GEO system has is that once a line of sight is established, then, provided the customer remains static the connection will not be interrupted by loss of signal. Access to the Iridium system (in near polar orbits) on the other hand does not vary significantly with latitude. Iridium's LEO constellation means that there is the probability of achieving a line of sight with at least one satellite from almost any open air location. The QoS will then depend on maintaining line of sight with that satellite as it moves across the sky and then achieving a good handover to another satellite.

14.3.3   Canyon effect

A particularly important effect that needs to be accounted for, that is unique to satellite QoS, is known as the canyon effect. LEO and MEO QoS is critically related to the direction of roads and the height of buildings, hills and trees and the way in which they interrupt the line of sight possibilities to satellites. Achievable QoS can vary considerably with time and as the customer moves along the street; hence, 'canyon effect'. Lines of trees along country roads can create the same effect.

14.3.4   Diversity

Diversity means that a variety of communication paths are available, that is, that for a satellite user, there is more than one satellite available for a ground terminal to communicate with at all times. At the physical level, diversity and multiple satellite visibility can be used to combat shadowing by buildings or terrain, to provide redundancy and increase the likelihood of a satellite always being in view of the terminal. Two basic implementations of satellite diversity are switched diversity and combined diversity. Switched diversity simply means that the ground terminal has a choice of multiple visible satellites with which it can communicate, and that after selecting a satellite, the terminal establishes a single duplex radio link with that satellite. Combined diversity is when the ground terminal communicates across multiple satellites simultaneously. This is also referred to as artificially introduced multipath. This is exploited in CDMA-based systems such as Globalstar, which recombine signals passed through more than one satellite at a ground station as a way of combating shadowing and reducing errors. This combined diversity can also be exploited to facilitate soft handovers. Ground-to-space diversity, across the air interface, can be exploited at various layers of the ISO network protocol stack. Physical diversity can be exploited in constellations without Inter Satellite Links (ISLs), for example, Globalstar's use of CDMA and power recombination of signals across multiple satellite transponders. It can also be exploited at the data-link layer, via TDMA management as in ICO.

14.3.5   Rain

As with all UHF and microwave systems satellite systems are vulnerable to the effects of rain interrupting the line of sight path. These effects are often very short term and unpredictable and there is a possibility that data systems will be seriously affected by packet loss. This would be due to the higher overall round trip delay meaning that detection and then recovery of lost packets takes longer than with terrestrial systems.

14.4   Latency

Latency is time delay between transmission and receipt of a telecommunications signal in telecommunications and computing systems. Latency has always been a particular issue for designers of communication systems that use geostationary satellites. This is because of the time required for a radio signal to travel to and from the satellite (GEO satellite orbit is at 35,787 km). Latency is also an issue for ground based packetised networks like the Internet. In the case of the Internet the latency is not because of distance but rather due to the amount of time spent in routing the packets at the routers.

Latency can be a major problem when there is a human interface to the Information, for example, a person making a phone call, or using a multimedia application, or watching TV. Latency along with bit error rate are the main determinants of the QoS.

For GEO satellite communications systems, the transmission latency is 320 ms (sometimes associated terrestrial network, framing, queuing, and on-board switching can add extra delays, making the end-to-end latency as high as 400 ms). This is approximately ten times higher than a point-to-point fibre optics connection across the continental United States. The latency might not affect bulk data transfer and broadcast-type applications, but it will hurt highly interactive applications that require extensive handshaking between two sites. Crucially, one of the major Internet transport protocols, TCP, requires such interaction.

Even for LEOs, flying 400–1000 miles from the earth's surface, some round-trip delay is inherent, although experts disagree on just how much.

The Internet Engineering Task Force, puts average round-trip (terrestrial) delay at about 100–150 ms; LEOs at 200 ms; and GEOs at about 600 ms. NASA has commissioned a study which gives corresponding minima and maxima of 42/446 ms for LEOs and 274/1068 ms for GEOs.

The NASA report concludes that only terrestrial services are suitable for interactive media, although LEOs come very close to making the grade.

This report re-enforces the opinion of experts who contend that latency-sensitive applications, such as voice, LAN-to-LAN interconnection, interactive video and gaming, will perform best when kept out of the sky altogether, especially for short hops. For longer transcontinental and intercontinental journeys, there is general agreement that LEOs will be quite competitive with terrestrial fibre optics.

An alternative view is that most LEO delays will be attributable to processing, not propagation, and that the use of high-speed laser-based switching – like that planned by Teledesic – will make LEO delay at least comparable to terrestrial delay.

Teledesic maintains that a transmission halfway around the world through Teledesic could be faster than one through a fibre optic cable, because although the arc is a bit longer for the satellite path, light travelling through fibre propagates at only around two-thirds the speed of light in a vacuum.

Teledesic expects typical round-trip latency to fall to between 40 and 100 ms, with a round trip of 5000 km, meaning latency of 150 ms, and a round trip of 500 km meaning latency of 20 ms. Motorola, too, contends that LEO delay will be comparable to terrestrial, and generally below the 100 ms threshold for voice.

Thus, whilst a satellite system typically does not outperform a fibre optic network, there are instances where the number of hops between routers and switches can be reduced by a satellite system. Fewer routing points in a network will result in lower latency. Latency in a well designed LEO system can be comparable with a poorly designed terrestrial fibre optics system but will usually be twice as large. However, because neither a LEO nor a MEO satellite stays in a fixed position relative to the surface of the earth, a constellation of many satellites is required to provide complete coverage, increasing the complexity of the system relative to GEO systems. Network management is a much harder problem because handoff, tracking and routing all need to be managed consistently. The advantage of simple topology is lost, and so is single-source broadcast/multicast capability.

Other than the problems regarding QoS there is a much debated issue of latency problems with TCP/IP data transfer. The TCP/IP protocol implementations have a buffer size limitation (around 4 k). When transferring data over satellites the bandwidth is limited since:

$$\text{Bandwidth delay product} = (\text{Satellite latency} + \text{Terrestrial latency})$$
$$\text{(i.e. buffer size)} \qquad * \text{Link bandwidth}$$

Thus the longer the delay (latency) the smaller the effective bandwidth. Teledesic claims that its LEO-based system with its smaller latency gives it a larger throughput than other satellite systems (particularly GEOs) for TCP/IP data connections. There are other potential solutions to the TCP/IP throughput problem when using GEOs including the use of large window TCP/IP. Teledesic claims that these implementations are not wide spread and, therefore, that they will have a considerable advantage over broadband GEO Satellites.

14.5   Jitter

GEOs typically maintain an orbit 35,787 km from the earth's surface and are tied to the earth's rotation, so that they appear to be fixed over a specific geographic area. Broadband LEOs will orbit only 500–1600 km from the surface and, therefore, a greater number of satellites, travelling in more complex orbital patterns, must be launched in order to cover the globe. For Teledesic that comes to 288 satellites; a downward revision from an even more ambitious start-up plan of 840.

One of the great unknowns about these LEO constellations is how well they will be able to handle variations in delay, otherwise known as variable latency or jitter. A low orbit satellite may only spend tens of seconds over a given geographical area, which means that a given transmission may be picked up and passed on by multiple satellites. For a TCP/IP system sequential data packets may be carried by different satellites and experience a different delay. This, in turn, can result in receipt of packets in the wrong order known as packet reordering.

The Internet Engineering Task Force maintains that this effect, due to the random and constantly changing distances between satellites in a LEO constellation, means that TCP's window duration measurement algorithms will be constantly opening and closing (rubber banding) resulting in a severe degradation in the available bandwidth.

For voice, video and IP multicast service applications, which are intolerant of jitter or packet reordering, GEOs are more effective. If the specific TCP implementation is very robust, and jitter and reordering are not significant issues, then LEOs will give lower end-to-end latency. If low end-to-end latency and minimal jitter are both required then the conservative tack would be to avoid use of satellite links.

One solution to the jitter effect would be to create larger re-sequencing memory buffers in ground stations. Transmission, then, would be delayed long enough so that the playback to the user is at a constant latency. Teledesic, in fact, believes that this is what should be done and says that it has already incorporated the cost of this buffering in its manufacturing cost estimates.

However, this buffering whilst benefiting voice and videoconferencing will not help the 75% of Internet traffic that consists of short Web queries and e-mail traffic. This is because this type of traffic is too short to load the buffers.

Even though re-sequencing buffers smooth out voice and video latency, they do add delay. If the jitter is modest the accompanying added delay should not be high enough to interfere with most interactive applications, including voice.

Another factor affecting jitter is the size of the satellite constellation. The smaller the constellation, the greater the jitter. This is because each satellite in a smaller constellation serves a larger ground footprint relative to the footprint served by a larger, denser constellation. Thus Motorola's Celestri, with 63 LEOS, would have had a greater magnitude of jitter than Teledesic with 288 satellites, assuming operation at similar altitudes.

A final and possibly quite serious issue is whether TCP will become confused if the jitter for a transmission exceeds the time required to send, receive and acknowledge the transmission. If variable latency exceeds the round-trip timing, TCP may interpret jitter as packet loss due to network congestion and not only begin a re-transmittal but also close down on bandwidth used. If this happens, data transfer rates would slow down significantly.

14.6   Interference

Satellite systems, both at the satellite and ground station ends of the link rely on the reception of very low level radio signals. Satellite-based receivers, in particular, can be subject to possible interference from ground-based systems transmitting in the same frequency band. Two major system types operating around the world and used primarily by the military can and do cause QoS problems. These are forward tropospheric scatter systems and radar systems.

14.6.1   Tropospheric forward scatter

Tropospheric forward scatter systems operate in some of the bands that are allocated to satellite systems and have been used by the military authorities of many countries to provide over-the-horizon highly reliable communication links between two locations. Although it has been agreed under the auspices of the World Radio Conference (WRC) of the ITU-R that tropospheric scatter systems occupying bands intended for satellite communications systems should be closed down and/or moved to other bands, in practice, this does not appear to be happening. Tropospheric scatter systems work by directing very high powered beams of VHF and UHF energy at a particular angle such that a small but well defined amount of the energy is reflected/scattered by the tropospheric turbulence back down to earth over the horizon. This phenomenon is repeated as the radio wave meets other turbulences in its path. The total received signal is an accumulation of the energy received from each of the turbulences. Unfortunately, for satellites, the great majority of the beamed radio energy passes straight through the troposphere and on through the ionosphere into space. Thus, a satellite aerial that happens to have a footprint that passes over a forward tropospheric scatter transmitting station will/may receive a high level of interference from that station. An added difficulty for satellite operators in predicting the effect that such stations might have on QoS is that since forward scatter tropospheric systems are military systems they are not well publicised and they also tend to transmit very short bursts of radio energy which means that they can have an inconsistent effect on the satellite system.

14.6.2   Radar

There is a very similar situation to that for tropospheric forward scatter systems that arises with respect to radar systems. The WRC has agreed that all radars in use in the satellite allocated bands should be closed down. In practice, this is happening slowly in some countries and not at all in others. Radar systems have characteristics where a high powered transmitted beam of microwave energy is directed in a sweeping movement across the sky (vertically and/or horizontally). The radar system relies on small amounts of reflected energy for its success. Inevitably, and unfortunately, the majority of the transmitted power goes straight on into space where it may hit a passing satellite interfering seriously with that satellite's ability to receive communications signals from its intended sources. Thus, radars can cause a reduction in QoS that will be very difficult to predict.

14.6.3   Satellite system response to interference

Satellite system receivers are designed to be highly sensitive and when they receive a high powered signal within the band that they are listening to, they react by rapidly closing down in order to protect themselves. Thus, the effect of interfering systems like tropospheric forward scatter and radar systems is to deny access to a particular satellite not only for the time that the satellite is within the offending beam but also for the time that it takes to recover from the close-down. The QoS effect experienced by users will be unpredictable and will depend on the availability and visibility of other satellites in the constellation at the time.

14.7   Handsets

Handsets for duplex use with GEO satellite systems have to use a directional aerial that can be lined up with the GEO satellite location over the equator. The further it is from the equator (i.e. at higher latitudes) the more important accurate lining up of the handset aerial is. Obtaining a good QoS is more difficult the higher the latitude. As smaller, higher transmitter power handsets have become available (current size is roughly that of a notebook computer) together with higher powered GEO satellites, the easier it has become to get higher quality connections through GEO systems such as Inmarsat.

Handsets for use with MEO and LEO systems use omnidirectional aerials similar to those in use on standard mobile phones. Although similar, the aerial design in practice has to be larger so as to have a higher aerial gain and be fed with higher power transmitter levels without raising the power directed at the user's head. LEOs move across the sky and hence directional aerials on a handset would be unmanageable in practice. LEO handset aerial design is critical to the QoS experienced by users. The health radiation problem is addressed by placing a shield around the aerial on the side facing the users head. Whilst large aerials are more conducive to better QoS, in reality the size of the aerial is usually driven by aesthetic considerations.

14.8   TCP/IP issues

14.8.1   Datacommunications

Historically, the economics of satellites has meant that they are only used for voice calls when other telecommunication systems are not available. The economics of broadband data distribution means that it is likely that the majority of satellite system capacity will be devoted to providing data access to remote users in locations not economically servable by terrestrial means. Thus, satellite system QoS issues, as they affect data transmission, are of particular interest.

TCP is the protocol used by the vast majority of Internet applications. The performance of TCP over long-delay networks will have a direct impact on the performance of Internet access using satellites. In particular, TCP-based solutions to the effects of greater latency and jitter on QoS are possible. Some of these are discussed below.

14.8.2 Window size

TCP flow control starts from the 'window size' concept of a TCP connection. It determines how much data can be outstanding (i.e. unacknowledged) in the network. In long-delay networks, there can be many unacknowledged segments. Theoretically, the amount of data that can be in a transit is given by the bandwidth delay product.

$$\left[\begin{array}{ll} \text{Bandwidth delay product} & = (\text{Satellite latency} + \text{Terrestrial latency}) \\ \quad (\text{i.e. buffer size}) & \quad * \text{Link bandwidth} \end{array} \right]$$

In practice, memory and operating system resources limit the window size. In the current TCP standard, the maximum window size in TCP is 64 kb. (Because of historical implementation issues concerned with signed and unsigned arithmetic, the practical window size is often limited to 32 kb.)

To maximise bandwidth utilisation in a satellite network, TCP needs a much larger window size. For example, on a satellite link with a round-trip delay of 800 ms and bandwidth of 1.54 Mbps, the theoretical optimal window size is 154 kb, or considerably more than a maximum window size of 32 or 64 kb.

A new TCP extension, or TCP-LW for 'large-window' has been defined to increase the maximum window size from 2^{16} to 2^{32}, allowing better utilisation of links with large bandwidth delay products. To obtain good TCP performance over satellite links, both sender and receiver must use a version of TCP that implements TCP-LW. Applications should also set the size of the send and receiver buffers to be bandwidth times delay.

14.8.3 Bandwidth adaptation

TCP adapts to the available bandwidth of the network by increasing its window size as congestion decreases and reducing the window size as it increases. The speed of the adaptation is proportional to the latency, or the round-trip time of the acknowledgement. In a satellite network with longer latency, bandwidth adaptation takes longer and, as a result, TCP congestion control is not as effective. Furthermore, it will take much longer for TCP's linear increase to recover the window size after a packet loss if a TCP 'large-window' extension is used.

14.8.4 Selective acknowledgement

The standard TCP acknowledgement scheme is cumulative. If a segment is lost, TCP senders will retransmit all data sent starting from the lost segment without regard to the successful transmission of later segments. TCP considers this lost segment as an indication of congestion and reduces its window size by half. Recently, the newly defined standard TCP-SACK (Selective ACKnowledge) allows the receiver to explicitly inform the sender of the loss. Consequently, a sender can retransmit the lost segments immediately rather than waiting for a timeout, reacting to supposed congestion, and multiplicatively decreasing its window. If lost segments are not caused by congestion, or the congestion is transient, throughput in TCP-SACK should be much better. This will be helpful in satellite networks because anything that triggers timeouts and window size reduction will force a lengthy recovery in TCP.

14.8.5   Slow start

When a TCP connection first starts up or is idle for a long time, it needs to quickly determine the available bandwidth on the network. It does so by starting with an initial window size of one segment (usually 512 bytes), then increasing the window size as packets are delivered successfully and acknowledgements arrive, until reaching the network saturation state (indicated by a packet drop). On the one hand, slow start avoids congesting the network before it has a good assessment of the available bandwidth; on the other hand, TCP bandwidth utilisation is suboptimal during the procedure. Therefore, the shorter TCP slow start lasts, the better performance it can achieve. The total time of a TCP slow start period is approximately $RTT * \log_2(B/MSS)$, where RTT is the round-trip time (twice the latency), B the available bandwidth, and MSS the TCP segment size. Although the growth is exponential, for high-bandwidth and long-delay networks (e.g. satellite links and terrestrial gigabit networks), this can take a significant amount of time.

14.8.6   Congestion avoidance

Recently, new techniques have been introduced in TCP to avoid congestion before it happens. The first approach, called Random Early Detection (RED) gateways requires each gateway to monitor its own queue length. When imminent congestion is detected the TCP sender is notified. By dropping a packet earlier than it would normally, RED sends an implicit notification of congestion. The sender is effectively notified by the timeout of this packet. The principle behind the RED approach is that a few earlier-than-usual drops may help avoid more packet drops later on. The TCP sender can then reduce its window size before serious congestion occurs.

Another approach is to have the TCP sender predict when congestion is about to occur and reduce its transmission window before intermediate routers drop packets (TCP Vegas). TCP can keep track of the minimum round-trip time seen during a transfer and use the most recently observed round-trip time to compute the data queued in the network. TCP can also keep track of the throughput before and after the congestion window changes to estimate the network congestion level. If estimates indicate that the number of packets queued in the network is rising, it reduces the congestion window. As it observes the number decreasing it increases the congestion window.

Although neither approach has been widely adopted, both hold promise for satellite networks. As we mentioned earlier, TCP congestion control responds to congestion slowly because of latency. If such congestion can be avoided before it happens, it is a big win for high-speed and long-delay networks.

14.8.7   TCP for transactions

Many TCP applications involve only simple communications between the client and the server. The interaction is called a transaction: a client sends a request to a server and the server replies. The Hypertext Transfer protocol (HTTP) for WWW browsing applications is a typical example of TCP with transactional behaviour. Under standard TCP, even a small transaction involving a single request segment and a reply must undergo TCP's three-way handshake in preparation for bidirectional data transfer. If the request is bigger than a segment, TCP must also undergo the slow start procedure. It is very inefficient to establish such a TCP connection, send and receive an insignificant amount of data, and then tear it down.

Transaction TCP, or T/TCP, is an extension to TCP designed to make such behaviour more efficient. T/TCP does this by bypassing the three-way handshake and slow start, using the cached state information from previous connections. Although T/TCP is designed mainly for short client–server interaction applications, it can be used to reduce the impact of latency on the beginning of a TCP connection. If slow start can be avoided, significant performance improvement can be achieved in a satellite-based network.

14.9   Tropospheric propagation

As the lowest region of the Earth's atmosphere, the troposphere extends from the Earth's surface to a height of slightly over seven miles. Virtually all weather phenomena occur in this region. Generally, the troposphere is characterised by a steady decrease in both temperature and pressure as height is increased. However, the many changes in weather phenomena cause variations in humidity and an uneven heating of the Earth's surface. As a result, the air in the troposphere is in constant motion. This motion causes small turbulences, or eddies, to be formed, as shown by the bouncing of aircraft entering turbulent areas of the atmosphere. These turbulences are most intense near the Earth's surface and gradually diminish with height. They have a refractive quality that permits the refracting or scattering of radio waves with short wavelengths. This scattering provides enhanced communications at higher frequencies.

Recall that in the relationship between frequency and wavelength, wavelength decreases as frequency increases and vice versa. Radio waves of frequencies below 30 MHz normally have wavelengths longer than the size of weather turbulences. These radio waves are, therefore, affected very little by the turbulences. On the other hand, as the frequency increases into the VHF range and above, the wavelengths decrease in size, to the point that they become subject to tropospheric scattering. The usable frequency range for tropospheric scattering is from about 100 MHz to 10 GHz.

14.9.1   Tropospheric scattering

When a radio wave passing through the troposphere meets a turbulence, it undergoes an abrupt change in velocity. This causes a small amount of the energy to be scattered in a forward direction and returned to Earth at distances beyond the horizon. This phenomenon is repeated as the radio wave meets other turbulences in its path. The total received signal is an accumulation of the energy received from each of the turbulences.

This scattering mode of propagation enables VHF and UHF signals to be transmitted far beyond the normal line of sight. To better understand how these signals are transmitted over greater distances, we must first consider the propagation characteristics of the space wave used in VHF and UHF line-of-sight communications. When the space wave is transmitted, it undergoes very little attenuation within the line-of-sight horizon. When it reaches the horizon, the wave is diffracted and follows the Earth's curvature. Beyond the horizon, the rate of attenuation increases very rapidly and signals soon become very weak and unusable.

Tropospheric scattering, on the other hand, provides a usable signal at distances beyond the point where the diffracted space wave drops to an unusable level. This is because of the height at which scattering takes place. The turbulence that causes the scattering can be visualised as a relay station located above the horizon; it receives the transmitted energy and then re-radiates it in a forward direction to some point beyond the line-of-sight distance. A high gain receiving antenna aimed toward this scattered energy can then capture it.

The magnitude of the received signal depends on the number of turbulences causing scatter in the desired direction and the gain of the receiving antenna. The scatter area used for tropospheric scatter is known as the *scatter volume*. The angle at which the receiving antenna must be aimed to capture the scattered energy is called the *scatter angle*. The scatter volume and scatter angle are shown in Figure 14.1.

The signal take-off angle (transmitting antenna's angle of radiation) determines the height of the scatter volume and the size of the scatter angle. A low signal take-off angle produces a low scatter volume, which in turn permits a receiving antenna that is aimed at a low angle to the scatter volume to capture the scattered energy.

As the signal take-off angle is increased, the height of the scatter volume is increased. When this occurs, the amount of received energy decreases. There are

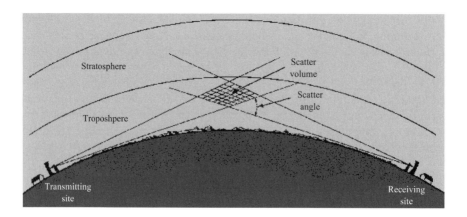

Figure 14.1   Tropospheric scattering propagation

two reasons for this: (1) scatter angle increases as the height of the scatter volume is increased; (2) the amount of turbulence decreases with height. As the distance between the transmitting and receiving antennas is increased, the height of the scatter volume must also be increased. The received signal level, therefore, decreases as circuit distance is increased.

The tropospheric region that contributes most strongly to tropospheric scatter propagation lies near the midpoint between the transmitting and receiving antennas and just above the radio horizon of the antennas.

Since tropospheric scatter depends on turbulence in the atmosphere, changes in atmospheric conditions have an effect on the strength of the received signal. Both daily and seasonal variations in signal strength occur as a result of changes in the atmosphere. These variations are called long-term fading.

In addition to long-term fading, the tropospheric scatter signal is often characterised by very rapid fading because of multipath propagation. Since the turbulent condition is constantly changing, the path lengths and individual signal levels are also changing, resulting in a rapidly changing signal. Although the signal level of the received signal is constantly changing, the average signal level is stable; therefore, no complete fade out occurs.

Another characteristic of a tropospheric scatter signal is its relatively low power level. Since very little of the scattered energy is re-radiated towards the receiver, the efficiency is very low and the signal level at the final receiver point is low. Initial input power must be high to compensate for the low efficiency in the scatter volume. This is accomplished by using high-power transmitters and high-gain antennas, which concentrate the transmitted power into a beam, thus increasing the intensity of energy of each turbulence in the volume. The receiver must also be very sensitive to detect the low-level signals.

14.9.2   Application of tropospheric scattering

Tropospheric scatter propagation is used for point-to-point communications. A correctly designed tropospheric scatter circuit will provide highly reliable service for distances ranging from 50–500 miles. Tropospheric scatter systems may be particularly useful for communications to locations in rugged terrain that are difficult to reach with other methods of propagation. One reason for this is that the tropospheric scatter circuit is not affected by ionospheric and auroral disturbances.

14.10   Satellite-to-satellite communication

A factor that influences the QoS offered by a satellite system is whether there are Inter Satellite Links (ISLs) within the constellation of satellites forming the system. This is because whether the signal passes through a series of satellites to reach the ground station closest to the user before returning to the terrestrial network as opposed to being relayed by only one satellite can make a considerable difference to the time delay (latency) between transmission and receipt of a signal. As examples of the different approaches Iridium's 66 satellites each have ISLs to four other satellites whilst the ICO system (ten working satellites) will not use ISLs but will return the signal back to the most convenient ground station. The different latency effects of these two systems can only be predicted using complex modelling techniques and are not addressed directly here.

14.11   Broadcast TV download systems

A unique aspect of satellite systems as they can be used in telecommunication systems involves the use of broadcast TV satellites to download broadband data to a user whilst the communication path in the reverse direction (user to data supplier) does not go via the satellite system but uses the terrestrial telecommunications system. These hybrid systems will have their own unique QoS characteristics, and can effectively be considered to be hybrid of completely satellite-based and completely terrestrial-based systems.

14.12   Summary

Satellite systems continue to remain an important segment in the design of communication networks. The over arching benefit of satellite systems is that they are able to provide universality of access. However, the universal availability is constrained by high unit cost and variability of the service experience. Satellite system designers continue to push the frontiers of design and deployment to improve service levels and reduce unit costs. Nevertheless, in general, satellite systems continue to require higher degree of user cooperation than terrestrial and mobile systems.

Web Sites

1 Inmarsat Satellite System **http://217.204.152.210/about_inm_satellite.cfm**
2 ICO System **http://www.ico.com/system/home.htm**
3 Unique Properties of Satellites to consider when designing telecommunication systems **http://www.interlinx.qc.ca/leehogle/satellite.html#considerations**
4 Radio Communication Agency **http://www.radio.gov.uk/**

Further reading

POSTEL, J.: 'Transmission control protocol,' Tech. Rep. RFC793, DARPA, September 1981
ALLMAN, M., HAYES, C., OSTERMANN, S. and KRUSE, H.: 'TCP performance over satellite links,' in Proceedings of the Fifth International Conference on *Telecommunications Systems*, March 1997

ALLMAN, M., GLOVER, D. R. and SANCHEZ, L. A.: 'Enhancing TCP over satellite channels using standard mechanisms,' Tech. Rep. draft-ietf-tcpsat-stand-mech-06, IETF Internet Draft, September 1998

MORGAN, W. L. and GORDO, G. D.: *Communications satellite handbook* (Wiley, New York, 1989)

JACOBSON, V., BRADEN, R. and BORMAN, D.: 'TCP extensions for high performance,' Tech. Rep. RFC1323, IETF, May 1992

STEVENS, W.: 'TCP slow start, congestion avoidance, fast retransmit, and fast recovery algorithms,' Tech. Rep. RFC2001, IETF, January 1997

Section IV

Customer impact

Service provider's offerings do have an impact on the customer. In this section, a selection of five topics, customer relationship management, numbering, billing, ergonomic considerations and requirements of those with special needs, have been used to illustrate how the service provider's offerings impact on the customer.

Chapter 15

Service surround and customer relationship management

15.1 Introduction

The stated aim of many public and private organisations is to provide world class service for their customers. Very few organisations actually achieve this goal. It is doubtful whether many of the organisations setting out in quest of this particular Holy Grail would actually recognise it when (or if) they find it.

There would almost certainly be a very wide variety of different answers proffered as to how such a state would be defined. All would undoubtedly include some notion of commercial advantage being gained as a result of their customers perceiving that they were getting better service than could be obtained from the competition. A few would probably point to revenues growing at a far faster rate than their nearest competitors. Yet more would point to all sorts of customer satisfaction statistics gathered by painstaking and expensive research work. The most knowledgeable organisations would point to the quality of their relationships with their customers and the high degree of loyalty and marketplace advocacy of their customers.

Organisations that provide world class customer experience do so in the well-founded belief that it gives them a very significant commercial advantage. Traditional customer care is a key component of the overall customer experience and allows an organisation to demonstrate to its client base the soft skills of its people resource.

However, it must be remembered that customers and suppliers consist of people. If the people interactions between customer and supplier are good then the relationship is likely to be good. Yet all too often the degree of trust in the relationship is poor, and customer loyalty is fragile or non-existent. The supplier often puts in place a so-called customer service organisation to manage this fickle relationship and thereby avoid subjecting customers to the worst excesses of the supplier organisation. Customer service advisers form a thin red line that currently mans (and womans) the call centre forts. These customer service call centres are harnessing all available technologies to enable the call centre operatives to hack the increasing volume and complexity of customer service enquiries. The inter-relationships between the suppliers' customer service organisation and the customer builds up a level of trust between customer and supplier. Such relationships are founded upon people-to-people interactions; clearly the better these fundamental interactions, the better will be the trust that is built up.

To a very great extent, customer service advisers who are properly skilled and customer empathetic can manage the overall relationship with the client. The skills and the personalities of the people in the customer service organisation is a key determinant of just how far trust can be built and to what extent customer expectations can be met. Those organisations which recruit (and manage to successfully retain) the best people clearly have a head start over those that are content to aim at mediocrity by recruiting less than the best. Retention of staff is a serious challenge, for the overall market place for call centre skilled people is expanding faster than suitable call centre agents can be selected and skilled.

The customer experience is shaped by how easy it is to contact the supplier organisation and by how available the supplier organisation is. However, having gained access to the supplier, if the relationship is to be furthered, it is vital to have to hand all the data that is relevant to the particular customer and the transactions in question. The problem of bringing the correct data to the customer service operative or the customer is non-trivial. Those organisations which solve this problem for their customers will be at great advantage relative to their competitors in the marketplace.

This chapter looks at these issues as part of QoS in telecommunications.

15.2 Customer expectations and requirements

Different customers have widely different expectations of their service supplier. Generalisation is dangerous and can only be done at a highly superficial level. Organisations like the Telecommunications Users Association and the Telecommunications Managers Association have over the years carried out many surveys on behalf of their members, and have lobbied network operators and suppliers to provide what their members want. In addition, individual suppliers have carried out their own research into what their customers want.

Below is a distillation of customers' expectations drawn from a wide variety of sources and the author's own experience. It is not in any way claimed to be a comprehensive statement of customer requirements that could be used to design the optimum customer experience in all marketplaces and over all customer products. Rather, it should be treated as a minimum checklist that service suppliers can disregard only at their peril.

When dealing with their supplier, on any service aspect, customers expect

- to be treated as individuals;
- to be valued and respected;
- to be dealt with by people who understand their business or personal needs;
- to have all requirements coordinated in a professional manner;
- to receive a flexible response to urgent and non-routine requests;
- to feel that they are a market of one.

When ordering new services customers expect

- understanding of their requirements, especially if that customer is not a telecommunications expert;
- realistic promised delivery dates and then delivery as promised;
- service installations carried out to a high technical quality;
- service delivery executed in a properly scheduled manner;
- some flexibility in responding to their requirements.

When seeking service restoration they expect

- full ownership by the supplier of resolution of faults;
- faults fully and logically explained in terms that the customer can understand;
- regular information updates on progress towards restoration;
- reported faults handled in a professional manner;
- formal handback of restored service to customer;
- maximum outage time to be guaranteed with compensation if exceeded;
- frequency of needing to seek service restoration to be very low.

During the billing cycle customers expect

- invoices presented in a clear easy to follow format;
- matching of the invoicing process to the customers normal accounting processes;
- billing and invoicing processes to be inherently accurate;
- fast access to supplementary data to support invoiced amounts;
- fast reconciliation and repayment of disputed amounts;
- complete integration between invoicing and payment tracking.

15.3   Customer perception

Customers' perception of service quality is shaped by many factors that are covered in great detail in Chapters 3 and 7. On each of these factors, the customer makes a conscious or sub-conscious value judgement. These can be grouped very broadly into

- factual experience of using the product or service;
- factual experience of dealing with the supplier organisation;
- less tangible 'impressions' from indirect sources.

These are shown in Figure 15.1.

All customers will weight these aspects differently. Most customers are likely to weight their direct experiences of using the product or service and dealing with the supplier most heavily. Thus, it is actual deeds, not advertising hype that will impress rational customers. However, customer perception will still vary over time and will be a function of the outcome of many customer-affecting events, both good and bad.

Over time, the feelings of a customer towards a supplier will become more and more positive if a whole series of events takes place with good outcomes for the customer. Confidence and trust is built little by little, and occurs over time as shown in Figure 15.2.

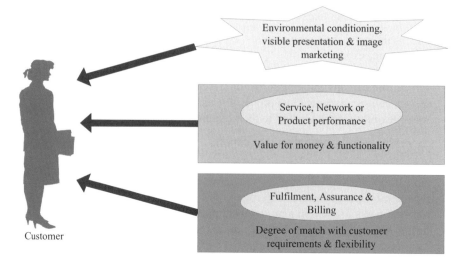

Figure 15.1   Customer perceptions of service quality

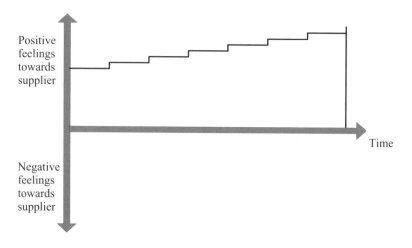

Figure 15.2   Customer perception following good events

Conversely, the feelings of a customer towards a supplier will become more and more negative as a sequence of events with poor outcomes for the customer progresses. Confidence and trust is destroyed with each negative happening. By comparison with the slow rate at which trust can be built up by a series of good events, the erosion of trust proceeds at a far faster rate, as shown in Figure 15.3.

Following a disaster situation, such as when a large customer loses all service and that customer's ability to conduct business with its own customer base is seriously compromised, customer perception plummets. The situation is bad, but can now swing either of two ways, as shown in Figure 15.4.

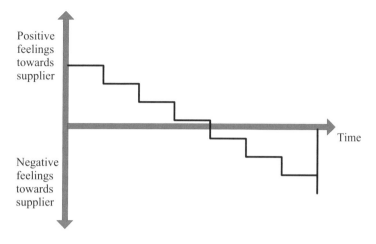

Figure 15.3   Customer perception following bad events

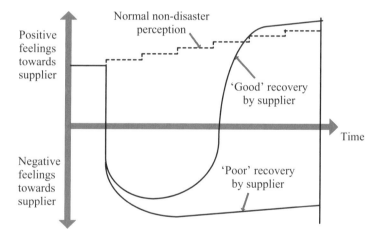

Figure 15.4   Customer perception following disaster

If the supplier pulls out all the stops and puts all of its resources (including top management) into solving the problem, then it is very likely that the customer will feel that the supplier can be trusted to put them first, thereby gaining respect in the eyes of the customer. Conversely, if the supplier exacerbates the problem by not moving resources to fix the problem for the particular customer, the supplier will come across as uncaring or arrogant. The customer at a stroke will have been converted into a potential defector.

Customer's problems actually present a glorious opportunity to impress customers, always provided, of course, that there are not too many such 'opportunities' per customer per year! All customers encounter a myriad of problems from a wide variety of sources in their normal lives. There is a great deal of positive goodwill generated with a customer when a problem is efficiently fixed. Perversely, this can actually create a better impression than if there was never a problem in the first instance. The reason for this is that if there is no problem, then the resultant blandness does not put the organisation within the customer's conscious value reference norms.

However, if an enquiry or a problem is resolved in a manner that really 'wows' the customer, customer perception rises and holds for a very long time at well above the original threshold. Even better than a 'wow' following a failure is a positive, proactive act. This could consist of a service centre operative calling a customer to report that a problem that had affected some aspect of the customer's service has now been fixed. Apologising for inconvenience, confirming with the customer that all is now well and enquiring whether the original event had caused serious impact to the customer will all greatly raise customer perception. But, it must be remembered that such 'wows' or proactive acts are only possible if the customer service infrastructure as a whole is highly effective and the product base is of high quality. Otherwise, the degree of basic quality shortfall will totally negate any good work done in this area.

15.4 Customer care

Customers purchase products and services from suppliers and expect to be able to utilise the product or service to meet their particular expectations. The skill, knowledge, experience and aptitude of any particular customer will govern the extent to which that customer will require more than just the supply of the basic product. Some customers will require a lot of advice and help on how to select and then use a particular product. Others will be sufficiently competent to be able to take care of some or all of these aspects themselves.

All customers will need and expect help when a product fails and has to be replaced or repaired. All customers will also expect to be billed accurately and to pay only for services that they have in fact utilised.

Service providers generally attempt to provide a high quality of service. However, because of the sheer scale and complexity of the way that services are provided, service guaranteed for 365 × 24 is not generally a realistic proposition. By way of illustration, the size of a typical incumbent Telco in Europe or North America is of the order shown in the table below:

Scale of Business for a Typical Incumbent Telco

Employees	100,000
Network assets	$40 Bn
Annual investment	$5 Bn
Turnover	$20 Bn
Customer Base	30 Millions
Annual orders processed	7 Millions
Annual faults processed	5 Millions
Annual bills issued	120 Millions
Product portfolio	1000 Items

This sheer scale means that service will normally be subject to many interactions between people, and once in a while these people will goof. In addition networks, systems and products will fail at infrequent intervals though usually with very high impact. Even concrete, glass and steel constructed buildings that are used to house equipment occasionally catch fire or suffer some other catastrophes!

To cater to the imperfections of the real world, service providers go to great lengths to build a service support infrastructure, or 'service surround', so as to ensure that customers enjoy the fullest possible utility from all products and services offered.

Notwithstanding that many organisations have eloquent mission statements to be the 'World's Number One Service Partner' etc., the purpose of service care is to leverage revenue, not to run the world's smoothest, best (and largest!) customer care operation. Furthermore, it is vital that the functions of service care are fully integrated with normal overall business processes, rather than being stand alone processes which exist in their own right.

Any organisation that does not carefully match the level of service surround that it provides to its customers with the best that is available in the marketplace will struggle to survive. All industries have to provide a commercially justified level of service surround for their customers, and telecommunications is no exception.

The match of how the supplier organisation interfaces with its customers determines the efficacy of the overall customer relationship. Although it is possible to provide some aspects of service surround by utilising technology, the vast majority of customers look towards people in the supplier organisation. These people understand the needs of their customers and are able to make things happen on their behalf.

15.5   Value add from service surround

In order to add value to a basic product or service, the customer service organisation must be able to be contacted easily by all users. In general, this will mean an easy to remember telephone number such as a 0800 number and appropriate call steering mechanisms to ensure that any call from a customer is directed to an appropriately skilled and empowered operative. E-trading and fax contact points may well reinforce such primary access where there is no need for real-time personal contact.

For order placement, provision and most billing enquiries, hours of availability can largely be matched to the hours that customers are actively trading with their customers, and for the most part this will be the normal business day. Not many customers will wish to place an order for a private circuit or router network at 4.00 a.m., but even here where customers are global it could be that orders are placed from within a different time zone into the single point of contact. Hitting the office automated voice answering service is not likely to gain favour with such global players! There is also a trend towards having to match customer's physical opening hours, in certain sectors such as the retail and financial services sectors.

The position on trouble handling, however, is much more demanding. With increasing globalisation, the trend towards 365×24 is inexorable. However, there is little point in having a 365 × 24 trouble/fault handling service centre if it is not backed up by service personnel actually available in the field to fix problems for 365 × 24.

In order to add value, the person dealing with the customer has to have all relevant details regarding the customer readily available. Indeed, larger customers will almost certainly demand a set of individuals to serve them who are dedicated totally to them.

Resource levels must be adequate to ensure that a rapid response can be made to all customer enquiries. There is little evidence to suggest that all calls handled in 15 seconds or better impresses customers one iota, but it is equally certain that 'ring tone no reply' for 15 minutes would be totally unacceptable.

Most organisations have a plethora of customers and have to decide on how much effort they will devote to each customer, largely on the basis of the importance of that customer to the organisation. Thus all customers are not equal. However, in organisations such as public utilities, notwithstanding that customers are not of equal importance, the regulatory regime dictates that no customer (or set of customers) can be discriminated against. Thus, it is vital, when segmenting the customer base, to ensure that all customers feel that the broad arrangements put in place for them are appropriate for their level of dealings with the service supplier and enable them to be dealt with on a 'market of one' basis.

It is possible to segment customers in a variety of ways. The following list of possible approaches is by no means exhaustive:

- by geography
- by products used
- by market sector
- by customer size
- by customer competence/expertise
- by customer power
- by customer marketplace leadership
- by customer profitability.

All of these have their uses and it is very much a case of selecting segmentation(s) that make customers feel that they are being treated appropriately. In practice, a lot of organisations segment along the following broad stratifications:

Large corporate and multinational companies
Each large customer will be given a unique access to a dedicated service team. The sheer volume of enquiries from such a customer dictates that a small team of individuals can be kept busy and add considerable value to the relationship with that customer.

Small and medium companies
These customers generally have relatively simple telecommunications needs, spread across very few sites. The volume of their enquiries is generally insufficient to establish a really good personal relationship between them and the organisation. Much can be done in terms of simple automation to allow small things such as automated fault

logging to take place, initiated by tone/voice response systems that can put a great deal of simple transactions beyond the need for human intervention.

Personal customers

Generally these customers are the most sensitive to price and the trick here is to provide a no-frills service at minimum cost that truly meets customer requirements. The scope for robotic and e-trading is quite phenomenal and much work remains to be done in this area.

Technically competent 'expert' customers

A lot of organisations and individuals are now very technically competent and where this is the case, there is a lot to be said for giving those customers direct access to people in the field or direct access to the service support systems to request changes, report faults etc. This is sometimes viewed as counter productive to the overall customer relationship, but given that these customers are 'expert', the only relationship that they want is with a corresponding 'expert'.

An organisation setting up a service surround for all types of customers would probably build an infrastructure as shown in Figure 15.5.

Customers gain access to customer service centres via published telephone numbers. Call centres can be accessed by any of the following methods:

- *Free call number* – The customer pays no charge with the entire cost of the call being met by the call centre provider.

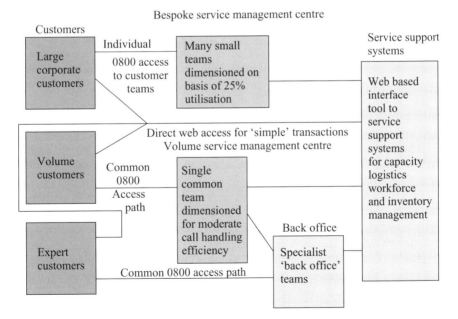

Figure 15.5   Generalised architecture of customer service centre infrastructure

- *Local call number* – The customer pays for the cost of a local call, with the call centre provider paying for the extra cost over and above this charge.
- *Standard telephone number* – The customer pays the full cost of the call.
- *Premium call number* – The customer pays for the call at above national call rates. The call centre provider receives a fee from the network operator for every call that it handles.

The type of access that is provided for individual customers is determined by purely commercial considerations. However, to indicate just how heavily organisations and individuals have become dependent upon freephone service access numbers it is worth looking at North America. In November 1998, on an average business day about 40% of the 260 million calls carried over AT&T's network were tollfree, adding up to some 24 billion calls per year. Since that date toll free calling has grown by about 30%. Thus the scope for getting customers to pay for calls to their service suppliers is probably a lot lower than it appears on face value.

The value of building relationships with customers and turning them into loyal customers and workplace advocates is enormous. Loyal customers generate additional sales and revenues. Marketplace advocates recommend your organisation's products and services and make it easier to capture new customers who are more readily persuaded to defect from other suppliers. The cost of selling to existing contented customers is very low.

However, if a company *believes* that customer loyalty and retention are key to its success (or even its existence) then it is vital that it corporately *acts* and behaves as though it does. If the company acts in this way, the existing sales force will be well geared to making regular contacts with the key opinion formers in its existing customer base and indeed, those customers will know exactly who to contact in the supplier to get what they want. Such relationships are formed, however, only after the passage of time and the outcome of many individual events has allowed confidence and trust to be built up.

Yet, many customers never stay with an organisation long enough for the organisation to have a sporting chance of turning them into loyal customers. Indeed, within particular service providers like mobile and ISP, high levels of churn appear to be the norm rather than the exception. The revenue cost of churn is fairly easy to quantify. If $x\%$ of your customers defect each year, than very broadly your total revenue is reduced by that factor.

To replace those $x\%$, it is necessary to recruit the same number of customers, but this means that heavy sales and marketing costs will need to be carried. The replacement $x\%$ are not likely to be cash positive until much further downstream, and in the meantime the haemorrhaging continues.

How big is x typically? It is believed that many mobile operators around the world suffer churn rates in excess of 20%. With annual revenue streams typically measured in billions of dollars, the lost revenue is several hundreds of millions of dollars. The problem is further compounded because the newly captured customer will probably have been attracted (bought?) from another supplier merely on the basis of price. As such, these customers will be quite likely to move again. If the churn rate amongst such customers is a factor of five or six higher than that for the more established customer base, then the newly recruited group of customers will probably never contribute to overall profitability at all!

Hanging onto customers can be likened to waging a war. It is not waged with guns and tanks but rather with more subtle psychological weapons aimed at winning the hearts and minds of customers. In any form of warfare it usually takes greatly superior forces to defeat an enemy who is well dug-in.

It is reckoned that in business terms, a competitor would have to deploy ten times the resource to prise a customer away from a supplier than that supplier deployed to retain the customer. Hanging onto existing customers must be the soundest strategy that any business can pursue and the loss of any customer should be regarded as an event that ought not to happen.

15.6 Service surround processes

There are many processes that need to be operated if the customer is to have a good experience. In describing such processes, it is very useful to use a model. The FAB process model is described in detail in much documentation emanating from the Telemanagement Forum, in particular document GB910 and is shown in Figure 15.6.

There are three components, fulfilment, assurance and billing and the model take its name from the acronym formed by the first letters of each of the components.

The fulfilment component covers all of the following processes:

- Pre-order enquiries;
- New order placement;
- Moves and changes order placement;
- Cessations;
- Delivery date negotiation;
- Pre-delivery dialogue and enquiries;
- Customer handover process.

The assurance component covers all of the following processes:

- Post-delivery enquiries;
- Failure and fault reporting;
- Complaint handling;
- Failure and fault management;
- Customer hand-back process.

The billing component covers all of the following processes:

- Customer inventory management;
- Invoicing;
- Invoice aggregation/disaggregation;
- Invoice query handling;
- Payment tracking;
- Payment reconciliation;
- Compensation payments.

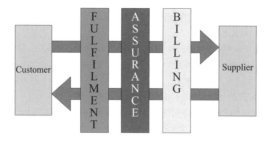

Figure 15.6   FAB model

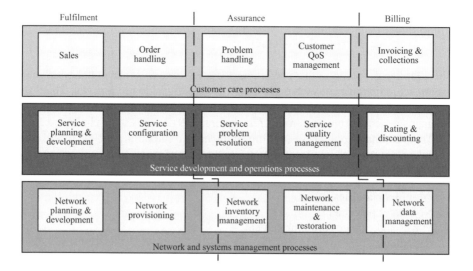

Figure 15.7   'FAB' high level process mapping

These three major process components in turn permeate across customer, service and network domains as shown in Figure 15.7 which is extracted from TeleManagement Forum 1999. Yet, even with such a model available, the detailed processes can be very difficult to understand and get to grips with as shown in Figure 15.8.

15.7   Service support systems design principles

The entire purpose of the service surround is to ensure that customers enjoy a high QOS.

The best way of not breaking promises to customers is to ensure that all major components of the promise to the customer can be checked off *before* the promise to the customer is made.

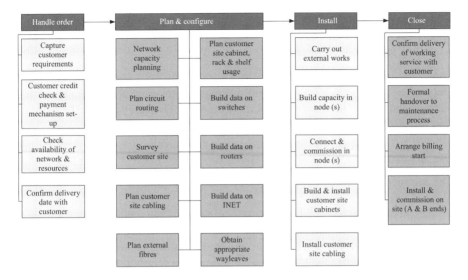

Figure 15.8   Typical generic provision process for a telecommunications operator

Take as an example, an order for telecommunications service that involves

- the provision of wiring and/or equipment at one or more of the customer's premises;
- the provision of capacity through or into the network;
- configuration or reconfiguration of the service platform for the customer;
- modification of the way that customer's invoices are currently aggregated and presented.

In order to be able to meet this set of requirements satisfactorily, it is necessary as an absolute minimum, to be able to answer the following questions:

- When can the CPE and/or wiring be delivered on-site to the customer?
- When can field staff with the right skill sets be available to fit the CPE and/or wiring on the customer's site?
- Is capacity available through or into the network? Can it be earmarked for this set of requirements or does it have to be provided as new capacity? If so, when will it be available?
- Can the service platform be configured as requested? If so, when?
- How does the customer inventory and invoicing process need to be modified so as to meet the customer's requirements? When can this be achieved?

Thus as a bare minimum, it is necessary for a customer service advisor to be able to view, interrogate and modify the following databases:

- customer database
- network capacity availability database
- network and equipment problems database

- service platform database
- logistics supply chain database
- billing database
- workforce availability database.

Only after there have been confident affirmatives to each of the sub-questions addressed to each of the database engines can a firm high quality promise be made to the customer:

> We will deliver this service on this date and fully meet your requirements.

The requirements for the high quality promise underpinning dataset is shown in Figure 15.9.

Yet, notwithstanding the need for generating such high quality promises to customers, there are very few organisations that have the luxury of a single support system that fully integrates all of the necessary databases together and allows a good promise to be made to customers. It is far more usual to rely upon the skills of the customer service advisers to negotiate between what the customer really wants and what the 'best risk' promise that can be made is.

For example, customer service advisers will normally have some knowledge of the broad availability of particular products in particular geographies. For instance, if there were a localised shortage of people skilled in fitting router equipment, local office notices might instruct customer service advisers not to negotiate delivery dates with customers until they have checked with the local workforce to ascertain people availability. Whilst such failures to be able to deal with problems on the spot will

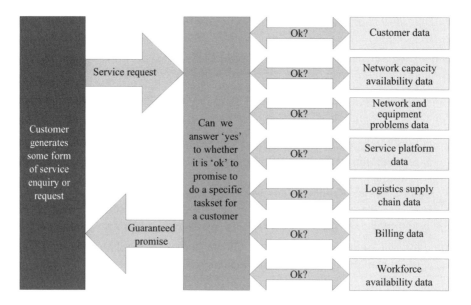

Figure 15.9   The high quality promise dataset

cause some dissatisfaction to customers, the alternative of accepting the order 'blind' with an 'agreed' delivery date that cannot be met is a source of far greater long term dissatisfaction to the customer.

Similar disjoints and disconnects can occur during service restoration and billing/payment enquiries. With regard to fault reports, there ought to be little to prevent a fault report being properly logged onto the system. After all, such a report is normally associated with a piece of customer inventory for which revenue is currently being received.

However, the fault management system may be totally lacking in detail of the results of diagnostic tests, or the expectation of when the particular aspect of the customer service can be restored. Most suppliers will only be able to issue a fault ticket number to the caller in these circumstances.

This will do little to give the customer confidence that the service will be restored within the specified maximum outage time. However, a good customer service adviser ought to be able to escalate the status of the problem as necessary and keep the customer contact properly informed. Whilst this does not amount to a 100% guarantee of a successful outcome, it partially mitigates against the lack of being able to give an adequately good promise to the customer and is far better than:

> Thank you. The fault has been logged and service will definitely be restored in the next four hours. If it isn't please phone this number and quote the fault reference number that I have just given you. Goodbye.

With regard to enquiries about billing and payment, the promise quality that is required here is to guarantee to the customer that the supplier understands and can track every penny that the customer has spent and will spend. Furthermore, it is vital to be able to demonstrate that of all the tariff packages that are open to the customer, the best tariff is the one that the customer is currently on. The real problem with billing enquiries is normally the inability to do this in near real-time, particularly if the customer account is large, and the invoice aggregation or disaggregation to match the customer's business requirements are somewhat complex. Here, again, the customer service adviser has a vital role to fill in bridging the expectation gap.

It must be remembered that for the most part, orders are negotiated, faults reported and bills queried in real-time while customers are on the telephone. As a result, response speeds from systems are crucial. It is vital that the operator can retrieve information from the system by at most three mouse clicks with an inter-click refresh rate on each of these of around 3–5 s. The achievement of such data retrieval speeds from the huge databases associated with large service provider organisations is an extremely challenging design parameter. All too often systems operate far too slowly.

The speed requirement is greatly exacerbated by the sheer scale of the undertaking. The support system data processing load is not trivial by any standards. For a typical large incumbent Telco, then, the following loadings are quite realistic:

- 1 million orders per year or about 50,000 per day;
- 0.5 million fault reports per year or about 25,000 per day;

- 0.25 million bill enquiries per year or about 12,500 per day;
- 5000 simultaneous users.

Furthermore, the rate of growth of business needs to be accommodated by a generous level of day one dimensioning otherwise throughput will very rapidly suffer.

Ideally, all data used in support of the customer experience should always be completely up to date, but in practice different databases are likely to be refreshed at different rates. It is important that time scales for decisions that are taken using particular data are always several orders of magnitude larger than the refresh rates. Thus, if a database were to be refreshed at five-minute intervals, it is likely that information about equipment, people or capacity availability seven days downstream would be very meaningful. However, a fault status database updated half-hourly would be very difficult to use as a basis for dialogue with customers over the expected maximum four hours duration of a fault.

Systems are not generally easy to use. Most major telecommunications service providers recruit operatives who are then trained to operate the complex systems that abound. This is unsatisfactory because there is usually a very high systems training overhead that can be anything from 2–10 weeks, whilst the skills of the operator are focused upon the hard skills of driving the systems and not upon communicating with customers using softer skills. Increasingly, it is these softer skills that are needed to build relationships with customers.

Only by bringing together the metrics from different disciplines to provide a holistic view of business activity and thereby reflecting the increasing complexity of today's business models can end-to-end business processes be properly supported.

To achieve this it is necessary to combine data from network and customer facing systems in an integrated BSS and OSS. This will give added benefits of being able to

- provide higher levels of customer service;
- enhance revenue assurance;
- enable more accurate service pricing;
- make better use of network assets;
- reduce cost of inefficient organisations and processes.

All too often, however, there is very little cross flow of information between the operations and customer care parts of the organisation as shown in Figure 15.10.

Even worse, systems and processes regularly reflect the structure of the organisation as shown in Figure 15.11. This means that systems reflect the demarcation lines within an organisation rather than the holistic needs of customers.

In order to break out of this situation, it is vital that the processes are re-engineered from the customer standpoint. All the end-to-end implications of how things need to be done must be fully understood. A good catalyst to do this will be to set up cross-functional teams to design the integrated systems architecture and implement processes workarounds during the phased migration to the new integrated systems.

Vendors have tended to produce software that does not address all areas of CRM, fulfilment, assurance and billing. This gives rise to a need to patch between various vendors' software and almost inevitably leads to a lowering of efficiency relative

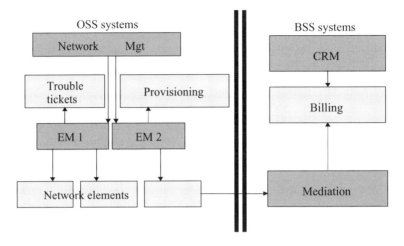

Figure 15.10   Systems lacking in cross flows

Customers

Network operations	Finance	Marketing	Customer service
Provisioning	Billing Ops	Product devpt	Contact centre
Routing	Revenue	Pricing	Order taking
QoS	assurance	Discounting	Bill queries
monitoring	Settlement	Packaging	Customer entry
Trouble tickets	Margin	Channels	Complaints
Capacity	management		
planning			

Figure 15.11   Systems reflect organisational structure

to software suites designed holistically from the outset. There is little 'one-stop' capability to choose from.

15.8   Impact of people

The words 'client' and 'customer' often tend to be used to describe entities that have neither gender nor persona. The 'client' is often viewed as an object that has to be successfully interacted with in a manner that allows the 'supplier' to maximise the amount of revenue flowing from 'client to supplier'. 'Supplier' is an equally vague term and is effectively a collective noun for all of the people, processes and systems that comprise the supply chain.

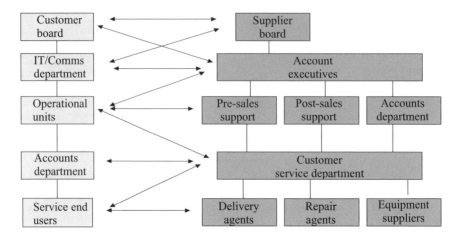

Figure 15.12   Customer/supplier interfaces

However, in actual fact, customers and clients consist of people. Suppliers consist of people. People form relationships with people. People buy from people. People supply goods to people.

Depending upon the complexity of the client, and the complexity of the supplier, there can be very many inter-personal relationships that collectively aggregate together to form the client–supplier relationship. This is shown in Figure 15.12.

When viewed in this way, it is quite obvious that the quality of the relationship as perceived by the people in the client organisation is chiefly dependant upon the skills, aptitude, attitudes and cultural values of the people in the supplier organisation with whom they have most dealings.

The provision of telecommunications services usually involves a large number of people. In order that clients can enjoy a relatively cohesive and integrated single point of entry into the supplier organisation, without needing to understand the intricacies of the supplier's organisation in great depth, most suppliers have built telephone call centres to handle enquiries and contacts from customers.

The customer service advisers who are resident within these customer service call centres are at the frontline in terms of managing and shaping the relationships with customers. They are in fact relationship managers in just as great a sense as is the customer account executive who is responsible for generating sales to particular customers. So what sort of people are these relationship managers, and what sort of skills do they need?

First, because of the complexity of the service support systems, it is necessary for these individuals to be highly skilled and competent in inputting data to, and extracting information from, these support systems. Without such skills, irrespective of how good or bad the customer perception of the experience is, nothing would actually get done. No order would get entered, no fault reports would be logged and no invoices would be tracked or checked.

It is important to recognise that to become competent on most support systems requires between two and ten weeks of intensive training. It is pointless giving such training to unsuitable people and so a good general intelligence level is also a prerequisite.

Rocket scientists and nuclear physicists are not required, but customer service advisers must have good general numeracy and literacy. Most supplier organisations look at candidates' general education attainments before recruiting them, or will set aptitude tests to ensure that candidates are competent in relevant areas.

In addition, vocational qualifications are increasingly being viewed as valuable by customer service organisations. In the UK, there is much formal training that takes place that leads to National Vocational Qualifications (NVQs) in customer service. Within the scope of such a formal framework, workplace experience is assessed by independent assessors with the aim being to coach people to attain the highest possible standard.

In addition to the hard skills above, customer service advisers need a range of soft skills to enhance the quality of the relationship. Customer service advisers need to be confident and assertive. They have to be willing to champion the customer's cause within the supplier organisation. This means that they have to build a network within the supplier organisation to get things done for customers. Passivity is no good; they need to be fixers, movers and shakers. Above all, the people working in the call centres need to like dealing with other people and have to be able to relate to the clients' point of view.

In this way, the customer service organisation is seen to be passionate about customer service. Such an organisation will usually be very effective at building a high quality relationship with customers who become increasingly loyal and turn into marketplace advocates for the supplier.

The greatest asset of any organisation is the ingenuity of its people resource. Yet people can only be creative if the environment is right. Creativity is a learning experience. Sometimes it is necessary to make mistakes in order to really learn. If the environment is one where calculated risk is only accepted if it succeeds, whilst failure is treated with total disdain, the people in the organisation will be placed in a situation where their individual talents cannot be fully and effectively harnessed.

Organisations often carry out recruitment to a very tight specification in an attempt to provide a high degree of uniformity in the service offered to customers. The impact of this is usually that each individual becomes a stereotyped 'clone', with an expectation that each individual will react to customer relationship situations in a particular and predictable way. Often such clones are recruited in the image of the service organisation manager. Such recruitment policies leave the manager surrounded by individuals of like mind who will rarely challenge the 'way things get done around here'.

Yet such challenges are absolutely vital if the organisation is to be able to embark upon the process of continuous (or quantum) improvements that are essential to keeping up in the marketplace. There is a tangible difference in atmosphere in an organisation where the customer service advisers are fully empowered to use their talents and creativity to ensure that the demands of customers are met. There is an exciting buzz about such a place. The people know they are doing a good job. The customers know that the service advisers enjoy their jobs and perform their tasks very well.

These positive vibes strongly reinforce the ongoing relationship building process. When things go wrong, the advisers don't hide behind expressions like 'Our equipment suppliers have let us down' ... or ... 'We can't do that because we didn't get adequate notice', ... or ... 'Our fitter phoned in sick'. They take it on the chin and fix it for customers whilst portraying an upbeat image to customers. When the chips are down the customer knows that everything humanly possible has been done by the human at the other end of the call to get a result for the customers. Such person-to-person honesty and professionalism is what builds and cements relationships.

Yet even in the best-managed supplier organisations containing very high calibre people, it is sometimes the case that best use is not made of the available skills and talents. Here, the problem is usually the sheer volume of low grade tasks that completely swamp the individuals. There is no time to think, no time to tidy up loose ends, no time to treat the client as a person. Questions like 'How are you today, Frank?' disappear from the dialogue.

The need for speed depersonalises relationships. Little can be done except to either have more people, or to automate the simplest tasks and keep the people available to deal at more leisure with the lesser proportion of more challenging customer enquiries. However, the impact that good people can make can be totally negated by poor customer experience prior to actually reaching the customer service adviser, whilst any weaknesses in people qualities will reinforce the poor impression. Thus, it is important that the customer service surround is designed holistically, so that all aspects properly combine to provide a high perceived quality customer service experience upon which the all important client–supplier relationship is totally dependent.

15.9   The pivotal position of the call centre

In the beginning, all trade was conducted face-to-face, either in premises owned or rented by the goods or services supplier, or on the customer's premises, or in exchanges and marketplaces that facilitated the coming together of customers and suppliers. As communications services developed, starting with humble traditional mail services, the possibility of a small proportion of non-time critical remote transactions was enabled. It is arguable that if mass telecommunications had not been invented, this state of affairs would have continued, with the great bulk of transactions still needing to involve a face-to-face physical inter-relationship.

However, following the advent of the telephone, customers started to use this latest technology gismo to communicate with their suppliers with great enthusiasm. Such contact with businesses by telephone was initially on an ad hoc basis and met with varying degrees of success and satisfaction across this interface.

As companies and organisations harnessed the power of the media to market their products and raise customer awareness and expectations, these ad hoc methods of engaging with customers seldom kept up with the expectations of those customers.

As the scale of business increased and the range of products and the geographies in which they were offered increased, it became increasingly difficult for customers to know which part of an organisation to make telephone contact with.

Widely varying customer satisfaction caused by dealing in a variety of different ways with the diverse parts of a supplying organisation was the norm. The telephone call centre was created to provide a single point of entry to the supplier organisation.

In its most basic form, it took the form of a switchboard operator with an internal telephone directory, routing callers to the appropriate point of enablement within the target organisation.

As the complexity of organisations increased, so did the size of their switchboards and the number of operatives required. Technology again came to the 'rescue' by allowing knowledgeable customers to contact the various parts of the organisation by means of DDI extensions thereby eliminating the need for huge increases in the numbers of telephone switchboard operators.

However, this meant that any customer (or anyone in any part of a customer's organisation) could dial into all parts of the supplier organisation with greatly varying success at reaching and enjoying a good customer experience with the correct point of enablement.

Rationalising the way in which the customer engages with the organisation by telephone and correctly positioning it alongside all of the other channels to market between the supplier and the customer has given rise to the development of the modern telephone call centre.

The role of a call centre is to enable telephone and computer technology to be used in a standardised manner in order that the people within the organisation can deliver a better experience for their customers.

Call centres are not simply a physical location where customer's calls are handled, but a set of functions that manage the many varied needs of customers. The people in the call centres act as the points of interface between the organisation and its customers, gate-keeping and interpreting the bi-directional flow of information.

A properly designed call centre will generally perform this interface function far better than the ad hoc methods of engaging the customer described earlier.

In addition there is the opportunity to focus upon enhancing the customer relationship and improving customer service, often allowing customers to make enquiries over extended opening hours whilst also facilitating faster completion of transactions and enquiries. Call centres radically improve the quality of contact with customers and maximise the possibility of conducting further business with the customer.

Scale economies are made possible by the concentration of sales, administrative and customer response activities in a small number of locations. The associated cost reductions can be very large indeed in comparison with those functions which remain highly distributed throughout the supplier organisation.

A first class customer experience relies on the optimum combination and management of the three key elements of process, people and technology. The call centre is a fundamental tool to implement this optimum winning combination and thereby enhances the quality of contact and builds up a relationship with the customer. The use of computer and telephony integration provides staff with customer records at

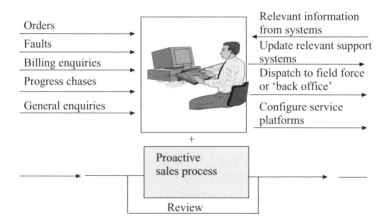

Orders
Faults
Billing enquiries
Progress chases
General enquiries

Relevant information from systems
Update relevant support systems
Dispatch to field force or 'back office'
Configure service platforms

+

Proactive sales process

Review

Figure 15.13   Functionality of generic customer service management

their fingertips to enable prompt response to a very wide range of queries and to enable those queries to be dealt with in a highly professional manner. Customer service quality is the key driver for call centres, with the success of a call centre being directly measurable by the impact upon quality for users of that centre. Figure 15.13 shows the general function fulfilled by customer service operatives.

Currently, in most organisations, operatives spend about 30% of their time communicating with customers, and about 70% of their time communicating internally within the organisation, or inputting data onto systems. Operatives are unable to build relationships with customers while they are communicating internally. The opportunity being missed, largely as a result of poor processes and support systems, is manifestly large.

Relationships will clearly benefit from unconstrained dialogues between individuals. To enable this, customer service centres should be dimensioned so that the service operatives can spend the appropriate amount of time building relationships with their customers. Unfortunately, many customer service centres are driven by 'economics' and are designed around getting good operator usage efficiency. Such over zealous design imposes a high degree of artificiality into customer relationships. Relationships being built in such an environment will generally be poorer. As well as giving the operative a comparatively small amount of time to dialogue with customers, such tight dimensioning will already have caused much customer dissatisfaction at having to queue for long periods waiting for the telephone call to be answered. The relationship between customer and supplier is often already at a low, before a single word is spoken.

Customer service call centres normally handle a very high proportion of quite simple transactions and a much smaller proportion of increasingly complex transactions. As customer service expectations have risen rapidly, so have the amounts of calls requesting very simple transactions. These have generally grown faster than service centre operatives have been recruited and it is a testimony to their skills and to improvements in the support systems that these increased volumes have been successfully met. However, the more complex transactions use a greatly disproportionate amount of service centre resources. Furthermore, the absolute complexity of these more complex transactions tends to be growing very rapidly. This will rapidly lead to a situation where more and more service centre operatives will be required unless something is done to streamline the complex transactions and offer alternatives such as web access to eliminate a high proportion of the simpler transactions. The ability to provide solutions to the complex transactions that leading customers wish to make is seen as a key differentiator in the marketplace. It is vital that resources can be freed up to handle these in a high quality way.

There are several stages in the life cycle of a call centre agent:

- recruitment and selection;
- training;
- coaching;
- effective working;
- investment;
- agent attrition.

These progression stages are related to each other as a people flow process as shown in Figure 15.14. There are several key features to this process. Although ingress to call centres occurs only at the initial stage of the flow with the intake of new people, egress occurs throughout the process, with people being lost for a variety of reasons.

People being lost to the organisation, for whatever reason, represents a cost of failure. Such costs can vary greatly. In well-managed environments these will be small, as poorly performing agents, whether employed directly or via agencies, will have their employment terminated by the appropriate process. However, in poorly managed environments there is a risk of large one-off costs resulting from industrial action or industrial tribunals.

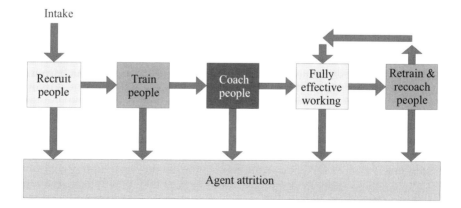

Figure 15.14   Call centre agent life cycle

Call centre costs vary widely according to the type of role that they perform. However, across a wide range of centres, the cost structure on an annualised basis averages out as shown in the table:

Cost component	Annualised cost
Agents' pay	40%
Managers' pay	12%
Property	10%
Training	21%
Telecommunications	8%
Systems	9%

15.10   Dimensioning of a call centre

There are very many parameters that need to be taken into account in designing and dimensioning a customer service call centre. The most important decision is whether or not all the resources are to be made available to all calling customers via a single queue, or whether the offered traffic is to be streamed into separate queues. From the point of view of dimensioning efficiency, a single queue is best but such arrangements are very unlikely to meet the requirements of the most demanding customers. Whatever the queue structure that is decided upon, from the mix of tasks that the agents will undertake it is necessary to calculate the time that it will take a typical agent to handle a call from a customer. This must include both direct time spent speaking to a customer, and indirect time spent completing the task offline, or 'wrapping-up'.

It is not the purpose of this book to produce a fundamental treatise on queuing theory. This has already been developed and documented by many eminent experts. Suffice it to say that queuing models exist that allow centres to be designed so that a given percentage of callers encounter a queuing delay less than a specified time [1, 2]. Typically this is set at around 95% of calls being answered within 15 s. Unfortunately, variability of demand throughout the day as shown in Figure 15.15 and/or major service affecting problems can and do conspire to mean that the original design parameters will often not be achieved.

Most call centres use some form of Integrated Voice Response (IVR) systems to allow callers to be steered towards the relevant part of the supplier organisation that can deal most effectively with their queries. Such systems can be used to help customers make a choice about whether to continue queuing or perhaps try later when it will hopefully be easier to get through. Messages like 'You are held in a queue and we expect to be able to answer your call in eight minutes' can easily be created by intelligent call control software. Callers can be prompted to key a certain number to leave a short message, or can be advised of a web-site alternative. Yet in spite of

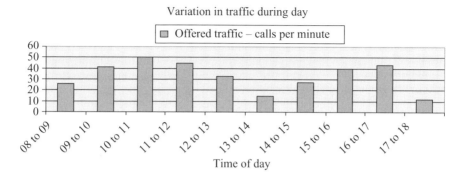

Figure 15.15   Traffic is not nice and smooth

such technologies being available, a lot of service providers only provide music on hold, interspersed with announcements like 'You are held in a queue and your call will be answered by the next available agent'. Both these infuriate callers intensely. None of the advanced IVR technology actually creates more capability to answer customer's telephone enquiries. It merely makes the process feel more acceptable to the end customer. It is therefore vital to provide as high a level of performance as can be economically justified in the first instance.

By far the best way of freeing off some agent time is to allow customers to interface directly with the supplier's service surround systems for a wide range of (relatively) simple tasks, such as problem reporting or order placement.

15.11   Evolution towards multimedia contact centres

For a long time to come, e-channels will co-exist with traditional voice channels. Customers will choose their preferred channel and will use it. Customers will expect uniform service from an organisation irrespective of how they are in contact with the organisation.

The most successful businesses into the future will deal in a way that suits the customer, not the organisation. It will be necessary to completely harmonise the information that is available via each channel. In particular, if a customer enters the organisation via an e-channel and then switches later to a traditional call centre channel, information relating to the holistic customer transaction set must be available. Ultimately, there needs to be a customer knowledgeable person dealing either actively or passively with the customer requirements, who can see all of the necessary information to enable the proper service to be provided for the customer.

By operating in this way, the organisation will harness the talents of its customers in combination with its own internal resources and so will have started off the relationship from a high point where the customer feels valued. Much can be done in this area and indeed there is the opportunity, as interfaces are developed, to actively involve more and more key users.

15.12   Customer relationship management

The overall goal of Customer Relationship Management (CRM) should be to create a seamless and personalised experience for customers (the market of one philosophy) and to monitor and continually enhance that experience in order to delight the customer. CRM harnesses deep understanding of the customer's needs and leverages that knowledge to increase sales and improve service, with the objective of increasing value for both customer and operator.

Definition of CRM is shrouded in confusion because it is a relatively recent management discipline and there is over-emphasis on the IT aspects rather than on the overall goal of building relationships with customers. CRM is the management of the whole framework of the customer relationship as shown in Figure 15.16. It is the macro picture and is not to be confused with so-called CRM software.

A CRM system is typically a hybrid of numerous software applications for both front office (ERM – enterprise resource management) and back office (ERP – enterprise resource planning) across multiple departments. These are integrated to ensure a more customer orientated and professional approach to customer contact. The sheer size of many operators means that full cross-departmental integration is a major challenge. CRM also facilitates a more proactive and dynamic approach to gaining knowledge on customers by enabling data on usage patterns to be analysed. This in turn can be used to predict future customer behaviour. Once this knowledge has been gained operators are far better positioned to generate more profit from their customers by ensuring that they stay loyal and pre-empting any move on their behalf to take their custom elsewhere.

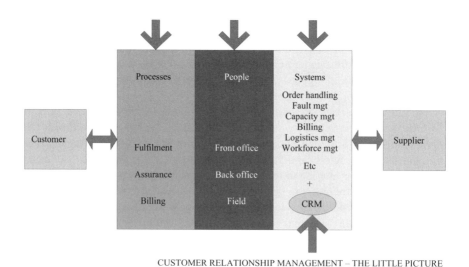

CUSTOMER RELATIONSHIP MANAGEMENT – THE LITTLE PICTURE

Figure 15.16   Customer relationship management – the big picture

CRM is about using information proactively rather than just reactively, understanding that customers are an essential business asset. It is a balance between delivering customer value and extracting customer revenue. CRM implementation is being driven by the emergence of new technologies that enable smaller segments of the market, down to the individual, to be precisely targeted and measured. In this way, the success of marketing campaigns can be ascertained. The key questions to be addressed by the CRM system are:

- Who are my profitable customers and how do I keep them?
- How do I increase and sustain the value of each individual customer?
- How can I manage levels of customer churn through servicing them more efficiently and effectively?
- How do I recruit new and profitable customers?

The supplier must answer these questions by having knowledge of the customer and knowledge of the supplier's business. The cost of serving particular customers, and those same customers' value to the business can only be arrived at by carefully manipulating the information from these sources. The roots of a successful implementation of CRM are in its alignment to business interests and the co-ordination of management towards that goal. To delight a customer, much data that already exists within the supplier organisation must be 'mined' [3].

It is vital that account management is involved in defining the functionality of the CRM software to ensure that the CRM application does not become just another technology burden for account managers with zero benefit for customers.

The end aim is an integrated CRM application that increases customer loyalty and revenues whilst reducing sales costs. It is vital that the customer's needs are understood and met.

15.13   Measurement of service quality

It is said that beauty is in the eye of the beholder. Nowhere is this statement truer than in the field of measuring service quality.

It is not how good you *think* you are that matters. It is not even how good you *actually* are that matters. What really matters is how good your client *perceives* you to be.

It follows then that carrying out detailed measurements of various aspects of service supply, service care and billing processes may not give insight into customer perception. The processes and sub-processes *are* vital to the actual performance of the organisation but usually such processes are defined around themes of $x\%$ achieved by a specified time, to $y\%$ accuracy.

Thus, it is quite possible for every stage of a process to be performed to specification, while customers affected by the tails of these probability distributions can be experiencing very poor service. The only effective way of measuring customer perception is to carry out a direct or indirect dialogue with the customer that teases out the information that is required. Clearly, the more intimate the relationship is, the more searching and meaningful the questions that can be asked.

When dealing with a lot of customers who generate relatively few transactions at relatively infrequent intervals, it is very hard to ask a series of in-depth survey questions. In this case, a sample of customers would be questioned about a particular type of event that was known to have occurred with them. Often such surveys are initiated by means of a fitter or faultsman leaving a pre-paid business envelope and short survey form with the customer as they depart from a customer's premises. The return from such drop-cards is usually at best about 20% and can often be considerably lower. Nevertheless, provided that the right questions are asked and enough cards dropped, some meaningful results can be obtained.

A variation on this is the telephone follow-up survey, whereby over the phone a skilled telephone researcher contacts the customer and poses a set of questions aimed at obtaining a view of how good that particular event was and how good the supplier was perceived to be, overall.

This post-event telephone survey tends to be a more risky affair in terms of data quality, particularly for the SME type customer. Often contact will be made with someone in the customer organisation who did not in any way experience the event in question. As indicated in Figure 15.17, it is vital that the correct individual is contacted.

For major customers generating a lot of transactions and providing the supplier with a large amount of revenue, it is both possible and economical to carry out a detailed survey. In its simplest form it can take the form of a research agency carrying out a face-to-face interview with the main player in the customer organisation to ascertain how the overall relationship is shaping. Although this can elicit much useful information and is (relatively) cheap to operate, it does nothing, directly, to build the relationship between the client and supplier organisations.

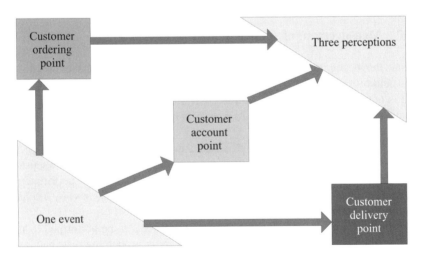

Figure 15.17   Problems with event driven customer service measures

A far better approach is to use the very people who are so instrumental in shaping the day-to-day client–supplier relationship to carry out a structured dialogue with the client aimed at measuring the quality of the actual relationship.

First the customer service manager asks his opposite member in the supplier organisation for (say) five names that he would like to be contacted. These people should all be people within the supplier organisation who could be expected to give authoritative views on the state of various aspects of the client–supplier relationship. The customer service manager then organises a meeting or series of meetings with these individuals and gets each individual to identify the key problem areas that the supplier needs to address if the relationship is to be built into one of complete trust and honesty.

The issues are recorded and are prioritised by all of the people contacted. A composite questionnaire is completed in conjunction with the customer service manager and for each key aspect identified, numeric assessments are made about the current importance, the current performance level and whether the performance is perceived to be improving or deteriorating. Verbatim comments are also captured. The survey is brought away and using various algorithms a customer confidence index can be computed. It cannot be stressed too strongly that by itself such an index is meaningless. Nevertheless, it might be possible to gain a view from a set of such indices of the relative satisfaction or otherwise of a whole range of customers.

However, the most important use of the client review is to produce an action plan for implementation by the supplier in order to address all of the issues which cause the current relationship to be less than perfect.

The action plan is signed-off by supplier and client. This signed off action plan is then worked upon by the supplier organisation and all relevant issues pertinent are addressed over a period of a few months. When the action plan has been successfully carried out, or, failing that, when *very significant* improvements have been made, those individuals in the customer organisation who were contacted originally are re-interviewed, and a review is carried out.

A revised list of top items to be fixed are identified and rated; a new action plan is constructed and actioned and the cycle is repeated over and over. This rolling relationship review as shown in Figure 15.18 is a very powerful way of building and reinforcing relationships with the largest customers. For most supplier organisations, the effort put in here will be highly leveraged because of the disproportionately large amounts of revenue that the largest customers generate, and because of the degree of influence that such customers have upon market place opinion and sentiment.

15.14 Summary

The QoS that customers experience is very largely determined by how well the supplier organisation meshes with the customer. The customer service surround needs to be considered as one of the key elements in delighting customers and in winning and retaining business. It must be designed as an integral part of the supplier's business,

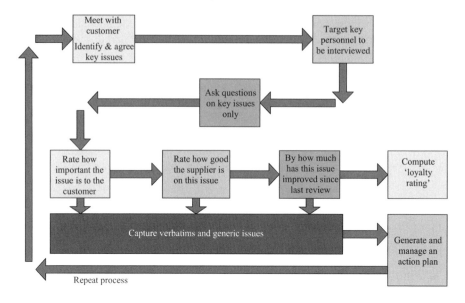

Figure 15.18   Rolling relationship review

not a costly and ad hoc adjunct. Every aspect of it needs to be viewed from the customer standpoint and not the organisational standpoint.

15.15   References

1   BUNDAY, B.D.: 'Basic queueing theory' (Edward Arnold, 1986)
2   BATEMAN, J.: 'Call centres – a social, economic and technical perspective', Institution Of British Telecommunications Engineers Structured Information Programme, Volume 34, April 2000
3   BRAMAR, M.A.: 'Knowledge discovery and data mining theory and practice', Institution Of Electrical Engineers, 1999

Chapter 16

Numbering and billing

16.1 Introduction

Numbering and billing are two relevant topics in the provision and management of any telecommunication service and contributing to their Quality of Service (QoS). These two topics are normally independent of technology and service. The issues are similar, though solutions may differ in different countries. However, the issues on numbering or addressing of IP-based services are slightly different from those on conventional services. In this chapter the principal issues of these two topics are identified and some broad guidelines for their resolutions attempted.

16.2 Numbering

16.2.1 Importance of numbering

A telecommunications numbering system is an extremely important aspect of the QoS given by a network. Far from being a mere artefact of bureaucratic minutiae, numbering may attract a high public profile and is capable of stimulating emotion, irate public opinion, letters to the press and questions in parliaments. Numbering impacts QoS in three ways:

- the friendliness of the user interface;
- the subjective 'look and feel' of a network conferred by numbering;
- the high costs of numbering volatility.

Numbering, once the sole preserve of the incumbent monopoly operator in each country, is nowadays typically managed by National Regulatory Authorities (NRAs), certainly strategically and, often, also at the level of day-to-day management.

Numbering is in some ways a paradoxical subject, where things are not always as they appear on a superficial first encounter. Although nothing in numbering is 'rocket science', it requires a surprisingly broad viewpoint to hold in mind all the relevant factors, of capacity, competitive fairness, long life, smooth migration, user-friendliness, tariff linkage and ease of implementation. It is not unusual to encounter people who are convinced of the obviousness of something that is, in reality, incorrect; should such a person be in a position of influence, then mistakes can be and indeed have been made. Many people experience numbering as an esoteric art, a perception not helped by a shortage of people able to articulate the key issues clearly and simply.

There is no standard reference on the subject of numbering. General-purpose works on telecommunications sometimes include a section on the subject, but these are usually too superficial to support a serious interest in the subject. A valuable overview may be found in [1].

16.2.2   The user interface

The user-friendliness of any network numbering system depends on the combined simplicity of operation of the *numbering plan* and the *dialling plan*. The numbering plan is the form of the numbers of telephones and services, while the dialling plan is the way that people use numbers to make calls. Most countries nowadays have a *national numbering plan*, where each number contains all the information needed to make a call. These systems may be regarded as much more user-friendly than the historical systems they typically replace, which often depended on exchange or city names and required people to consult lists of dialling codes when making an unfamiliar call, or a familiar call from an unfamiliar location.

National numbering plans are of two types, *unitary* and *two-part numbering systems*. In a unitary system, the same digits are dialled regardless of location, as in the eight-digit systems of Denmark and Norway. Two-part numbers are composites of an *area code* and *local number*. The entire (two-part) number has to be dialled for the general call, while the local part only need be dialled for a call to a number having the same area code. Market research in the UK and elsewhere has established that people like the abbreviated in-city dialling facility, as it can substantially reduce the average number of digits a user has to dial. With fixed line services, the local option is usually permissible rather than mandatory, so catering for a user who for some reason dials the entire national number when a local option was available. The detailed structure of a typical two-part national number is shown in Figure 16.1, though the number of digits in each part of the number will vary according to different countries' formats.

Most countries' national numbering plans include an invariant initial digit, known as the *prefix* and usually zero. This is the indicator to the exchange that distinguishes a local from a full number. Countries not having a prefix in the numbering plan usually have it in the dialling plan instead, for example in Russia callers must dial 8 before an area code, and in the USA callers must dial 1 before an area code for a long-distance call.

The user-friendliness of a system depends to some extent on its subjective resonance with users' perceptions and expectations, for example, in the size of numbering areas and the amount of information inherent in numbers. Often there is a design trade-off, for example, where an increase of structural information in numbers reduces the efficiency of numbering capacity utilisation, and so leads to numbers being longer

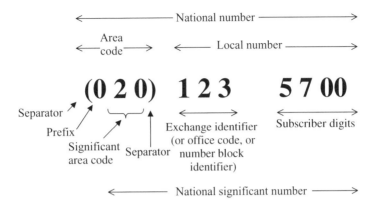

Figure 16.1   Anatomy of a two-part national number where there is a prefix

(or having to change more often) than would otherwise hold. These aspects introduce a degree of 'softness' into the science of numbering, and allow different countries to develop different solutions without the one being more 'right' than the other.

16.2.3   Numbering volatility

16.2.3.1   The costs of volatility

The costs of number change are considerable, so a network that aims for high quality of service will avoid, or at least minimise, change, and ensure that necessary change is accomplished in the most user-friendly way possible. The costs inflicted on users by number change arise through the need to reprogramme repertory dialling devices, advise callers, reprint stationery and repaint vans. It is difficult to quantify these exactly, though Oftel's consultants produced the figures of £2 per line for residential users and £200 per line for business users [2]. The costs to operators of a major number change may run into tens and even hundreds of millions of pounds. The direct costs of reprogramming the network to handle new numbers may, surprisingly, be a smaller element in these costs, which also include extensive publicity, and the identification and modification of the hundreds of management and customer care databases that may contain telephone numbers and use them as access keys.

16.2.3.2   User-friendly numbering changes

Given that numbering change may from time to time be unavoidable under growth and technical innovation, much can be done to relieve the burden that changes place on users. A number change may be, in order of reducing inconvenience, that is, increasing quality:

- a random number change;
- a semi-logical number change;
- a logical number change with 'step change';
- a logical number change with periods of parallel running.

A random change takes place without any transformational logic, for example 200 goes to 31604, 201 goes to 77918 etc., so finding a changed number requires a fresh directory look-up. A logical number change follows a simple rule such as 'prefix by 2', and applies without exception over a large radius such as a whole country or city. Semi-logical changes have a logic that varies with different numbers, for example 'prefix 2*xxxx* numbers with 5 and 4*xxxx* numbers with 7'. A step change applies at one instant in time, at which the old numbers cease and the new ones start, while parallel running allows both old and new to run together for a defined period. Parallel running allows old numbers to be trapped to a recorded announcement for a further period. If parallel running is to be possible, numbering capacity must be reserved and kept free in the form of spare initial digits or initial pairs of digits to provide the space into which numbers can move when they change. Prior reservation of the initial patterns 2, 52 and 74 would have been necessary to support the migrations in the examples above.

The costs of volatility quoted above apply when there is a logical number change with parallel running; the different figures of £8 per line for residential users and £3000 per line for business users [3] were attributed by those same consultants to random number changes.

When numbers change in a user-friendly way as described above, it is inevitable that features of the older numbers are carried forward, leaving their footprints in the new system. The numbering system of almost every country, therefore, is of inferior quality to the one it would have had, had it been able to switch off its network and design from a clean start.

16.2.3.3   Strategic development of numbering

Numbering system managers should develop a long-term strategy for the development of numbering in their country. This will provide a framework where they will plan in advance how they will handle exhaustions of numbers, both in specific cities and in the national system generally. This gives number changes a degree of coherence, which must result in a higher quality system than one that treats each change in a piecemeal way and possibly surprises users with a new format, structure or way of doing things every time. Apart from harming conceptual comfort, ad hoc handling of number change frequently complicates numbering systems and blocks expansion paths, raising the overall level of volatility. Two factors commonly obstruct the formation of sound numbering strategy. First, it can be very difficult to argue for sensible strategy in cultures that habitually prefer 'quick fix' solutions with unfortunate results. Second, it is not unknown for persons in a position of influence to inject idiosyncratic numbering features that may seem clear or pleasing, at least to them.

16.2.4   Numbering system design

16.2.4.1   Stability

The need for a high level of numbering stability is as important as it has ever been. The old British Post Office set a target of thirty-year stability for telephone numbers and was mostly successful with this until the 1980s, when growth pressure proved hard to manage. Stability may indeed be more important now that numbers are embedded in computer applications, where a change may trigger unexpected malfunctions. Sadly, however, an aspiration for longevity conflicts with other QoS objectives, listed below, forcing numbering designers to a balanced judgement:

- minimisation of number length;
- a single number format;
- the amount of user information conveyed by numbers.

16.2.4.2   Numbering capacity management

A national numbering system with n significant digits (note that prefixes do not change and so do not count as a significant digit) provides a total basic supply of 10^n different numbers. The UK's ten-digit system, introduced in 1995, provides, for example, ten thousand million numbers, and as this represents a multiple of about 170 relative to the national population, it is fair to question why so many digits are needed there and in other countries.

The demand for numbers over a thirty-year planning horizon cannot be obtained by extrapolation of past trends, but needs to be estimated demographically. Oftel's consultants [2] obtained a saturation requirement for the UK of 390 million numbers, and this figure, nearly seven per head of population, is possibly a useful planning benchmark for other developed markets. Numbers were first classified into ordinary geographic telephone numbers (fixed lines), personal and mobile numbers and, finally, information services. The potential for ordinary numbers was set at three per worker and two per household (a total of 145 million), whilst the ultimate demand for mobile and personal numbers was estimated at one to each person and three to each worker (also adding up to 145 million). The estimate of 100 million for information services is arbitrary.

An n-digit national numbering system cannot supply the whole of its 10^n basic capacity for use as working numbers, since certain inescapable loss mechanisms make as much as 60% of this unavailable. First, most countries reserve a couple of initial digits for uses apart from normal numbers, typically 0 for the prefix and 1 for special service codes, removing 20% of capacity. Next, it is necessary to reserve blocks of numbers to provide migration paths for logical number change, and taking as a sound planning standard two prefixing opportunities per initial digit range absorbs another 20%. Finally, because number change takes time, typically two years, and because current growth rates are in the region of 10% per annum, another 20% must be deleted to arrive at the 'eleventh hour' threshold point at which, under growth, capacity must be regarded as exhausted and reorganisation set in hand. This is not all, since there are further loss mechanisms that are, however, amenable to relief by management and design choices.

Churn loss arises because vacated numbers, typically running at 5–10% per annum for fixed services and much more for mobile services, cannot be reallocated immediately to prevent nuisance calls to a new user of the number. Operators have discretion over how long they keep numbers free. *Granularity loss* arises if blocks of numbers are allocated to individual areas and cities in larger units than needed. Many operators have traditionally assigned 10,000-line (four-digit) blocks to specific exchanges; when the average exchange size is well below this, then capacity will be wasted, perhaps in the order of 50% or worse. Operators such as BT are now minimising this loss by allocating by 1000-line blocks, while in the USA the procedure known as *number pooling* is used. This process 'flattens' numbering somewhat, attacking the user-friendliness of numbering in the eyes of those who like to see a clear geographical logic in city numbers (68*xxxx* in this suburb, 21*xxxx* in the downtown etc.). A final loss mechanism, *structural loss*, arises from numbering structures such as service digits and area codes. When a numbering system is divided into equal spaces, then the whole system needs reorganising when any *one* of the compartments exhausts, effectively gearing available capacity on the most intensively used partition. The UK's historic area code system, admittedly a fairly extreme case, gave for example equal million number blocks not only to large cities like Leeds, Bristol and Newcastle but also to many much smaller areas with a mean fill in 1990 of 30,000 numbers, effectively locking out nearly 95% of capacity when exhaustion in these large cities was triggering a need for system reorganisation.

Aggregating these loss mechanisms shows that a numbering system may be doing well if it can use 10% of its basic 10^n capacity. This leads to a conclusion that a national numbering system needs to be dimensioned to a basic capacity of at least 70 times the national population to have a chance of meeting the design lifetime of 30 years. Numbering is not, of course, an exact science, so this figure should be taken more as an indicative scale factor rather than as a firm parameter. The situation of any particular country may be more or less demanding than this.

16.2.4.3   Number length

For user-friendliness, the shorter the numbers are, the better. This conflicts with the requirement for system longevity, leading to the balance above of about 70 times population in a developed market. Of course, since fractional digits are not available, the chosen number length will have need to be on the topside, not the underside of this target.

16.2.4.4   Number formats

User comfort is best when there is a single well-known number format such as in the USA. This allows users to know 'what a number looks like', and simplifies the design of computer databases that hold and use numbers. Unfortunately, this conflicts with the longevity or shortness of length objectives for a numbering scheme, since a fixed format may make it much harder to deploy efficiently the basic 10^n capacity of a numbering scheme. As a result, many countries make the trade-off of having two formats. Taking the Netherlands as an example, 30 cities have 0*ab* area codes with seven-digit numbers, while the remaining 111 areas have 0*abc* area codes with six-figure local numbers. A system with a more complex array of formats should be regarded as of lower quality.

Multiple formats improve the utilisation of numbering capacity. To understand this, it is necessary to realise that a digit once allocated, say the area code 01 for the capital, cannot simultaneously be used for other codes such as 012, 013, 014 etc. Because there are only ten first (significant) digits, the code 01 takes 10% of the numbering range regardless of the number length in use, while a code such as 021 takes 1%, 0254 takes 0.1% and so on. Using different area code formats makes it possible to distribute numbering capacity where it is most needed, and avoids the absurdity of giving a block size big enough for the largest city everywhere. One method of mitigating the capacity loss inherent in a fixed format scheme is to design the numbering areas to *make* them more evenly populated with demand for telephone numbers. This is basically the pattern followed in the USA, and results in widely varying geographic area sizes, for example, being relatively compact in a city like New York but going on for hundreds of miles in the mountain states. Many European countries have not favoured this option because they wanted to align numbering areas with charging zones, thus ensuring that a within-area call is always a local call.

16.2.4.5   Numbering structures

An important aspect of the user-friendliness of a numbering system is the amount of information that is inherent in a telephone number. For this reason, many numbering systems provide a structure that allows people to inspect a number rapidly and know, for example,

- the charge rate for the call;
- the service type (fixed, mobile, Freephone, premium, personal number etc.);
- the geographical location.

These structures add to the quality of a numbering system, unless, of course, the information being conveyed is itself subject to volatility and to going out of date. As we have seen above, however, structures may compromise the ability of a system to utilise its capacity efficiently, conflicting with the quality objective of keeping numbers as short as possible.

Geographical information in numbers may apply at various levels. The early digits of a local number may show the specific exchange within a city or area. An area code indicates a part of the country with a greater or lesser fineness of resolution. Some countries, like Germany and Austria, have a *regional area code* system that structures codes in successive stages of decimal subdivision. The first digit gives a broad zone, the second a region within the zone, and so on for as many digits (up to four in Germany) as are needed to define individual exchanges. Others have a more approximate and less formal regional structure, and some (the UK and USA) have apparently random assignments, based on obsolete historical criteria. Despite the intellectual appeal of regional logics, it appears that the actual values of area codes do not matter to many people, who will quickly learn to recognise their own and neighbouring codes. In the past, some structures may have been conditioned by implementation considerations, that they were easier to handle with mechanical switching systems.

Current trends in numbering system design are tending to raise the visibility of service type (and hence charge class) as a structural feature, but to dilute the geographical information. This may occur through the adoption of larger numbering areas, and of flatter, smaller granularities of assignment of local capacity within areas. The latter is regarded as a grievous loss by some sections of the user community, though perhaps not by a majority.

16.2.4.6   Area coding systems

Any system using two-part national numbering faces a basic design decision as to the size of the areas to be used for coding. This needs a great deal of care, because systems once established are not easy to unwind. The choice of a small coding area

- implies fewer digits are needed for dialling local (within area) calls;
- imposes the need to dial an area code on many, everyday and quite nearby calls;
- gives a finer geographical resolution to the area code;
- allows coding areas to align with charge zones;
- may impose a high structural loss on the numbering scheme as a whole, unless combined with multiple formatting.

On the other hand, the choice of a large coding area

- implies that more digits are needed for dialling local (within area) calls;
- relieves the need to dial an area code for a lot more calls;
- gives low geographical resolution to the area code;
- possibly also reduces the geographical information in local numbers;
- breaks charging linkage with area codes, putting some within-area calls into a higher, long-distance price bracket;
- mitigates structural loss in the numbering scheme by allowing an area's number capacity to be concentrated in the urban centres;
- permits further gains in numbering utilisation if rural coding areas are allowed to spread and take in more population.

It is arguable that many European countries' two-part numbering areas, designed around a charge linkage or to fit the routing boundaries (switching centre catchment areas) of a historic analogue network, have become too small for users' modern, mobile lifestyles. People find themselves having to dial an area code for many nearby calls that they regard as everyday and routine, and certainly do not in any sense perceive as 'long distance' calls.

16.2.4.7   Granularity and implementation

A numbering system must be capable of implementation by the switches of all the operators in a network. While historic mechanical exchanges may have made implementation a dominant issue, modern software-controlled systems use look-up table techniques and can implement any reasonable numbering system. Nonetheless, it is still necessary to set ground rules about how many digits of a national number need to be analysed to determine, first, the charge rate and, second, the routing point in the destination network. These rules will imply a granularity that is in effect an implementation constraint. In the UK, for example, no more than five digits need to be analysed to determine the charge (100,000-number blocks), and six for the operator routing (10,000-line blocks). An operator can choose different internal network rules, but cannot ask other operators to apply further digit analysis than stated in the national system conventions when processing calls to its network.

16.2.5 Case histories

16.2.5.1 The UK

The principal features of the current UK numbering system are shown in Table 16.1. The UK commenced with a historic nine-figure system with prefix at the introduction of long-distance dialling in 1958. This system was basically of good quality and user-friendliness, although there was high structural loss in the area coding scheme. Management of the system suffered from loss of coherence in the 1980s, when at least six different methods of increasing capacity were tried in various places. This resulted in excess volatility for some places. London's area code split (from a unified 01 to an inner 071 and outer 081 area) in 1990 doubled capacity, though this provided only ten years' relief at the 7% growth rate and was possibly a mistake, chosen for its being a 'quick fix' solution.

Exhaustion of area codes made migration to a ten-figure system inevitable. A reorganisation in 1995 employed the (significant) prefix digit 1 to transform old geographic area codes 0*abc* into 01*abc*. The ten-figure extension paved the way to a fundamental reconstruction, giving the UK a strategic direction for handling future

Table 16.1   UK numbering system at a glance

No of digits	10
Prefix	Yes, 0
Type	Two-part
Dialling plan	Two-part
Recent migration	The ten-figure system replaced a historic nine-digit system in 1995, while an opportunity was taken to reconstruct the geographical area code structure and formats in 2000
Coding areas	Two systems are in operation. Most areas are historic coding areas, which are small and align in principle with the 638 charging areas, though exceptions are common. A reconstructed system of ten (or possibly more) regional super-areas, such as London and Northern Ireland, was introduced in 2000
Principal formats	(02*a*) *bcde fghi* – for the new regional areas (01*ab*) *cde fghi* – for large cities under the historic area system (01*abc*) *defghi* – for most locations under the historic area system At least two other legacy formats also persist
Special features	The first significant area code digit, the 'S' digit, denotes the broad service class, as shown in Table 16.2
Mobile numbers	These have distinctive area codes under 07…
Special service numbers	These have distinctive area codes under 08… and 09…

Table 16.2   UK service digit (S-digit) allocation

Digit	Service
0	Blocked for international prefix 00
1	Historic geographic area code system
2	New geographic area code system (for use when old areas exhaust)
3	Reserved for further geographic area code expansion
4, 5, 6	Free for further use and for evolution
7	Find-me-anywhere services: mobile, paging, personal numbers
8	Special charge services up to national trunk charge rate: 0800 and other Freephone, local rate, national trunk call rate
9	Premium charge services costing more than national trunk call rate

change while migrating to a higher-quality system. The area code space was parti-
tioned using the first significant digit, or 'S' digit, to denote service type as shown in
Table 16.2. A new geographic S = 2 range was designated to host a new system of
regional super-area codes with eight-digit local numbers, and in 2000 the first of these
were applied to give capacity relief in six areas including London. The simplicity of
the new system will be complicated for many years, however, by the persistence of
older codes and formats, since number change can only be entertained when actually
necessary. Prevention of volatility is more important than ensuring the intellectual
coherence of the system.

16.2.5.2   North America

The United States shares with most of North America the *North American Numbering
Plan* (NANP). Developed in 1947, this has provided a stable, fixed format, numbering
plan that has become ingrained in the minds (and databases) of the region. Figure 16.2
shows the NANP format, and Table 16.3. outlines the principal features. The very
success of the system poses a present-day problem, since although there is strong
demand pressure it faces a very high barrier to migrate to a different format. As a
result, cities are fragmenting into a growing patchwork of separated numbering areas,
while many users are losing the local option of within-city dialling without area code.

The NPA codes in the 1947 plan had 0 or 1 as the middle digit; this provided for a
maximum of 152 area codes, given reservation of 0 and 1 as initial digits, and of the
$x00$ 'generic area codes' for special purposes. Exclusion of these patterns for CSOD
codes reduced the stock of these to 640. The ability to distinguish NPA and CSOD
codes by numerical value dispensed in theory with the need for a prefix, although the
prefix 1 entered through a complex dialling plan (see Table 16.3).

Numbering exhaustion pressure is currently met by two methods. The first is
to split coding areas, so preserving the format but breaking the numbering unity
of previous areas. This process has been repeated in the largest cities, so that, for
example, Chicago now has five area codes (312, 630, 708, 773 and 847) where

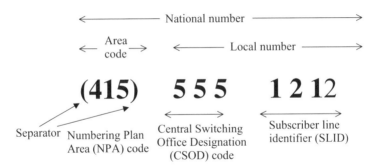

Figure 16.2   Anatomy of a North American Numbering Plan (NANP) number

Table 16.3   North American numbering system at a glance

No of digits	10
Prefix	None
Type	Two-part
Dialling plan	Fundamentally two-part. A complex dialling plan requires a prefix 1 whenever the call needs to enter the long-distance (toll) switching layer. This may be with or without an area code, and a user must look up what is required for a particular destination. Recent developments have imposed ten-figure dialling of local calls in many communities
Recent migration	A change was made in 1995 to the permissible values of area (NPA) codes and exchange identification (CSOD) codes, making many more of either type available for use
Coding areas	Basically population-homogenous and of very different topographical size
Principal formats	One universal format as shown in Figure 16.2
Mobile numbers	These are frequently allocated under geographical area codes, looking like 'normal' numbers
Special service numbers	These have distinct area codes, such as 800 or 888

not long ago there was only 312. The second method was the introduction in 1995 of interchangeable NPA and CSOD codes (this is visible in the Chicago example), which removed their distinctiveness and so made roughly eight hundred codes available for both. This has forced many areas to require ten-digit dialling of local calls, which is not a comfortable evolution of the dialling plan. The strategic development of the NANP may focus on using the second (middle) digit of the area code for service indication, and for eventual migration to an eleven or twelve-digit system when that becomes necessary.

16.2.5.3   France and Australia

The principal features of the French numbering system are shown in Table 16.4. France has five large coding areas and has, unusually for a country of its size, a unitary numbering and dialling plan. Although a prefix is not necessary in a unitary system, France has retained its prefix and given it a possibly unique significance as a digit that can be varied for carrier selection.

Australia provides an example of a country that has combined a number length extension with evolution from a patchwork of coding areas to a smaller set of much larger regional zones. The principal features of the Australian numbering system are shown in Table 16.5.

Table 16.4   The French numbering system at a glance

No of digits	9
Prefix	Yes, 0
Type	Unitary
Dialling plan	The ten digit number is dialled for all calls regardless of location
Recent migration	The ten-figure unitary system was introduced in 1997, replacing a bi-zonal national eight-figure system from 1985
Coding areas	Five large areas, for Paris and four provincial quadrants of the country
Principal formats	One universal format 0a bc de fg hi
Special features	The prefix may be replaced by another digit (or pattern) to signify carrier selection. This is possibly a unique feature
Mobile numbers	These are distinct and follow the standard format under 06
Special service numbers	These are distinct and of the form 08…

Table 16.5   The Australian numbering system at a glance

No of digits	9
Prefix	Yes, 0
Type	Two-part
Dialling plan	Two-part
Recent migration	The nine-figure system replaced historic eight-digit numbers area-by-area over the period 1996–97
Coding areas	Four large areas
Principal formats	Single format for geographical numbers (0a) bcde fghi
Mobile numbers	These have distinctive area codes beginning 04…
Special service numbers	These have distinctive formats beginning 18… and 19…

16.2.6   Some contemporary issues

16.2.6.1   Numbering for competition

The advent of competitive supply in many national markets has raised the issue of how the numbering system should serve new entrants' customers. It is very important that there should be no numbering barriers to competition, whether by there not being enough numbers, or through there being more and less desirable numbers that might confer competitive benefit or inflict disadvantage on particular operators. Three possible solutions are as follows, of which the last is by far the most usual choice.

- Separate numbering: each operator has an independent system.
- Partly-integrated numbering: each operator obtains a block in a national scheme.
- Fully-integrated numbering: each operator obtains blocks for each city or service where it wishes to operate, on a level playing field basis with the incumbent former monopolist.

The first is a non-starter and of theoretical interest only, as it degrades the quality of national numbering by breaking its coherence. The second involves estimating players' market shares and so capping them, probably seriously raising the overall volatility of numbering. The third allocates blocks of numbering to each operator just as for the incumbents' own exchanges. The new entrants' numbers are only identifiable as such by people taking detailed interest in such things. The advent of competition amplifies the demand for numbers and has accelerated volatility, hastening the number changes seen from the 1980s in many countries. There can be no doubt, therefore, that competition has imposed costs in the sphere of numbering.

16.2.6.2   Number portability

The need to change number when changing operator is a user barrier that restricts the ability of a new entrant to capture customers, and so hinders the development of competition. National regulatory authorities are therefore mandating number portability, which allows someone to keep their number when moving to another serving operator. Number portability cuts across the normal routing of numbers by digit analysis to a destination exchange, and so incurs costs by the technical solutions necessary to recognise ported numbers and handle them exceptionally. In return for these costs, the industry and user community benefit from the greater efficiency brought about by greater competition. Recent European Directives lay down that number portability should be treated as a consumer right, and a national network that provides it may rightly be regarded as a higher quality network than one that does not. Most countries have set different priorities for different classes of number portability, usually in this order:

- 'ordinary' fixed-line numbers, having an area code and geographic locality;
- 'non-geographic' numbers such as Freephone (800), Premium Rate and other number translation services;
- mobile numbers;
- personal 'find me anywhere' numbering services.

16.2.6.3   Individual number allocation and trading

Number portability raises the philosophical and practical issue of who owns a number. The traditional view placed ownership firmly with the operator, but portability raises the possibility that a *user* might be able to ask an operator to provide service on a number of *his or her* choice, subject only to conformance with the national numbering conventions and, of course, the number not being already in use. This is known as Individual Number Allocation (INA). This facility is especially though not exclusively relevant to the number translation services, where users often seek memorable patterns such as 10 20 30, FLOWERS or 747747. Still under discussion in many countries, INA is available for non-geographic numbers at the time of writing in Germany and the USA. Without it, someone wanting a particular number would have to contract with the range-owning operator of the number, and then contemplate porting to the chosen operator thereafter. The freedom of choice represented by INA is a quality feature in a national number market.

Memorable numbers are sometimes referred to as 'Golden numbers' or 'Coveted numbers', this terminology implying an economic value. Many countries are discussing the possibility of having a trading framework for such numbers. This would supply the missing market, making possible the economically most efficient allocation of these numbers.

16.2.6.4   Alternative methods of addressing

We live in a time of rapid convergence of the domains of information, communications and entertainment technologies. The telephony numbering framework, governed by ITU standard E164 [4] and to which we have so far exclusively referred, is not the only one employed in the wider, combined field. Other broadband and data services do not make use of the E164 numbering environment, nor is it at all obvious that they should. Others include

- physical addresses, for example, for TV channels;
- physical addresses obtained via a name server, for example, digital TV channel location through an electronic programme navigator;
- Internet addressing (IP addresses);
- Internet addressing via name servers using the domain name system;
- closed user group addressing, for example, virtual circuit numbers for a switched virtual circuit environment;
- portal addressing, via menus or links provided by a portal owner or service provider;
- proprietary addressing environments, for example, within computer networks.

Services that typically need dedicated terminals and traditionally have not used E164 telephone numbers to access them, such as Internet pages and broadcast television, are unlikely to start using them. Factors that might drive the adoption of E164 telephony numbers as the address 'handles' for people and the new services are as follows:

- Any service accessing or being accessed from terminals in the telephony domain, for example, an ordinary telephone, data modem or mobile phone, is likely to have an E164 number.
- Services likely to involve unified access methods by a variety of terminal types that include ordinary and mobile telephones are likely to have E164 numbers. Examples are unified messaging and personal numbering services.
- A broadband service having an imaged or adapted form that can be accessed from ordinary, non-broadband terminals, is likely to have a E164 number presence.

16.2.6.5 The Internet addressing system

The addressing systems used by the Internet for accessing web pages, e-mail destinations and the like are important and will continue to be so with advancing convergence of computing and communications. This section examines these addressing frameworks, contrasting them with the traditional telephony numbering framework governed by Reference 4.

The E164 numbering framework is serially hierarchical, where an international number consists first of a country code, then a national significant number, which is itself made up of an area code, possibly an exchange identifier, and finally the subscriber number. Two-part dialling plans place the user at a location-dependent default addressing point: if no area code is dialled, the local city is assumed; should no country code be dialled, the local country is assumed. Prefixes, both national and international, serve to move the user's point of reference up the hierarchy. While one of the Internet addressing frameworks is hierarchical, there is no sense of a default position and so complete addresses are always required.

E164 numbers are used both in the user interface for direct input by humans, and for the routing of traffic by switches. The Internet does not work this way, but has distinct user-visible and routing addressing systems. The logical (and user visible) naming system is known as the Internet Domain Name System (DNS), while normally only machines handle the Internet Protocol (IP) routing addresses. Domain Name Servers provide the translation interface between the two.

An IP address applies to each service, host, router and user terminal on the global Internet, and has 32 bits (4 bytes), typically represented by the decimal values of each byte, for example, 192.168.45.230. Regional registries allocate IP addresses under the ultimate control of the Internet Assigned Numbers Authority (IANA) [5]. IP addresses possess no hierarchical routing information. Internet routers develop a large database of the routings for frequently encountered addresses, and have recourse to master routers possessing a complete set. A current issue with IP addresses is that the 32-bit format does not provide anything like enough addressing space to meet forecast worldwide demand, and evolution from the present Internet Protocol Version 4 (IPv4) to Version 6 (IPv6), which provides a 128-bit addressing space, is under active discussion. It is likely that the new addressing space will incorporate some hierarchy to ease the routing function.

Domain names are alphanumeric and hierarchical, and are designed to be user friendly, for example IBM has *ibm.com* and Netherlands Railways has *ns.nl*. Whereas E164 numbers have fields in decreasing order of importance, domain names are in increasing order of hierarchical significance. The various types of domain are

- *country code Top Level Domain* (ccTLD), defining a country, for example, *.uk*, *.nl*;
- *generic domain*, defining a type of user, for example, *.com* for companies, *.org* for non-profit organisations;
- *user domain*, for example, *ibm*, *aol*.

The worldwide generic domain names are registered by the Internet Corporation for Assigned Names and Numbers (ICANN) [6]. User domain names are registered via a number of *registrars*, who are private companies operating in competition with one another; for a list, see Reference 7. Registration is subject to the name not being already in use, and to any restrictions that may attach to the use of a particular generic domain name, for example, *.museum*. Country code top level domains and the registration of user domain names under them are administered independently in the separate countries, which may or may not use generic sub-domains, and which may have different levels of restriction on who may register. An applicant for registration must provide the IP routing address of the computer that serves the domain name. This is then sent out to the network of Domain Name Servers, which perform the domain name to IP address translation.

Recent problems with domain names have included *cybersquatting* and *cyber-piracy*, which include the abusive registration of domain names by users intent neither on using nor paying for them, and the registration of names in which the registrant has no real interest or rights, but has hopes of extracting a payment for releasing the registration to an organisation with valid interest in the name. Recent developments in the registration system and codes of practice for registrars have helped curb the worst abuses of the system.

The attractiveness of e-mail addresses based on the domain system, for example *john.smith@xxx.com* where 'xxx' is a company domain name, has led some to speculate whether Internet addressing might in time displace the E164 numerical addressing environment by virtue of its apparently greater user-friendliness. However, a little reflection will convince that naming at this level of beauty is hardly scaleable, and there remains the problem of inputting domain names at simple telephony terminals.

16.3 Billing

16.3.1 Introduction

The interaction between network services, the customer service systems surrounding those services, the customers themselves and the billing systems are now examined. It is not intended that this section will describe in detail how billing systems should be structured. Rather, it aims to show how aspects of billing impact upon the overall

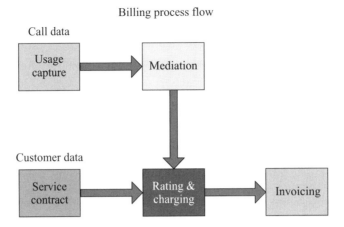

Figure 16.3   Billing process flow

QoS that customers' experience. Billing is as complex, in many ways, as the networks that they support. Billing systems have traditionally been viewed as an overhead on operations but today are increasingly being viewed as essential marketing tools. Figure 16.3 represents the major elements of the billing process.

16.3.2   The function of billing

Telecommunications service suppliers and information service providers are fundamentally companies that own a database of potential customers, are able to support a range of telecommunications services that they wish to market and sell to those customers and are capable of accurately billing those customers for the services actually used by them.

Modern billing systems are largely software-based systems and are often generically referred to as Billing Operating Support Systems (BOSS). Systems that are closely associated with billing but are not usually defined as being part of BOSS include facilities management, fraud detection, trouble reporting, revenue assurance and financial systems. The various processes normally forming part of a typical billing system are

- the service order and data capture process;
- mediation and message processing;
- rating and charging;
- invoice production.

The function of billing is to accurately track and amass all relevant charges for a customer in a form that is acceptable to the customer, for the customer to approve and then subsequently make payment. Thus accuracy is the yardstick by which the QoS of billing is judged.

An increasingly important function of billing is as a marketing and customer management tool for the operator/service provider. It is through the billing systems that the operator/service provider gains intimate knowledge of the customers usage patterns and preferences. The recognition of this key role of billing has led to the investment of huge sums in designing ever more sophisticated systems. From the customer's perspective, the bill has come to represent a lot more than a simple invoice of yester year. The personal customer uses the content and breakdowns provided in the bill to analyse his own usage patterns of the range of services he procures to assess whether he is getting the best deal in the market place. The business customer does the same but also uses the bill to support his own business processes, for example, operating internal charging regimes. Thus the content and the presentation of the bill has become a fundamental issue of quality for the customer.

16.3.3   The service order process and inventory management

Accurate billing of a customer for his use of telecommunications services depends on the existence of an accurate inventory of both the customers' equipment and of his service and usage profile of network services.

The accuracy of these records impacts on the customers' perception of overall QoS. Initial accuracy of the inventory record depends on the quality of the service order process. Subsequently maintenance of accurate records depends upon the capture of changes as they occur and then a regular audit of records against reality.

Customer usage of network services is captured by the network switches in a circuit switched network in the form of Call Detail Records (CDRs). These are held in the detail necessary to support the charging algorithm involved. For example, it is common for a full CDR of time, length and destination of call to be held in European switches (in conventional circuit switched networks) because all calls are charged for by length of call; whereas in North America where it is common for local calls to be 'free' a CDR for a local call is not always generated or held by the switch.

The trend towards the provision of packet switching networks, such as Internet Protocol (IP) networks makes the use of 'simple' CDRs impossible. This is because IP network elements do not generate 'Billing Friendly' usage information. Instead they generate large volumes of piecemeal usage information, across many disparate network elements. Data Reduction Units (DRUs) close to the network and Remote Gathering Agents (RGAs) running on IP Application Servers (Web Hosts, Video on Demand servers, Mail Servers, Voice over IP Gatekeepers etc.), allow the real-time collection of vast volumes of network data. Once collected, this data needs to be reduced down to meaningful billing information using filtering, aggregation and correlation tools. This billing information can then be represented in the industry standard IP detail record (IPDR) format.

Both for circuit switched networks (CDRs) and for IP networks (IPDRs) the overall volume of transactions that need to be tracked by the inventory management system will be very large, in the order of tens of millions of events per year for a large supplier. The sheer scale of the task of inventory management is enormous and the possibility of error can be significant.

16.3.4 Mediation and message processing

The mediation function is primarily the collection of raw network usage data (e.g. CDRs and IPDRs) and the transformation of this data into billable business information that can be applied to the rating and charging information at the time of invoice production allowing the calculation of the correct bill for each billable event. The most common use of this business information is of course for Retail and Whole-sale Billing. However it may also be distributed to other Operational and Business Support Systems (OSS and BSS) in any required format, for fraud detection, traffic analysis, network planning and marketing.

A mediation platform should be able to accept any event/record format that the network may generate and be able to generate accurate information rich records in any format required by OSS or BSS systems. A key mediation objective is to shield the OSS/BSS systems from any changes in the network elements.

16.3.5 Rating and charging

Rating and charging involves the application of the various rating algorithms to the information provided by the inventory and mediation systems at the rates that form a part of the service contract with the customer.

Examples of different charging algorithms range

(a) from the fairly standard one of the charging of a fixed monthly rental together with per call charges at different rates by time of day;
(b) through to a fixed monthly charge for unlimited use typical of some ISPs;
(c) and at the other end of the spectrum a per call charge without any monthly fixed charge typical of some virtual network service operators.

An effective mediation processor will allow a service provider to change from one algorithm to another without having to make any changes to the data capture arrangements.

The flexibility afforded by a modern billing system in which the mediation and charging and rating functions are separated allows suppliers to offer a wide range of discount structures even combining charges for calls made by the same customer through different networks into the same bill. This facility underlies the provision of what are known as convergent services.

Pre-payment options have provided an increasingly common method for payment of services. This option results in significant increase in complexity in the end-to-end billing process involving real time passage of CDRs through the mediation processor and onto the charging and billing functions. The effect of any errors in this process becomes immediately visible to the customer.

As IP-based services become more prevalent, charging for services based on speed and effective bandwidth and even on volume of data delivered (e.g. per Gigabyte) and nature of content will become more usual.

16.3.6   Invoice production

The most common form of invoicing the customer is still the production of a printed invoice produced monthly or even quarterly and posted to the customer. The subject of invoice formats, their content and how the invoice may be used as an effective customer care and marketing tool are the subject of much study and research. The impact on customers of inaccurate invoices is significant and forms the basis of many if not most customer complaints. Improving the quality of presentation and clarity of the printed invoice can have a disproportionately positive effect on the customers' perception of quality.

Most network operators are making plans to move towards online Internet billing so as to eliminate the costs and delays inherent in a paper-based system. Online billing will allow the provision of much more detailed levels of information to the customer and should improve the quality of the billing experience.

16.3.7   Wireless services billing

There are some unique issues of billing associated with wireless services, including roaming administration and settlement, global roaming and the impact of multiple technical standards. There are also issues to do with whether calling or called party pays in different jurisdictions. Most calls are still between mobile and fixed networks or vice versa and the inter-network billing and regulatory rules governing fixed to mobile charges can be confusing to customers. This area of confusion should be reduced as convergent systems allow the integration of customers' fixed and mobile bills.

16.4   Summary

Telecommunications numbering is a strong element in network QoS because it provides 'look and feel' and forms part of the user interface. Numbering volatility can inflict great user cost. In this chapter, the principal design factors, a few case studies and future directions were explored.

Components of the billing process for providing telecommunications services may be grouped under usage capture (with emphasis on accuracy), accurate inventory management, mediation function (preparation of network usage data into billable information), rating and charging, and invoice production. Accurate billing and presenting this to the customer in an acceptable and pleasing format is one of the dimensions of QoS in telecommunications. This chapter dwells on the key issues and addressing of these issues.

16.5   References

1   BUCKLEY, J. F.: 'Telecommunications numbering', *IEE Electronics and Communications Engineering Journal*, 1994, **6** (3)

2 'Numbering for Telephony Services into the 21st Century', Oftel consultative document, July 1989
3 'Numbering for Telephony Services into the 21st Century', a report to Oftel by Ovum Ltd, 1991
4 'Numbering Plan for the ISDN Era', ITU(T) Recommendation E164
5 www.iana.org
6 www.icann.org
7 www.internic.net/regist.html

Chapter 17

Ergonomic considerations in the design of products and services

17.1 Introduction

Ergonomic considerations are another dimension in the QoS as experienced by the user who is its ultimate judge. Telecommunications is delivering an ever-widening array of services to the end-user. As the services become more sophisticated the requirement to deliver QoS becomes more critical and demands consideration of a wider array of issues. To examine these issues this chapter looks at QoS from the perspective of the end user and focuses not on the user of basic telephony services but on the user of emergent broadband services capable of delivering multimedia services.

From the end user's perspective telecommunications has changed dramatically. A service that has traditionally offered two-way voice communications can now offer data transfer, pictures and video etc. These capabilities are now being used for many different applications; for communication, entertainment, access to information services, e-commerce etc. Users may access these services from business premises, their home or on the move. Mobility and using services from anywhere are key issues and terms like m-commerce and m-learning are beginning to be widely used.

These developments raise many questions about the human–machine interface necessary to offer quality services. To offer these services, the familiar telephone handset may now require complex keypads, joysticks and other input devices. The presentation of images and video to the user will require high-resolution displays, and other sensory modalities such as speech synthesis and speech recognition may also be used.

Attaining QoS for the end user of these services varies with the nature of the application, the characteristics of the user, the location of the user etc. The aim of this chapter is twofold; first, to explore these issues by looking at the way four application domains have treated the service to be delivered and second to consider the design process that is necessary to create QoS for a diverse array of applications. Initially, however, some pertinent frameworks and concepts will be introduced which will inform the analysis of the four application areas.

17.2   Frameworks for understanding the response of the end user

Ergonomics (or Human Factors) has a long tradition of studying how people respond to products and services. The following propositions summarise some of the major findings.

17.2.1   Fitness for purpose

Initial (and continued) usage will depend on the end user's judgement of the 'fitness for purpose' of the service, that is, it must have perceived utility or benefit for the user as compared with other ways of obtaining the service. There are important distinctions to be made between buying a service and using one. The features that encourage buying include perceived utility but are unlikely to include many service qualities. This is especially true of usability (see Section 17.2.3) which is particularly important for subsequent usage.

17.2.2   Media and fitness for purpose

Some media and forms of interaction are better suited for some purposes than others. As a result, the way a service is offered should depend on the nature of the application.

There is a large volume of human factors data about the human issues of each medium of interaction. Some examples are as follows. Sound is good for alarms because of its non-directional attention getting capability but it is not good for large volumes of information because short-term memory limitations mean that only a small amount can be attended to in a short period. Graphs are very useful for showing trends in information but tables of numeric data are better if precise figures are required. A voice commentary can focus attention and add explanation to video or an animation.

Thus, human factors information can be useful in establishing which medium or which mixture of media should be used in a particular application. Guidelines on the properties of each medium and how they interact in multimedia applications are available, for example, the LUSI guidelines [1], and an international standard for the human factors of interactive multimedia systems also exists in ISO 14915, 2000 [2].

It is important to note that these guidelines can only be applied when the task and context of the user is known, for example, does the user need to detect general trends to complete the task or are precise figures needed? Another way of stating this requirement is that the mix of media has to be fit for the purpose of the user.

17.2.3   Usability

It is not sufficient for a service to be fit for purpose. It must also meet a range of quality criteria including reliability, confidentiality, safety etc. Of special importance to end users is the usability of the service, that is, ease of use and ease of learning. Issues of this kind may manifest themselves in many ways; the user may, for example, not be able to navigate a system and may not find what they are looking for, they may not understand commands and may make errors etc.

17.2.4   The determinants of usability

No product or service can be universally usable. According to Part 11 of the multipart IS0 standard 9241 [3], usability is defined by three factors.

17.2.4.1   Characteristics of the user

An interface designed for an experienced user may need to be very different from that for a beginner. The beginner may, for example, need much more structure and more 'fail safe' mechanisms than the expert. The old may also need different interfaces from the young. An issue of increasing concern, leading to legislation in some countries, is the need for 'inclusive design' to ensure that people, no matter what physical disabilities they may have, can all have access to services.

17.2.4.2   Characteristics of the task

The task to be undertaken defines the process to be undertaken, the information to be accessed etc. and services may be more usable for some tasks than others. A complex unstructured and uncertain task involving information search may, for example, tax the usability of a service much more than a simple task that is often repeated.

17.2.4.3   The environment

A service may be quite usable in a quiet office. Try the same service on a noisy train or in the High Street and it may be much more difficult.

Together (17.2.4.1–17.2.4.3), these factors are known as 'the context of use' and they determine the range of conditions a service must meet if quality standards for usability are to be achieved.

17.2.5   Discretionary users and implicit cost–benefit assessments

Most users have some degree of discretion or choice about the way they use services. Members of the public have high levels of discretion and even people in a working setting have a lot of control over the extent to which they use the various facilities in services. Research suggests people use an implicit form of cost–benefit assessment to determine the choices they make where benefit is fitness for purpose and utility and cost may involve effort to use (usability) as well as financial considerations. Many studies have shown the way people restrict their use of multi-functional services to the few facilities they find most useful and usable [4].

17.2.6   The TAM model

In 1993 Davis [5] published the Technology Acceptance Model (TAM) to explain the factors which define an end user's attitude to a technology service. The attitude is

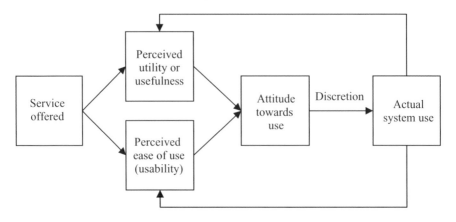

Figure 17.1   Factors affecting end user behaviour (after Davis [5])

a product of the end user's perception of the utility or usefulness of the service and the perception of its ease of use or usability. High utility makes it worth battling with poor usability but where there is marginal utility (or a good alternative) poor usability may result in non-use of the service.

Figure 17.1 is a modified version of the original TAM framework. The initial attitude towards use is translated into actual system use according to the amount of discretion the user has. For example, when a user is free to use or not use a system it may take high-perceived utility to overcome doubts about usability. A person at work who is required to use a service may have no alternative but to persevere with poor usability. Actual system use provides feedback for the user which changes perceptions of utility and of usability. Unfortunately, all too often, it is the use of the system that reveals the usability problems and reduces the range and frequency of usage.

In the review of the four applications that follows, evidence of user behaviour will be reviewed in terms of this framework to ask whether users are getting benefit from the services and whether they meet adequate standards of usability.

17.3   Case studies of multimedia applications

In each of the four cases reviewed below multimedia applications have been developed to serve specific domains and there is evidence available from evaluation studies of how users have responded to them. An example of what Christie [6] calls Type A forms of communication is where the system acts as a mediator for real-time communication between human beings. Three other examples are called Type B forms of communication; when the user interacts with stored information held on a computer database. The Type A example is telemedicine. The Type B examples are route guidance systems, teleshopping and electronic journals.

17.3.1 Telemedicine

Broadband systems can provide multimedia communication channels between people allowing speech, pictures etc. to be transmitted. These possibilities give rise to videophone and teleconference applications etc. One particular form of these applications which is exciting attention is telemedicine. In many circumstances it is difficult to get patients to hospitals or clinics where specialist medical practitioners can examine them. Telemedicine systems can provide video-conferencing facilities in the location of both the patient and the medical specialist and can enable the medical specialist to examine the patient at a distance and decide on appropriate treatment.

Telemedicine then is a particular application of teleconferencing in which video transmission allows people in different locations, in this case the patient and the specialist, to see one another and to communicate via an audio channel. In telemedicine, a local doctor or carer can also focus the camera on particular parts of the patient's body to help the specialist diagnose the patient's condition. Other data, for example, heart rate or medical records, can also be transmitted.

Harker and Eason have reported on studies carried out primarily by using scenario-based evaluations [7, 8]. In this process a paper-based description of the technical system is developed and embedded in an application scenario which shows how it would be used in practice by stakeholders. These scenarios are then studied by stakeholders and they are provided with criteria for assessing them, for example, relevance to the task, ease-of-use, organisational implications etc. The relevant stakeholders in this case (i.e. medical staff, carers, emergency services etc.) concluded that telemedicine was a service of great potential because it could bring expert medical attention to a patient in remote locations. However, they felt that the current applications relied heavily on a standard video connection as the basis for medical diagnosis and communication. The medical specialists felt this may not give sufficiently good quality information and images to allow them to make good diagnoses and that different illnesses required different kinds of information. Some illnesses, as a consequence, are more easily diagnosed using the video communication facilities of telemedicine than others and the range of equipment needed to provide a general purpose telemedical service might be considerable. The collection and transmission of information would be in the hands of the local carers of the patients and they were concerned that their skills as technicians may be very limited. They may also be expected to transmit good information under difficult conditions. The experience of all of the stakeholders indicated that medical staff tended to rely heavily on speech communication with the patients, local doctors, and rescue staff etc. and many of these systems, having placed the emphasis upon video communication, did not seem to have good quality sound channels. Similar findings have been reported by Veinott *et al.* [9] in other video conferencing applications. It is often the case that the video channel is of less importance to good communication than other media, especially the audio channel.

As a result of these perceived difficulties staff felt they might find themselves working under difficult circumstances in telemedicine sessions. This led them to a wider set of questions. Who is responsible if the wrong diagnosis is made or the wrong treatment given? Is it the 'remote' specialist or the 'local' doctor or carer?

If the local person, who is in a position to appreciate the wider context of the patient, is unhappy with the advice of the specialist, can they disregard it? What happens to patient privacy and the confidentiality of records when they are being broadcast on a telemedicine network?

These and other issues are already leading to changes in the way this technology is being applied in practice. There is a movement to build specific applications to facilitate telemedical services within particular medical specialisations where the technology can be geared to the transmission of the critical information required. Second, each application requires the development of a framework of use which ensures each person understands the rights and obligations attendant upon their involvement in telemedicine. The requirement to create a new organisational framework is often a necessity when a new technical system is used across work roles and has a profound effect upon the way in which people cooperate [10]. It is thus the case that a successful application depends upon the QoS of the technical system and the development of organisational frameworks which establish the rights and obligations of agents involved in the use of the service.

17.3.2   Route guidance systems

When one drives to a destination not visited before one has to navigate new roads and junctions. Often the person has to follow a map whilst driving. An application of telecommunications which has very exciting potential is to provide in-vehicle, real-time route guidance. If the system knows where one wants to go and where one is now it can help us find our way to our destination. If the system also has access to the current status of local roads, for example, where there is congestion, road works or delays because of an accident, it can help us navigate around problems. Satellite tracking means it is possible to locate where a vehicle is at any point in time and the provision of up-to-date traffic information is becoming available in many parts of the world. All these services can now be provided and they should provide high utility for drivers. But what kind of 'front-end' is required to provide a usable service?

The first applications of this kind made use of multimedia opportunities by providing road maps of the immediate vicinity of the driver with indications of the destination, the current location of the vehicle and of road works and congestion. This information was presented on a visual display screen positioned adjacent to the steering wheel. The design rationale was to use the power of current display technology to provide the driver with a familiar road map where it could be studied whilst driving and with all the 'value added' information inserted on the map.

A number of off-road trials of this kind of system have been undertaken in driving simulators (see Burnett and Joyner [11]). They show that drivers spend between 14–33% of the time they are in motion looking at the display of the map. This is a very high percentage of time not devoted to watching the road. The drivers reported that it was difficult to read the complex map and that they struggled with the dual tasks of map reading and driving.

Adopting the ergonomic framework presented earlier, we can see that there are a number of context of use problems. This is an example where a secondary task for the user (map reading) is interfering with the performance of the primary safety critical task (driving). The driving environment in which the map reading task is being undertaken means the information has to be quickly and easily assimilated. It may be possible for the driver to study a fully detailed map whilst stationary but a different interface is required when the vehicle is in motion if the driving task is to be undertaken in safety. This is an example where the service provides high utility but has severe usability problems.

These findings have led to a number of alternative design proposals including a Head Up Display (HUD) of the information on the vehicle windscreen so it can be read without the driver diverting his or her gaze from the road. Alternatively, the information can be spoken to the user which allows attention to remain on the road. This solution has to be very carefully timed so that navigation information is given just in time for the driver to make the necessary manoeuvres. Navigating a road system is, however, primarily a visual task and another solution is to offer a 'map' display with minimum clutter which can be easily assimilated. This map is usually limited to the next junction the driver is approaching with a sound commentary of the action to be taken. Early evaluation of the mixed, limited display/audio guidance approach suggests it is likely to prove both safe and effective.

This example demonstrates that there are a number of ways different media can be used to serve a particular application and that they have very different effects upon human performance. In this case the designers followed a common pattern; they started by using the technology to replicate the 'paper-based' display, that is, the road map, and only considered the role of the other media when this was shown to be dangerous. In this instance, a mixture of media is appropriate so that the visual channel is not diverted from the primary task of driving. It should also be noted how sensitive quality issues are to particular aspects of the task and the environment. The timing of the delivery of voice instructions, for example, is absolutely critical. The timing of an instruction on a screen is not so critical because it persists but its clarity and simplicity are vital if it is to be assimilated with minimum distraction.

17.3.3   Teleshopping for the elderly

One application of telecommunications which is receiving a lot of attention is the provision of a database of goods for sale which users can study remotely and place their order. This is the equivalent of ordering from a catalogue but with the advantage of online ordering. E-commerce and m-commerce have become major ambitions of many companies but have proved very difficult. Many of the issues that lay behind the collapse of major 'dot.com' companies were in the end user experience of electronic shopping. A number of surveys, for example, De Kare-Silver [12], show that the quality of service necessary to give consumers the confidence to shop electronically has to cover

- the design of usable sites (without too many flashy attention grabbing advertisements);
- personal privacy (so that user information is not used for purposes for which they have not given permission);

- payment security;
- accurate and timely delivery of products and services;
- ease of cancellation and return of goods.

There are general lessons here for the delivery of successful electronic shopping but it should be noted, following the earlier framework, that meeting these requirements will vary with the characteristics of the consumer, the products and services being offered and the environment in which the interaction is taking place. It may, for example, be possible to deliver complex information about a range of products to a home-based PC. It is not so easy to deliver the same information to a mobile phone.

To illustrate the effects of consumer characteristics consider the delivery of teleshopping to the elderly and the disabled.

If a satisfactory way of delivering this kind of application can be found it could be of great value to the elderly and the disabled who have limited mobility because it will enable them to buy goods without travelling and visiting shops. In a study in Sweden by Kaulio and Karlsson [13] such a service was tested. Elderly people were offered equipment with a simple interface which enabled them to place a weekly order with their local supermarket who delivered the order to their homes. The designers recognised that elderly people would have difficulty with a complex and sophisticated interface and deliberately kept the form of interaction simple. The system had a menu structure which enabled the user to locate a text description of each product and its price. To place an order the user checked a box against the relevant item on the menu and indicated the quantity required. The service was limited to normal supermarket products, again to keep it simple and familiar.

After a trial period the elderly users reported that this was a valuable service in principle but they had many problems with it. They were often surprised and disappointed by their order when they received it and, on this measure, the system was not wholly successful. They often got products they did not intend to get, or the wrong size or quantity, and the quality of fresh food was often not of a standard they wanted. They had many difficulties with the usability of the service. The users were not experienced equipment users and they found the system complex to navigate to find the products they wanted. They often did not understand the designer's classification of products and in consequence often looked in the wrong part of the menu structure. They had difficulty recognising products from the text description because they were more familiar with properties such as the size and colour of the packaging than the formal text description of the product. They also felt uncomfortable ordering many kinds of products this way. They usually chose perishable products such as meat, fruit and vegetables, for example, by inspecting it for quality, ripeness etc. and this was not possible from a text description. As a consequence they were often disappointed with the quality of perishable items.

This study and other teleshopping trials show that the shopping experience is complex and people use many cues and factors in making their decisions. As a consequence, when they teleshop, users need access to many kinds of information to come to a decision with confidence. A picture rather than a text description may enable them to make regular orders of familiar products with standard properties in this way. Other products ranging from vegetables to clothes and holidays need much more information, perhaps including a rich sensory experience of the product.

These findings suggest that the power of multimedia could well be used to enhance the opportunities the user has to explore the properties of products – to see products, to examine clothing, to take virtual reality tours of tourist resorts etc. The design issues are likely to be complex because different products require different kinds of representation. This example demonstrates that the task characteristics of 'shopping' can be quite diverse and different task types may require different system solutions. Different user types will also define what kind of service will be required.

The assumption that a simple text-based interface would be the most usable for this user population proved to be unfounded. This elderly population was unfamiliar with computing equipment and did not find this simple interface easy to use because it was not shopping as they knew it. They needed an interface which represented the product characteristics which they used to make purchasing decisions. Whether a multimedia system can be created that has these multimedia capabilities and is readily usable by the elderly is a design problem worthy of serious attention.

17.3.4   Electronic journals

The knowledge developed by scientists and scholars has traditionally been published in journals issued regularly in printed form to libraries and members of the relevant community. Full text articles in electronic form can now be delivered to users wherever they are. Initial thinking in this domain considered there were two main advantages for users. First, they would be able to read the articles on screen. Second, the journal service and the articles could offer added value. Articles could, for example, adopt multimedia forms not possible with printed materials. They could, for example, include video, sound, animation etc. to show the results of studies, explain theories etc. Hypertext links could be developed to allow users to search on names and concepts in the text, references etc. In short, the electronic journal could be a revolution in scholarly publishing.

A recently completed study is a three-year trial of electronic journals [14] to test user responses to these ideas. In this study 17 publishers collaborated to deliver 49 journals to 13 universities in the United Kingdom. The journals were organised into four discipline areas; two science subject areas and two social science subject areas. The journals and their articles were provided as the electronic equivalent of the printed journals. Many additional features were provided including a wide variety of links between articles and services, search engines, alerting functions, communication facilities and, in a few cases, multimedia content for articles.

Three thousand staff and students of the 13 universities registered as users and over nine-hundred became repeat users of this service, that is, they returned to it repeatedly after their first use of it. A lot of effort went into the design of the 'value added' special features for linking services, search engines, using multimedia etc. The views expressed by users in interviews and questionnaires however, demonstrated that these were not the features that mattered. They made little use of them and placed them low on rating scales of desirable properties of electronic journal services. What they did value was having available at their personal workstation a cluster of journals relevant to their work which were up-to-date, easy to access and easy to find, that is, had the utility of ease of access and were easy to use. A set of journals well matched to the interests of the user had the highly valued property of being a manageable way of keeping up-to-date with the area of scholarly activity. All usage of the journals was monitored and the usage log showed that few users studied articles in detail on the screen although they did use the full text version to check for relevance. When an article was found to be relevant it was usually printed because of the convenience of the paper form for detailed study, annotation, etc. One of the reasons users did not rate multimedia a priority was that it could not be printed.

This study shows that what were perceived to be the most obvious benefits of electronic journals were not the immediate priority for users. They defined their task requirements in a different way. They want a manageable set of full text relevant journals delivered to their workstation. This service is of considerable benefit when compared with the effort required to visit the library, study individual journals or use comprehensive indexing services before seeking particular articles. When they do find an interesting article, they want to be able to print a copy because, for detailed study, paper is much more usable than continuous text on a display screen.

The technical specification for a service that is fit for purpose in this case is, therefore, quite different from the initial ideas. The revised specification calls for mechanisms for defining a limited range of relevant journals (different for each user) that can be easily browsed and from which full text articles can be printed. It may be that, in time, multimedia and hypertext links embedded in the full text of the article will become important but only after a manageable and relevant journal service has been established.

17.4   Lessons for the design of multimedia products

Although the four case studies are of different applications in different contexts, their development histories display similar patterns in three respects. First the initial concept of the service to be provided in each case turned out to be, at worst, wrong and, at best, over simple. In some cases, for example, telemedicine and electronic journals, the belief in the value of new media, such as video conferencing and animation had to be tempered by the realisation that other forms of media had important roles to play. In other cases, for example, route guidance, the obvious solution of copying an existing medium of representation, that is, the road map, proved to have severe disadvantages. In teleshopping trying to avoid advanced forms of interaction in order to keep the system simple led to an inadequate solution.

The second similarity is that the successful service seems likely, in each case, to be one in which quality characteristics map specific characteristics of the tasks to be performed, for example, the critical information needed to diagnose different forms of injury and illness. In some cases there are critical issues associated with the users, for example, the elderly in teleshopping, or arising from the context, for example, the safety critical context of the use of route guidance. In some cases, for example, telemedicine, there is a wider organisational context (responsibility, privacy etc.) which has important ramifications for the design of a successful system. The case studies are all examples where the service has to be closely coupled to the task-user-context of the work practice it is seeking to support if it is to be effective. It is necessary to develop a well articulated account of the user task world or the 'user experience' before service characteristics can be defined.

The third similarity is the way in which the development of these products came to be sensitive to the requirements of the user settings. In each case it was by examining how a particular form of service did perform or might perform in support of the users. In the route guidance example the lessons came primarily from the use of prototype systems in driving simulators. In the electronic journal and teleshopping examples field trials provided the basis for understanding user reactions. In the telemedicine example the findings presented were the result of providing would-be users with a scenario (a paper-based version of the system and a context of use which they could evaluate). Although the form of user evaluation differed, the result in each case was that the strengths and weaknesses of the potential service were revealed. In each case this led to a revised service design which was focused more precisely on the QoS required by the users.

17.5   Towards a user-centred approach for the creation of end user applications

What are the implications of these case studies for the process by which QoS is delivered to end users? If the process is not sensitive to the particular user requirements of the application it is likely to fail so the important question is how user issues are addressed in the development process. An important point to note is that the end user makes few distinctions between the various agencies that deliver the service. It matters little that the equipment is designed by one company, the content of the service is the responsibility of another and it is delivered by yet another organisation. In a successful application these distinctions are seamless to the end user and come together to deliver an integrated, quality service.

How is such a service to be created? The practice at the moment appears to be primarily one of delivering the service, discovering what works and does not work for the end user and modifying future services accordingly. Unfortunately, such a process can lead to much wasted investment and the provision of unsuccessful applications can create negative attitudes amongst end user communities.

To target applications more successfully a user-centred approach is required in which the design of all aspects of the service is informed by a rich knowledge of the relevant 'user experience'. One approach is to engage in studies of the users in the settings in which they undertake the activities a future application may serve. This might be interpreted as a need for task analyses to demonstrate how users engage in relevant tasks, for example, driving or shopping. Since the cultural and organisational setting in which the tasks are undertaken is often important, it might be a call for ethnographic studies [15] in which the shared understandings and practices of the 'microworlds' of users is explored. These approaches were not fully used in any of the cases described in this chapter and so it is difficult to know whether they would have led to a vision of the user world which would have guided designers away from initial, dysfunctional, multimedia solutions towards the kind of services that were ultimately found to be effective. They do, however, have one major drawback; they examine task behaviour and work practice as it currently exists. If a new product is to offer real utility it is likely to produce a fundamental change in task behaviour and work practice. From seeing a patient in a consulting room, for example, the consultant will see the patient on a visual display and travelling for either will be unnecessary. Using a teleshopping system is very different from visiting a shop. How best are people to operate in these new settings and what kind of system will best help them? Whilst analyses of the existing situation will undoubtedly provide an informative base for new product development it will not provide evidence of the future user experience. To achieve that, users need some early exposure to potential new applications so that their reactions can guide the development. As the case studies illustrate, this evaluation can be undertaken in a number of ways, for example, through the testing of prototypes or scenarios and through the evaluation of field trials. All are likely to yield valuable information to guide product development.

In recent years there have been many efforts to define a systems development process in which the user is the central focus. This has culminated in an international standard on human-centred design ISO13407 [16]. Figure 17.2 is based on the principles of the standard. The instantiation presented here is to render it relevant to the

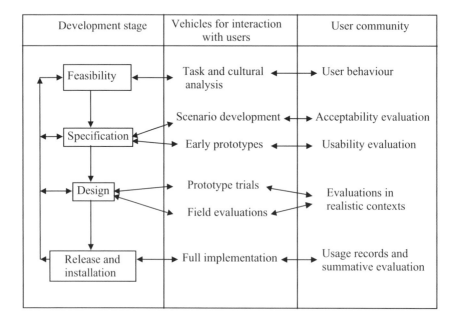

Figure 17.2   A user-centred development process

development of telecommunications applications to provide quality service to end users. This approach envisages development as a number of stages with major iterative loops. At each stage the development team work directly with members of the potential user community. The central column of Figure 17.2 identifies the 'vehicles' by which the development team and the user community interact to ensure a quality product or service is being created.

The initial step is to study the current user behaviour relevant to a product or service which may be feasible. These studies may have a specific task focus, for example, navigation whilst driving or a wider cultural perspective, for example, the communication patterns of teenagers which might be supported by advances in mobile telephony. These studies should both develop an understanding of the 'user experience' and provide evidence of the potential utility of future products. If this leads to a decision to proceed with development, the specification of the product or service needs to be based on a more detailed understanding of user requirements. To some extent these can be derived from 'user experience' studies but, as indicated above, there is the additional need to understand user requirements in the new setting that would result from use of the service. Two ways are advocated for making user studies of this, as yet non-existent, situation. The first is to build an early 'throw away' prototype of the service for the express purpose of obtaining user responses and thereby being able to specify more precisely the functionality and qualities of the service that will be required. Ideally, the early prototype has as many properties of the real service as possible, for example, users can interact with it and obtain services from it, and constitutes an integration of the contributions of the various providers, for example, hardware, content, carrier etc. To obtain good feedback from users the prototype should ideally perform no better and no worse than the expected performance of the ultimate service. Prototype services that will permit some level of user engagement are particularly useful for assessing usability issues especially when the prototype can be mounted in a realistic task environment as was the case in the tests of the prototype navigation aids in a car simulator.

There are two limitations to early prototype testing. The first is a practical consideration; it may not be possible to develop a working prototype early enough to inform product specification. The second limitation is that the prototype may not help users specify important requirements that arise from the wider organisational and cultural setting. To overcome these limitations more use is being made of scenario-based design [17]. In this approach a description of the envisaged product and the way it would be used is developed often using paper-based storyboard techniques. The users are asked to engage with this future scenario, perhaps playing the roles of the characters in the story. They are then able to evaluate the experience; what utility it would have for them, what problems they envisage etc. These techniques are particularly useful for identifying 'non-functional' qualities of service, for example, needs for privacy, confidentiality, reliability, availability etc. The issues revealed in the telemedicine study were based on this kind of analysis. Some examples of scenarios are provided by Eason [18].

Following the specification process the design phase will ideally create more comprehensive and working prototypes which users can test in ever more realistic settings. This should culminate in full scale trials of the service in situations where the users are making realistic use of the system. The feedback from these prototype trials tends to lead to iterative design of the service and as the trials become more realistic, identify all the issues to be addressed in implementation, for example, manuals, training, local support, organisational changes etc. Many of these issues were noted in the telemedicine example.

The final phase is implementation. This is a time when there is much valuable information to be obtained about the new user experience from field evaluations. One important source of information is usage records which are often available but rarely fully exploited. The usage records in the electronic journals project were vital in developing an understanding of user behaviour with this new service, revealing as they did, for example, that very few full text papers were read on screen and search engines were rarely used etc. Data from usage records are vital to indicate what is being used but they need backing up by 'summative evaluations' in which users can be asked 'why' questions. All of this information is in fact the analysis of the new user experience which provides the basis for the specification of the next generation product and begins the development cycle once again.

17.6   Summary

Broadband communications offer many exciting possibilities for new services for all kinds of users. This potential can, however, lead designers to believe that being inventive with the technology will be sufficient to create services which users will find valuable. The evidence of the case studies is that technology alone is not sufficient; users are looking for a form of technology which is fit for the purposes they have and has service qualities appropriate to the context of use, for example, is usable, acceptable etc. In the design process by which these services are created it is necessary to develop an understanding of the users' world and to test emerging forms of the service in order to ensure it will work at all levels; the form of interaction is usable, the mix of media is appropriate to the user's purpose, the system can operate within the organisational framework etc. If we can achieve a user-centred design process with these properties at all stages we can realise the enormous potential of these systems without going through a succession of failures before we identify a successful way of delivering the service.

17.7   References

1   CLARKE, A. M. (Ed.): 'Human factors guidelines for designers of telecommunications services for non-expert users' (HUSAT Research Institute, Loughborough University, UK, 1996)
2   ISO 14915 (2000) Multimedia User Interface Design, International Standards Organisation, Geneva, Switzerland
3   ISO 9241 (1998) Ergonomics of office work with VDTs – Part II: Guidance on Usability, International Standards Organisation, Geneva, Switzerland

4   EASON, K. D.: 'Towards the experimental study of usability', *Behaviour and Information Technology*, 1984, **3**(2), pp. 133–43

5   DAVIS, F. D.: 'User acceptance of information technology: system characteristics, user perceptions and behavioural impacts', *International Journal of Man–Machine Studies*, 1993, **38**(3), pp. 475–87

6   CHRISTIE, B. (Ed.): 'Human factors of information technology in the office', (Wiley, Chichester, 1985)

7   HARKER, S. D. and EASON, K. D.: 'The use of scenarios for organisational requirements generation', Proceedings of HICSS-32 Conference, Maui, Hawaii, 5–8 January 1999

8   EASON, K. D., HARKER, S. D. P. and OLPHERT, C. W.: 'Representing socio-technical systems options in the development of new forms of work organisation', *European Journal of Work and Organizational Psychology*, 1996, **5**(3), pp. 399–420

9   VEINOTT, E. S., OLSON, J., OLSON, G. M. and FU, X.: 'Video helps remote work: Speakers who need to negotiate common ground benefit from seeing each other'. CHI '99 Conference Proceedings (ACM Press, 1999) pp. 302–09

10 EASON, K. D.: 'Division of labour and the design of systems for computer support for co-operative work', *Journal of Information Technology*, 1996, **11**, pp. 39–50

11 BURNETT, G. and JOYNER, S.: 'Route Guidance Systems: Getting it Right from the Driver's Perspective', *The Journal of Navigation*, 1996, **49**(2), pp. 169–77

12 DE KARE-SILVER, M.: 'e-Shock 2000: The Electronic Shopping Revolution' (Macmillan Business Press, Basingstoke, 2000)

13 KAULIO, M. A. and KARLSSON, I. C. M.: 'Triangulation strategies in user requirements investigations: a case study on the development of an IT-mediated service', *Behaviour and Information Technology*, 1998, **17**(2), pp. 103–12

14 EASON, K. D., YU, L. and HARKER, S. D. P.: 'The use and usefulness of functions in electronic journals: the experience of the SuperJournal Project', *Program*, 2000, **34**(1), pp. 1–28

15 NARDI, B. A., 'The use of ethnographic methods in design and evaluation' in M. G. HELANDER, T. K. LANDAUER and P. V. PRABHU, (Eds): 'Handbook of human–computer Interaction' (North Holland, Amsterdam, 1997) pp. 361–6

16 ISO 13407 (1999) Human Centred Design Process for Interactive Systems, International Standards Organisation, Geneva, Switzerland

17 CARROLL, J. (Ed.): 'Scenario-Based Design' (Wiley, Chichester, 1995)

18 EASON, K. D.: 'People and computers: emerging work practice in the information age', in P. B. WARR, (Ed.): 'Psychology at work' (Penguin, 2002, chap. 4) pp. 77–99

Chapter 18

Telecommunication services for people with disabilities

18.1 Introduction

A surprisingly high proportion of the world's population suffer from some form of physical or mental disability, and for many of these people this disability has a major impact on their daily lives. For them the Quality of Service (QoS) measures traditionally used for telecoms services are of little value: their key question is 'has this service been designed in a way which allows me to use it at all?' This chapter is intended to provide a basic appreciation of the needs of this sizeable subset of the overall Telco market, and point to ways in which telecoms operators can assess how effectively they are serving it.

The initial sections review the available data on the size and nature of the disability issue, and look at some of the Legislative and Regulatory provisions with which Telcos may need to comply. It then looks at the way in which people with some of the principal types of disability can be helped or hindered by the way in which Telco services are offered to them. A later section considers the implications of disability issues for Telco's internal processes.

A huge range of studies have been carried out across the world in recent years into ways of helping people with disabilities make best use of telecoms systems, and the references cited can only provide an initial introduction to the associated literature.

18.2 Who are the disabled?

The UK Disability Discrimination Act 1995 [1] states that:

> a person has a disability if he has a physical or mental impairment which has a substantial and long-term adverse effect on his ability to carry out normal day-to-day activities.

This definition is consistent with the general spirit of legislation in many other countries, but does not lend itself too well to precise statistical measurement. Consequently the majority of published statistics relating to disability issues should be treated with a degree of circumspection. However, it is beyond doubt that huge numbers of people around the world have to live with a significant level of disability. Starting with the US, the area of Microsoft's Web site devoted to their commitment to accessibility for people with disabilities [2] states that about one in five Americans, or roughly 54 million, has a disability. About one in ten, or roughly 26 million, has a severe disability.

An alternative snapshot is provided by the Spring 2000 Labour Force Survey for Great Britain, an extract from which is shown in Table 18.1. This shows that in Great Britain there are about 6.6 million people of working age with one or more forms of significant disability. This compares with approximately 28.7 million people of working age who are not disabled; suggesting that about 19% of the British work force has one or more forms of disability.

However, comparing some of the individual figures within Table 18.1 with estimates from alternative sources illustrates the difficulties involved in preparing and using these estimates. Taking hearing problems as an example, the Royal National Institute for the Deaf estimate [3] that there are about 8.7 million deaf and hard of hearing people in the UK. Of these fractionally under 8 million have either mild deafness, a hearing loss in the better ear of between 25 and 40 dB, or moderate deafness, a hearing loss in the better ear of between 40 and 70 dB. These numbers are rising as the number of people over 60 increases. The remaining 673,000 people are severely or profoundly deaf, of whom 420,000 cannot hear well enough to use a voice telephone even with an amplifying device. But the Labour Force Survey data shown in Table 18.1 only identifies about 123,000 people in Great Britain of working age for whom hearing loss is their major disability; and in 1998 only 190,000 people in England (which for these purposes represents about 84% of the UK) were registered with social services departments as deaf or hard of hearing.

Similarly the Royal National Institute for the Blind quote estimates [4] which suggest that in 1996 just over 1 million people in the UK had a sufficiently serious visual impairment to be registerable if they so wished; and on 31.3.1997 the number of Registered Blind and Partially Sighted people in the UK was 348,000. Again neither of these estimates tallies very well with the 118,000 identified in the Labour Force Survey as having difficulty in seeing.

These wide variations in estimates of the scale of the disability problem reflect the genuine difficulties of assessing conditions that exhibit a wide range of degrees of severity. It is also well known that the numbers of people registering their condition with social services departments and other government bodies invariably underestimate the true scale of the problem. In general, statistics from Governmental sources will usually understate the disability problem rather than overstate it.

It should be noted that the prevalence of most forms of disability increases with increasing age. There are many people who do not classify themselves as having a disability, but nevertheless, through normal ageing effects, experience some of the

Table 18.1   Employment and unemployment rates by the type of main disability

	Number with this as main disability (% of all disabled)	Number in employment and employment rate (% of total)	Number on state benefits and not in work (% of total)
All long-term disabled in GB	*6,449,000* / *18%*	*2,996,000* / *46%*	*2,626,000* / *41%*
Problems with arms, hands	409,000 / 6%	192,000 / 47%	164,000 / 40%
Problems with legs, feet	706,000 / 11%	295,000 / 42%	328,000 / 46%
Problems with back, neck	1,225,000 / 19%	572,000 / 47%	524,000 / 43%
Difficulty in seeing	113,000 / 2%	57,000 / 51%	47,000 / 42%
Difficulty in hearing	123,000 / 2%	80,000 / 65%	27,000 / 22%
Speech impediment	10,000 / *	*	*
Skin conditions, allergies	123,000 / 2%	76,000 / 62%	30,000 / 25%
Chest, breathing problems	926,000 / 14%	554,000 / 60%	225,000 / 24%
Heart, blood pressure	738,000 / 11%	335,000 / 45%	292,000 / 40%
Stomach, liver, kidney, digestion	310,000 / 5%	150,000 / 48%	111,000 / 36%
Diabetes	279,000 / 4%	182,000 / 65%	61,000 / 22%
Mental illness	502,000 / 8%	81,000 / 16%	369,000 / 73%
Epilepsy	135,000 / 2%	57,000 / 42%	63,000 / 47%
Learning difficulties	143,000 / 2%	41,000 / 29%	90,000 / 63%
Progressive illness n.e.c.	244,000 / 4%	86,000 / 35%	131,000 / 54%
Other problems, disabilities	443,000 / 7%	227,000 / 51%	149,000 / 34%

GB – Great Britain (Estimates from the Winter 1999/2000 Labour Force Survey for Great Britain)

* = fewer than 10,000/less than 0.5%; estimate not shown. Base: All people of working age (men 16–64, women 16–59)

problems of disability. For example, the great majority of people over the age of 70 have a significant hearing loss.

18.3   The disabled as a market opportunity for Telcos

The large numbers of people affected by disability issues suggests that the provision of services to them should be seen by Telcos and CPE suppliers as a major market opportunity. As an example, the UK Disability Rights Commission has recently stated 'There are 8.5 million disabled people in Britain with a combined annual spending power of £40 billion.' Similarly there are few families who do not contain a member with a disability problem. Telcos who are seen to be unresponsive to their needs risk indirectly alienating a sizeable proportion of their market.

Unfortunately, few Telcos appear to have taken positive steps to develop what is *prima facie* a major market opportunity. There are two major reasons for this.

- People with disabilities are disproportionately likely to be unemployed, or in low wage occupations, or to be elderly people living on pensions. In most countries people with disabilities experience an unemployment rate that is two to three times higher than that of their able bodied contemporaries. Consequently they tend to have relatively low disposable incomes, making them an unattractive market sector. A study in Scotland which analysed household incomes into five bands found that only 6% of households containing a disabled person fell into the highest of the five income bands, compared to 24% of households without a disabled member [5].
- Because there are so many forms of disability [Table 18.1] and so many degrees of impairment within each disability group, the apparently large overall market for services to support people with disabilities has to be disaggregated into a large number of very small niche markets. These tend to have low volumes and high development costs, making them a relatively unattractive market, especially from the point of view of a large organisation such as a typical Telco.

For these reasons market forces have rarely supplied an acceptable range of products and services to people with disabilities, and most countries have developed legislative and regulatory solutions to the issue.

18.4   Legislative and regulatory provisions

Managers responsible for QoS issues for their Telco must familiarise themselves with the specific requirements of their local legislators and regulators. In principle legislative requirements should have the greatest importance, but in practice the regulators in several major countries were mandating various forms of support for people with disabilities several years before explicit disability legislation was enacted. Thus a typical Telco will have to comply with a mixture of obligations, some stemming from statute law and others mandated by their regulators.

In the US the dominant laws are the Americans with Disabilities Act and the US Telecommunications Act 1996. However, in addition to overseeing the application of these Acts in the field of telecommunications, the Federal Communications Commission [6] has historically adopted measures of its own to encourage or enforce the provision of services for people with disabilities.

A similar pattern exists in the UK, where the Office of Telecommunications made various forms of disability support mandatory on British Telecommunications plc via the Conditions of its Operating Licence several years before the Disability Discrimination Act (DDA) 1995 [1] was enacted. Part III of the DDA 1995 places an obligation on service providers (in all service industries, including telecommunications) to make 'reasonable adjustments' to their premises and services to allow them to be accessed by people with disabilities on a broadly equal footing to able bodied people. This specific part of the Act only came into effect from October 2000, and at the time of writing (spring 2002) there is limited evidence as to how it will work in practice. Consequently, the requirements defined by the Office of Telecommunications continue to play a major role.

It may be noted that in the UK the formal obligations imposed by the Office of Communications have been regarded as a subset of the wider Universal Service Obligations, and thence are restricted to British Telecommunications plc and to Kingston Communications, as a consequence of their roles as the legacy operators prior to privatisation. Many people would regard it as inequitable that the dozens of other suppliers operating in the UK telecommunications market are untouched by these requirements, and it is probable that other licensed operators will have to accept an increasing level of responsibility in this area.

The European Commission makes reference to the needs of people with disabilities within the recent EU Universal Service Directive [7]. This specifies that:

> Member States should take suitable measures in order to guarantee access to and affordability of all publicly available telephone services at a fixed location for disabled users and users with special social needs. Specific measures for disabled users could include, as appropriate, making available public text telephones or equivalent measures for deaf or speech impaired people, providing services such as directory enquiry services or equivalent measures free of charge for blind or partially sighted people, and providing itemised bills in alternative format on request for blind or partially sighted people. Specific measures may also need to be taken to enable disabled users and users with special social needs to access emergency services '112' and to give them a similar possibility to choose between different operators or service providers as other consumers.

As can be seen from this EU Directive, mandatory requirements have historically tended to focus on voice traffic, and thence on the needs of people with hearing problems. Typical national regulations would call for the provision of a Text Relay Service (see below), and the provision of inductive couplers on public payphones. Some countries require the provision of public payphones mounted at a height that makes them accessible to wheelchair users. It is not uncommon for there to be a requirement for free access to directory enquiry services for some categories of disabled users, where this is provided as a chargeable service. With the increasing role of telecoms networks in the transmission of visual information it is probable that additional requirements will emerge to ensure that visually impaired users have at least some access to vision based services.

It should be noted that in many countries the mandatory requirements for adaptation of services to make them accessible for people with disabilities stop at the 'Network' boundary. In these cases the market for the provision of terminal apparatus is unregulated, and there are few or no legal pressures to secure the provision of suitably adapted terminal equipment. A welcome exception to this generalisation is provided via Section 255 of the US Telecommunications Act 1996 [8].

18.5   Telco services as seen by some of the key disability types

This section looks at the way in which Telco services tend to be perceived by people with various forms of disability. By consciously considering these perspectives, Telcos can develop ways to manage their services so as to minimise the negative impacts on disabled people and maximise the positive impacts.

18.5.1   *People with hearing problems*

The basic product of Telcos has traditionally been the transfer of speech from point to point, so it is logical to start by looking at the way their services are viewed by people with hearing problems.

The great majority of customers with hearing problems have a relatively low degree of loss, for example 25–40 dB, and their difficulties can be substantially alleviated by provision of suitably modified CPE. This includes phones with an in-built amplifier to boost the sound level presented to the user's ear, and phones incorporating an inductive coupler in the handset. This generates a localised magnetic field mimicking the audio output, which can be detected by a matching pick-up coil that is built into the majority of modern hearing aids. This is of particular value in situations of high background noise. The unit cost of incorporating an inductive coupler in a handset at the design stage is so trivial that it would be reasonable to expect that all commercial designs of CPE would include this facility, but sadly this is not the case. Different countries have different mandatory requirements for the provision of handsets with inductive coupling: for example, their use may be mandatory in public payphones, in motorway emergency phones, and in emergency phones in lifts in high rise buildings. In some countries the regulator will specify a maximum value for the amount of gain which can be provided in amplifying handsets: this is to reduce the risk of acoustic shock if a person with normal hearing inadvertently picks up a handset working in amplifying mode.

A network based facility which can have an unexpected beneficial effect for customers with hearing problems is Calling Line Identification, for example the Caller Display service offered by BT. This provides the called party with a visual display of the calling number, which makes it hugely easier for hearing impaired customers to know who is ringing them, and to confirm the number to ring back if necessary.

Unfortunately, experience in the UK is that a high proportion of calls are sent over the network with either Number Withheld or Number Unavailable codes, which seriously reduces the value of CLI facilities to the hearing impaired. This is particularly the case for calls originating on PBXs and from Call Centres, and Telcos have a role to play in educating such users in the importance of not automatically setting their systems to Number Withheld!

The growth of the Call Centre industry, and the associated way in which firms in many industries look to their advertised Call Centre numbers as the preferred (or only) route through which their customers contact them clearly pose serious problems for hearing impaired customers. Telcos who provide or facilitate Call Centre systems should develop codes of practice which oblige the end users of the Call Centres to ensure that they always advertise alternative access routes for hearing impaired customers.

For people with no usable hearing an alternative method of communication is via Textphones – sometimes referred to as Minicoms. These combine a QWERTY keyboard and a simple alphanumeric display with a low speed modem to allow material to be typed at one end and read at the other. A typical modern Textphone is illustrated in Figure 18.1. The modem operates over a normal PSTN connection, either through a plug and socket network connection or (especially for portable models) through a voice coupler. In one or two countries there has been limited deployment of public payphones incorporating a Textphone unit. Unfortunately, this is rarely economic

Figure 18.1   A typical Textphone. Model shown is the Uniphone 1150, courtesy of Teletec international

other than in locations where it is known that there are high concentrations of hearing impaired users, for example, special schools and Deaf Clubs. However a recent development in the UK points to a possible solution to this problem. In the first half of 2002 BT is aiming to install an initial tranche of 2000 text payphones at high usage payphone sights across the UK. These units are designed to offer e-mail and SMS services to the general public as well as offering access to text relay services such as Typetalk (see below) and the ability to make calls direct to other textphones. This multi-user capability should significantly improve the economic viability as seen by the Telco.

The majority of Textphones currently operating in Europe use the V21 protocol, and it is therefore possible to load suitable software onto a modem equipped PC to allow it to act as a Textphone. One drawback to the use of V21 is that Group 3 fax machines also use V21 during the call set-up phase, so it can be difficult for auto-answer equipment to distinguish reliably between a fax call and a Textphone call.

To allow Textphone users to talk with friends and businesses who are only equipped with normal voice instruments Text Relay Services have been developed in several countries. They use specially trained operators equipped with PCs adapted to operate as a Textphone who relay speech between the Textphone user and the other party to the call, who talks to the operator via a standard PSTN call terminating on the operator's headset. In the UK this service is operated by the Royal National Institute for the Deaf under the Typetalk brand [9]. The substantial costs involved are met by BT as part of its Universal Service Obligation. Services of this type provide an invaluable lifeline for the deaf, and are frequently heavily used. If the Text Relay Service is engineered to high standards of availability (e.g. duplicated call handling centres) then it can be used to provide Textphone users with access to the emergency services normally accessed via codes such as 112 or 999.

The perceived value of text relay services to hearing impaired users can be inferred from some operational statistics for the UK Typetalk service as in late 2001. It was then employing approximately 500 operators to handle roughly 35,000 calls a week to or from hearing impaired users.

A recent innovation within the UK has been the development of the Text Direct service, which provides customers with easier access to Typetalk operators. Previously the deaf caller had first of all to log on to the Typetalk service, and give the call details to the operator, who then attempted to establish the call on their behalf. This involved a significant amount of effort and delay for both the caller and the operator, even if the operator then had to tell the caller that the called line was busy! With the Text Direct service the caller dials the required destination number direct from their textphone, preceded by a prefix code that tells the BT network that it is a textphone originated call. If the network then detects that the answer signal has come from another textphone it simply lets the call proceed as a direct textphone-to-textphone connection, but if the called party is a voice line the network arranges to bring a Typetalk operator into circuit. This substantially improves the customer experience for the Textphone user, and reduces the amount of operator time wasted in trying to establish calls that fail to mature due to No Reply or Engaged conditions. A key QoS issue is to ensure that the operator is associated with the call as quickly as possible following receipt of the called subscriber answer condition. Failure to achieve this can cause various problems, for example, if the called line is working in Answering Machine mode the Typetalk operator may miss the first few seconds of the message.

Another positive recent development has been the ability of modern telecoms networks to support Videophones. A combination of reliable medium speed network links (e.g. an ISDN2 'B' channel) and modern image compression techniques developed under the auspices of MPEG allow networks to economically handle remote videophone systems that will support Sign Language conversations. An obvious use is to allow two Sign Language users to talk directly to each other via suitable Videophones. A less obvious but potentially very valuable service is to provide remote access over the telecoms networks to Sign Language interpreters. This avoids the need for (extremely scarce) Sign Language interpreters to waste time in travelling to be physically present at the meetings or interviews at which their services are needed, perhaps only for a few minutes.

On the other hand it is very difficult to regard the explosion in use of Mobile phones as being a positive development as seen by people with hearing problems. The main issue is the poor standard of sound quality delivered over the great majority of commonly used Mobile phones, especially in conditions where the radio signal is marginal. The increasing use of Enhanced Full-rate Codecs slightly improves the transmission quality, but this is normally still well below fixed link standards. Perhaps more importantly, Mobile phones are frequently seen and marketed as fashion accessories, with constant pressure on designers to produce ever smaller handsets. As a result most Mobile phones flout many of the basic rules of handset design, with the microphone typically positioned a long way from the speaker's lips, and with no attempt to shield the microphone from surrounding environmental noise. Virtually all forms of hearing problems leave the user acutely sensitive to any form of noise interfering with the signal, so calls originating on most Mobile phones give serious audibility problems to many people with hearing problems.

A further unwelcome feature of some Mobile phone handsets is their ability to generate interference for adjacent hearing aid users. These effects have been reported with both GSM and DECT systems. It is important that the designers of these systems do all they can to eliminate these effects. To the extent that the Bluetooth system uses a digital frame structure that has many similarities to GSM and DECT, there must be a concern that the emerging generation of Bluetooth equipped devices may also generate hearing aid interference if this is not properly controlled at the design stage.

For some hearing impaired users the availability of Text Messaging services provides a welcome additional method of communication, but for some users the minuscule keypads fitted on many Mobile handsets are a deterrent to extensive use of this option.

18.5.2 *People with vision problems*

The traditional focus on voice-based services means that telecommunications systems are relatively accessible to people with vision problems. Indeed the trend to conducting business over the phone via Call Centres rather than via correspondence is a welcome development for many people with vision problems. This does however presuppose

Figure 18.2   The Big Button phone

that they are equipped with CPE that allows them to use the phone with relative ease. For many people an instrument equipped with a large and clearly marked keypad (e.g. the BT Big Button phone, see Figure 18.2) will suffice, for other users a Braille keyboard is required. A simple aid with zero manufacturing cost is to ensure that the button at the centre of the keypad (usually digit 5) is provided with a small moulded pip to allow location by touch. Sadly, there are many instruments in volume production that do not provide this elementary adaptation.

Telcos will need to ensure that their customer literature is available in alternative formats suitable for use by people with vision problems. At a minimum they must ensure that customer service agreements and bills are readily available in both large print and Braille formats.

However, the main area impacting on people with vision problems is probably the explosive growth in the Internet. Anyone without ready Internet access is increasingly at risk of being disadvantaged and marginalised in the emerging Internet-based society, so the way in which the Internet has evolved as a predominantly visual medium poses a major problem for people with vision problems. A wide range of devices and products have been developed in recent years which attempt to make Internet pages accessible to people with vision problems, see Reference 10 for an extensive treatment of this subject. It is less clear at present how effectively these products are reaching the people who need them, and Telcos can help increase Internet usage in their customer base by ensuring that their customers are made aware of these products. Simple and uncluttered Web page design is also a great help to visually impaired users – and to the large numbers of elderly people who do not classify themselves as being blind, but do nevertheless have significant vision problems. The use of appropriate fonts and text sizes is a low technology but effective aid in these cases, see Reference 10 for suitable guidelines.

A large number of people have quite high levels of optical resolution, but are colour blind. For these people the way in which many Web sites use colours chosen more or less at random, or because 'it looks pretty', is very unhelpful. Guidelines are available on how to select colour combinations that maximise readability by people with colour perception problems [11]: at a minimum, Telcos should ensure that these guidelines are applied to their own Web sites.

18.5.3   People with voice projection problems

There are a significant number of people with some form of speech disability, who have great difficulty in using the standard telephone service. Typically people who have a speech disability cannot communicate by telephone because the public, friends or even family members cannot readily understand their speech. This is also sometimes the case for people with cerebral palsy, multiple sclerosis, muscular dystrophy, Parkinson's disease or others who are coping with limitations in the aftermath of stroke or traumatic brain injury. Those who stutter or had a laryngectomy may also have difficulty being understood.

A recent development in America is tackling this problem through the introduction of a new service called Speech to Speech Relay (STS). The service is mandated by the Federal Communications Commission and enables people with a speech disability to use their own voice, voice prosthesis or communication device to make a phone call. The service became available nationally in the USA from 1 March 2001 [12].

In general, STS can be used by anyone with a speech disability or anyone who wishes to call someone with a speech disability. STS calls can also be made by people, or to people, who use a TTY, or other TRS-communication modes such as Voice Carry Over (VCO), Hearing Carry Over (HCO), or to another person with a speech disability.

18.5.4   People with mobility and dexterity problems

Telecoms systems are generally helpful to people with mobility and dexterity problems, in that they allow them to access information and services from their own homes that they would otherwise have to make difficult journeys to access. Many products and services which have evolved to support Teleworking are well suited to people in this category, and the ability of the Internet to deliver a huge range of information and support services to the immobile customer's armchair is equally welcome.

However, commonly available CPE is rarely designed with the needs of these people in mind. People who have difficulty in gripping things are often unable to hold standard telephone handsets, and the small buttons used on many forms of CPE (often combined with a poorly defined 'collapse' action) are a hazard for people with arthritis or coordination problems. Various forms of adapted CPE are available to help people with these problems, the key issue is to ensure that the right information on these products is reaching the right people. In the longer term it is likely that an increasing range of voice actuated equipment will become available to help these users.

Wheelchair users will often have difficulty in accessing kiosk mounted payphones. Many of them have heavy doors that are difficult to open by wheelchair users – or by people who have difficulty in gripping things. Sir Giles Gilbert Scott's classic 1920s design for the UK General Post Office falls into this category, as do derivative designs still in use around the world. Other designs have doorways that are too narrow to allow wheelchair access, and even when the user is inside the kiosk the payphone instrument is usually mounted at a height and angle optimised for someone standing, and is therefore unusable from a wheelchair. Most major Telcos have developed alternative housings that mount the instrument at a height that suits wheelchair users, but their deployment is usually fairly limited. Figure 18.3 reproduces a photograph taken in the early 1990s in Germany and illustrates one solution to the problem, that is, a pairing of a 'classic' kiosk design with a wheelchair accessible one.

Whilst Entry phone systems rarely form part of the regulated telecoms network it may be noted that they are as inhospitable to wheelchair users as they are to people with hearing problems. They are usually mounted at a height that is difficult to reach from a wheelchair, and their sound quality makes them virtually unusable to hearing impaired users. To compound the felony, Entry phones rarely have the bright colours needed to help vision impaired customers locate them, nor are they usually equipped with Braille symbols.

Figure 18.3   Payphone accessible by wheel chair

18.6   The implications of disability issues for Telco's internal processes

18.6.1   Customer service and product launch processes

Ideally all Telcos should seek to emulate Microsoft, which in 1995 made a public commitment to people with disabilities:

> Microsoft Corporation recognizes its responsibility to develop products and information technologies that are accessible and useable by all people, including those with disabilities.
>   We will devote the time and resources necessary to ensure that all users enjoy access to our products, technologies, and services.
>   It is the responsibility of everyone at Microsoft to deliver on this commitment.

Apple also has a similar public policy on accessibility to their products. See References 2 and 13 for further information on Microsoft and Apple policies.

Even when a Telco prefers to stop short of making a holistic statement of this type, the examples in the preceding sections show that quite minor features of the way a telecommunications product or service is designed can have a major impact on its accessibility to people with disabilities. A positive corollary to this is that quite minor changes to the way a product is designed or presented to the customer can dramatically improve its accessibility to people with disabilities.

Most Telcos use a formal Business Process Management system, so it should be feasible to include a loop within the Product Design and Launch process to systematically assess how easy or difficult the product is to use by customers from the major disability groups. (This is not an expensive or high technology process, all that is required is to issue some of the design staff with ear plugs or frosted glass spectacles for a few days!) On the other hand it will often point the way to simple improvements that can be implemented at negligible cost. For example, to have an internal rule that any form of customer publicity material which quotes the Telco's contact phone numbers must also include fax and textphone numbers will dramatically improve the company's QoS score as perceived by customers with hearing difficulties. Similarly the printing of customer service information in (for example) dark blue print on a light blue background may appeal to the artistic sensibilities of some Telco marketing departments, but renders the material unusable by many of the visually impaired customers they are trying to reach.

Major Telcos may also consider it appropriate to take a proactive approach to drawing the attention of their disabled customers to the wide range of specially adapted equipment that exists to assist them. A good example of this approach is the BT booklet *A guide for older and disabled people*, which includes details of many items of equipment available to UK customers from suppliers other than BT [14].

Most Telcos will aim to offer high standards of repair services to all their customers, and may therefore consider that it is not necessary to offer special standards of repair service to customers with disabilities. This should, however, be a decision that is taken consciously rather than by default, and in some cases it may be appropriate to offer enhanced standards of repair services to specified groups of especially vulnerable customers. It is also important to check that customers with disabilities have effective access to the repair service. For example, automated fault reception systems using automated voice prompts etc. can be an unintentional but serious barrier to customers with hearing problems.

18.6.2   Employee management processes

The major thrust of this chapter has been the need for Telcos to systematically review the way their services are perceived and used by customers who have some form of disability. However, it is important to note that many Telcos are major employers, and should therefore expect to have a significant number of people with disabilities amongst their employee base. The Spring 2000 UK Labour Force survey indicates that nearly a fifth of people of working age have some form of disability. Various studies have shown that these people are two to three times more likely to be unemployed than their able-bodied peers. Telcos should thence have employee management processes that respond positively to the needs of job applicants who have an existing disability, and of employees who become disabled during their employment. Rather than viewing this negatively as an obligation imposed by local employment law, Telcos should view this issue positively as an opportunity to ensure that their teams are representative of the customer base they seek to serve. It is also an opportunity to showcase the way their products can be used to help and empower the disabled.

Telcos should ensure that their recruitment processes demonstrably give a fair opportunity to applicants with any form of disability. The key to success is to focus on the applicant's abilities and identify roles in which these can be exploited, and only then look for ways of working round the disability aspect.

18.7   International collaboration and standards

The global nature of most disability issues makes this an obvious field for international collaboration in the development of solutions to these problems, and in the establishment of relevant technical standards. A comprehensive review of this rapidly changing field is beyond the scope of this chapter, so we list a few initial references below:

- The European Commission has sponsored numerous research and development projects, for example the 5 year COST 219bis project which examined many issues relating to access to telecoms services by people with disabilities. Its report was published in late 2001, with an associated conference at Leuven. A proposal is being developed for a follow on study which would focus on access to mobile telecom services [15, 16].
- The European Technical Standards Institute (ETSI) makes allowance for disability issues in several of its standards [17].
- In the USA there is a Center for Applied Special Technology (CAST), founded in 1984. It is an educational, not-for-profit organisation that uses technology to expand opportunities for all people, including those with disabilities. One of its

more interesting outputs is an online tool called 'Bobby' which automatically assesses the ease with which people with disabilities will be able to use a given design of Web site. Bobby can be accessed at the address given in Reference 18.

- The World Wide Web Consortium maintains a discrete Web Content Accessibility Guidelines Working Group, which produce guidelines on how to make Web sites as accessible as possible to people with disabilities [19].

18.8   Summary

The following are the key issues in the provision and management of telecommunication services for those with special needs. These issues have a QoS dimension for reasons discussed in this chapter.

- What data is available on the incidence of various forms of disability in the market(s) served by your Telco?
- What requirements have your local regulator or legislature specified for provision of services to people with disabilities?
- Does your Telco have an explicit policy statement covering these issues? Who is responsible for implementing it? How regularly is compliance assessed, and in what ways?
- What specific services, products, or tariff packages are offered for use by people with disabilities?
- What arrangements exist to ensure that all major products and services are checked for accessibility by the key disability groups, both at the design stage and periodically throughout their in-service life?
- Have any specific requirements been defined for maintenance of exchange lines or ancillary equipment known to be used by people with disabilities?
- Does the Telco's marketing and customer service literature give clear information on contact methods that can be used by customers who would have difficulty using the standard methods? Do Web sites comply with guidelines for accessibility to people with disabilities?
- What proportion of your Telco's employees have disabilities? Are they satisfied with the way the disability issues associated with their employment are managed?
- Does your Telco have defined arrangements to monitor and participate in the activities of national and international study groups looking at disability issues, and to implement the results?

18.9   References

1   The text of the UK Disability Discrimination Act 1995 can be found at: http://www.hmso.gov.uk/acts/acts1995/1995050.htm
2   Microsoft Accessibility: Technology for Everyone. See: http://www.microsoft.com/enable/guides/default.htm

3   A variety of statistics relating to the incidence of hearing problems in the UK can be found at: http://www.rnid.org.uk

4   A variety of statistics relating to the incidence of vision problems in the UK can be found at: http://www.rnib.org.uk

5   Scottish Household Survey, 1999

6   A wealth of information on the US FCC's approach to provision of Telecoms services on people with Disabilities can be found at: http://www.fcc.gov/cib/dro/

7   European Commission: Common position adopted by the Council with a view to the adoption of Directive of the European Parliament and of the Council on universal service and users' rights relating to electronic communications networks and services (Universal Service Directive) Document 10421/01

8   Information on the US Telecommunications Act 1996 can be found via [e.g.] http://www.fcc.gov/telecom.html

9   Further information on the RNID/BT Typetalk service can be found at www.typetalk.org

10   For a comprehensive discussion of the issues involved in interfacing computer based systems with people with Disabilities see Nicolle, C., and Abascal, J. (Eds) 'Inclusive design guidelines for HCI' (Taylor and Francis, 2001)

11   RIGDEN, C.: 'The Eye of the Beholder – designing for colour blind users', *Journal of the Institution of British Telecommunications Engineers*, **17** (4), 1999, pp. 291–5

12   For more information on STS see: http://www.stsnews.com/Pages/WhatisSpeech toSpeech.html

13   Information on Apple's accessibility policy can be found at: http://www.apple. com/disability/

14   'A guide for older and disabled people' published annually by British Telecommunications plc, and accessible online via: http://www.bt. com/aged_disabled/index.jsp

15   Information on the work of the 1996–2001 COST 219bis study group on the impact of Telecoms systems on people with disabilities [and of the preceding COST 219 study] can be found at: http://www.stakes.fi/cost219/

16   The possible scope of the proposed follow on study COST 219ter can be seen at http://www.stakes.fi/cost219/COST219ter8.htm

17   Information on the coverage of disability issues within ETSI standards can be found via: http://www.etsi.org/sitemap/home.htm

18   For details of 'Bobby' see: http://www.cast.org/bobby/

19   W3C Web Content Accessibility Guidelines Working Group: see http://www.w3.org/WAI/GL/

External drivers

The service provider's mission on QoS has to be realistically tempered with the external drivers. The customer's requirements are an obvious one. Others chosen to illustrate this aspect are the role of regulation and the standards bodies. The interaction between the customer and regulator with the provider has to be studied on a country-by-country basis whereas that with the standards body is perhaps more uniform with all countries. This section addresses the key issues in the relationship between these drivers and the providers.

Chapter 19

Role of consumer and user groups

19.1   Introduction

User groups exist to represent the users' interests in many industries and trades. User groups representing the users of telecommunications services have evolved, noticeably in the recent past. For other utilities, such as water, electricity and gas, which have reached maturity, the current principal issues are related to distribution, cost of supply and perhaps also quality of the commodities. Telecommunications offers additional challenges to representatives of users on issues such as quality of services being developed, worldwide compatibility and interworking of services and greater interaction between the service providers and the terminal equipment manufacturers for better use of network capabilities. With the prospect of rapid and continued growth in the range and use of telecommunications services and depth of technology there is increased scope for user groups to influence various bodies involved in the provision of services. User groups could play an exciting role in the field of Quality of Service (QoS), provided they take the initiative to contribute effectively. They could contribute towards improved quality (including value for money), coherence in service applications, user-friendliness of the service–person interface, cost of service etc. These groups are in a unique position of being aware, first hand, of their member's needs on all aspects of a telecommunication service.

   Existing telecommunications user groups are involved in a wide range of activities. Typical activities are: addressing issues on tariffs, provision of service at the required time, providing efficient customer care and QoS. These issues are usually taken up with the individual service providers. The more sophisticated user groups deal with wider issues, such as numbering, provision and coverage of new services (e.g. mobile and IP supported services) and regulatory matters. Some groups have taken the initiative on matters of QoS. However, it is the possible future role of these bodies that is given more attention in this chapter. Before exploring these possibilities some of the principal concerns of three segments of the user population are explored. Finally some guidelines to assist the formation of a new user group are also given.

19.2   Principal issues of users

19.2.1   General

For the purpose of analyses of the principal issues facing the users, the users may be divided into three broad segments: the business users, residential users and those referred to as 'special interest' groups. The latter comprise the disabled, the low income group and anyone with special needs. The key concerns of these groups are first identified before suggesting how user groups could contribute in addressing these issues. Those with special needs are dealt with separately in Chapter 18.

19.2.2   Business users

In recent years, quality has become a major area of concern for all business users [1]. In the search for competitiveness, some manufacturers have cut back on factory testing and have sacrificed experienced customer-support staff, so that a growing problem among users is not just a lack of quality, but also a lack of confidence in manufacturers and their products. To an extent this concern has been projected at service providers as they too cut down on their staff strengths. At the same time, technological advances such as digitalisation, common-channel signalling and IP backbone concentrate traffic onto fewer high-volume routes, so that even a minor hardware failure or software bug can result in a widespread loss or degradation of services.

In the past, telecommunications and later informatics, was an important but subsidiary element of most businesses. But their survival has come to rely more and more on information and the corporate network has become a major resource that ensures the flow, integrity and availability of all information. Users recognise that even small deteriorations in network quality can profoundly damage their business; hence, the need to monitor and assure the quality of the telecommunications services they use. In addition, their customers in turn demand high quality services and hence sophisticated internal monitoring is needed to guarantee their own performance levels.

Until recently, quality was regarded mainly as a technical issue, something to be 'built into' the design and manufacturing process, 'assured' by network and system design and dimensioning and 'maintained' by network monitoring and management systems and by competent and alert maintenance teams. The ITU and ISO were seen as the major guardians of telecommunications and informatics service quality, which tended to remain an esoteric affair and the domain of 'quality experts': the user 'called-up' the appropriate standards within the purchasing process and the vendor 'assured conformance' within those standards.

For many years, however, it has been recognised, that quality cannot be 'assured' by standards alone and exigent equipment users, such as military and government agencies, have always demanded vendor certification and regular inspection of the production process to create confidence. Manufacturers have instituted 'total quality' programmes and service providers have implemented sophisticated network management systems to shorten outage times and improve service quality. User-oriented quality assurance programmes have become marketing tools, widely advertised and exploited for product and vendor differentiation.

Business organisations may emphasise different quality indicators because they have different business interests or priorities and some organisations may be more sensitive to certain quality degradations than others. Bit-error rates and outage rates may be tolerated by some users, but may be totally unacceptable to others, depending on their business application of communications.

Often users may have to interact with several different organisations to obtain the services they require, particularly, if the service crosses national boundaries. The users' perception of quality then reflects the 'poorest performer' rather than the 'average performer', that is, 'the weakest link in the chain'. It must be in everyone's interest to eradicate such problems, since perceived quality of telecommunications services is globally damaged by arbitrary and often avoidable regional differences.

Furthermore, the users' perception of quality often goes far beyond technical issues. Although the service provider and the service user are almost certainly organisations, people act as 'agents' for those organisations and interact with one another in the search for quality. Therefore, these 'agents' play roles, both within their own organisations and at service provider/user meetings. While customers, who are bill-paying users, are particularly aware of costs and will tend to focus on individual cost–benefit relationships, providers will tend more to lean towards general technical oriented solutions. Also, subtle interpersonal factors are called into play, such as language and cultural differences that participants may find hard to identify. Customers often find that such problems are much more difficult to resolve than straightforward technical issues. The increasing interdependence and complex interactions encountered in international telecommunications make ad hoc solutions unworkable in the long term. Customers need to do all they can to ensure the quality of the services and equipment they lease or purchase. The prime aim of quality monitoring by users nowadays is not to demonstrate lack of quality to a vendor, but to maintain internal and external, national and international performance of the services used.

In all cases, the measurement of QoS is of fundamental importance. However, it gives rise to serious problems, especially when the results need to be understood, agreed and interpreted by many different actors, including notably the service providers and the service users.

The user groups have expressed concerns on the dominance of major suppliers. These suppliers run the risk of exhibiting intellectual arrogance towards business customers arising from the technical know-how they possess for the provision of new services.

User groups are in the best position to identify the special needs of particular segments of the industry (perhaps by pooling their collective requirements) in addition to dealing with requirements on a user-by-user basis with a provider. The collective voice would strengthen and complement their representation in meeting their needs.

Caution should be exercised by service providers when taking into consideration independently researched findings. There have been cases of inaccuracy in the data. It would be in the interests of providers to establish the credibility of the findings before any action is undertaken.

19.2.3   Residential users

User groups have a significant role in representing the interests of residential users in a variety of contexts. Already apparent are national and international consultative forums on different aspects of policy, mostly associated with regulators.

Several factors combine to suggest that this role will grow:

- Telecommunications is pervading everyday life, with society as a whole moving towards a new 'information society' in which telecommunications plays a vital part.
- The industry as a whole has a strong interest in ensuring that what it offers matches market needs, particularly in the presence of competition.
- Regulators are increasingly conscious of their duties towards consumers.

However, necessary resources for this important job may not yet be in place. At present, residential consumer groups specifically for telecommunications or IT are not numerous. Usually the task falls to a generalist consumer body, which of course has to put forward consumer views on every service and good on the market, or to a telecommunications user group, which tend to be dominated by business users.

Consumer representatives call for a dedicated independent group with adequate expertise and resources to advance the consumer interest in the information society at national level. The resources for such a body are unlikely to be found from membership subscriptions; if it is thought to be in the public interest, it will need public funding. This may be indirect industry funding, for example, from licence fees or an industry levy.

The issues with which such a group will be concerned cannot be expected to fall neatly into any preconceived categories. Users do not perceive the same distinctions as an industry manager, for example, between telecommunications carriage and content, or between quality and pricing issues. Examples of forward-looking consumer concerns found in a recent European survey include [2]

- misuse of personal data;
- inadequate complaints and redress systems;
- insecure personal billing accounts;
- inadequate price indications;
- unclear contracts;
- unclear bills;
- unwanted calls;
- service content harmful or illegal.

At a more immediate level users are already concerned from personal experience about system reliability, security and responsiveness.

User groups are most effective when they can combine the jobs of policy-level input with the day-to-day handling of actual users' problems. This way a feedback loop is formed whereby individual complaints can be seen as part of a larger picture, undesirable developments dealt with at an early stage and a positive contribution made to a more user-friendly future.

19.3 Possible contribution by user groups for the future

19.3.1 General

Due to the specialised nature of needs of various segments of the telecommunication user population, it would be advantageous if these needs were separately identified. In certain countries, for example, in the UK, many user groups exist. For example the National Health Service Telecommunications User Group (NHSTUG) deals with the requirements of the telecommunications service needs unique to hospitals throughout the UK. Other user groups are the UK ISDN User's Forum, Communication Management Association (CMA), Telecommunications User's Association (TUA) and International Telecommunications User Group (INTUG). A European body considering user's interests is ANEC. Brief background notes on INTUG, CMA, TUA and ANEC are given in an Annex at the end of this chapter. Further user group details are given by Macpherson [3]. As the user groups have ready access to member's requirements on quality they are in a better position to ascertain an unbiased and unpressurised response from their members than is possible by service providers. This accessibility, when exploited by the user groups could be used to the advantage of their own members, if necessary, through the regulators. If the user groups take on such responsibility it could put them in a powerful negotiating position.

In the non-business sector the following organisations undertake interests of the residential users; 'Consumer Association' publishes customers' perception of various consumer goods on a monthly basis in the UK. Occasionally this includes telephone quality. Similar organisations exist in other countries, for example, Consumentenbond in the Netherlands. There is also, in the UK, an industry-funded body called Independent Committee for the Supervision of Standards of Telephone Information Services (ICSTIS). Other bodies are National Consumer Council and the Public Utilities Access Forum. Their web sites give further information on their activities. These bodies can collectively influence the service provider. The following are examples of how user groups could provide this support to their members.

19.3.2 Common set of definitions of parameters to express QoS

The lack of an agreed set of definitions for QoS is a disadvantage. These disadvantages affect all users but particularly the business users. The published QoS performance data by the service providers are often couched in terms that mask the real needs of the customer. For example, a recent survey in the UK showed that 95% of the respondents quoted a requirement for repair time of around 1.5 hours or less. However, in the UK the published performance for repair is expressed, by most of the service providers, as $x\%$ of faults cleared in 'y' hours [4]. This does not tell the reader what proportion of the customer's repair needs are satisfied. It merely states how well the service provider's target has been met. In the current age when new services and applications are available more frequently than before, for example, Internet-based services and mobile services, it is particularly important that the user groups identify the real concerns of users and attempt to ensure these concerns are input to the standards and regulatory bodies.

User groups could define their own sets of performance parameters. These could be presented to standardising bodies for inclusion as parameters on which delivered performance ought to be published. The service providers could have their own sets of parameters for their own internal management purposes in addition to these. The initiative to make this happen could come from the user groups.

19.3.3   Standards bodies and user groups

User groups are eligible for membership in international standardising forums such as the ITU and ETSI. Whilst the majority of the membership is from the service and network providers it is feasible for the user groups to initiate the adoption of standards that will be of benefit to them. Such standards ought to be complementary to those that represent the interest of the service and network providers. User groups could play a useful role in identifying new services, studying their implications and contributing towards setting quality standards in collaboration with other industry organisations. An example is the participation of user groups in ETSI. Absence of such participation might result in a poorer set of contributions, dominated by the service provider's or the network provider's interests.

19.3.4   User groups and equipment manufacturers

Members of larger user groups who are sometimes Multinational Companies (MNCs) are aware of the capabilities of equipment manufacturers. Some of these MNCs are in a unique position to influence the development of equipment for measurements on performance parameters. Even though the final development of performance-measuring devices would normally await the specification by standardising bodies and contract specification by service providers, nevertheless user groups have an influence in the development of measuring equipment. It is an opportunity for the user groups to take the initiative to influence an increasing number of service applications and the resulting increase in performance parameters.

19.3.5   User groups and regulators

One of the prime functions of a regulator is to ensure that the interests of users are met fairly by the service providers. To fulfil this function user groups have a responsibility to 'educate' and advise the regulator on quality matters. In the European Union the principal regulator is the European Commission, which usually seeks advice from user groups, service providers and independent consultants before a 'Directive' is issued. Some of the directives which have a QoS content are listed in Chapter 20. Rather than wait for the regulator to seek out the opinions of user groups, a forward thinking group could formulate policies for the best interests of their members for consideration by the regulator.

19.3.6   User groups and industry organisations

Various telecommunications industry organisations exist which cover a range of activities. These range from education, seminars and conferences, studies into new service uses etc. The knowledge of user's quality needs, obtained by the user groups from their members, could be a useful input to these activities.

User groups could influence other industry organisations on matters related to QoS. Such organisations include the government ministries responsible for telecommunications and universities. These organisations could benefit from an understanding of the user needs in order to formulate policy and carry out fundamental research into quality.

19.3.7   Service providers, network providers and user groups

Perhaps the most obvious party to whom the user could relate and provide support is the service provider. The user group could give the provider up-to-date information on what the members expect on QoS. User groups could advise their members on realistic levels of quality that could be achieved by the service providers. In many countries where user groups exist, dialogue between these and the service providers already take place; however, there appears to be scope for a greater involvement in the form of regular feedback to the providers on the levels of quality expected by the members, quality criteria to be defined for new services and selection of criteria for delivered quality on which statistics are regularly published. The relationship between service and network providers and user groups must be on a partnership basis for maximum benefit to all parties. To an extent, the service and network providers must concern themselves with the implications of commercially sensitive information imparted to user groups becoming common knowledge. However, in the current business climate it is considered beneficial to all parties to increase the level of cooperation among these parties to a mutually acceptable level.

19.3.8   User groups and quality related tariff

It is only right that the quality delivered by the service providers bears more than a resemblance to that offered or promised. It is also desirable to move towards an era where quality delivered will be linked to the tariff increases or tariff penalties imposed by the regulator. This point is further discussed in Chapter 20. The user groups could play a useful role in determining their member's opinions. Members have a first-hand knowledge of the revenue lost or at stake to them if the promised quality is not maintained by the service provider. Support may thus be provided to the regulator in establishing the formula for the tariff penalties as well as rewards based on the quality delivered.

19.4   Guidelines for the formation of user groups

The following steps outline the principal steps to be considered in the formation of a new user group.

1   *Identify specialist areas*: User groups function best when the group is a representative number of a uniquely identifiable number of users. For example, the

publishing industry have a unique set of QoS needs that could be different from those of a high street retailer. It would be most useful if such segments of the population are identified. Alternatively, a large user-based group may be formed with sub-groups for different industries.

2 *Nominate representatives*: Representatives who hold offices in various categories of administrative and technical functions are needed to run the user group. Such representations may reflect the industries.

3 *Formulate terms of reference*: A set of rules or procedures for the running of the group to cover all its activities ought to be agreed on and published.

4 *Legal status to be established*: It would be necessary to establish the status of the user group in the eyes of the law. This would give it credence and clout in whatever action it deems necessary and wishes to carry out.

5 *Meetings*: A programme of meetings and activities are to be organised with a clear objective. The output is to be fed to a body that will act on the information.

6 *Alliances*: It would be helpful if the user group forms alliances with international user groups or regional user groups. Additionally its relationships with standards bodies and any other body with whom it could work constructively for the benefit of its members ought to be considered and implemented.

7 *Publicity*: The user group ought to make plans for its publicity both in terms of activities and its achievements. This may be used to increase its membership and its effectiveness in its mission and goal.

19.5   Summary

With increased sophistication in telecommunications, the role of the user groups could grow, from that of merely looking after the interests of their members to that of support and partnership, to service providers, regulators and the industry in general, in a more positive way. Contributions could be made by user groups towards the QoS parameters for Internet-based services and mobile services. Absence of such influence could result in the network provider and service provider imposing their concepts of quality on the customers with the possibility of not optimising customer needs with the network capabilities. Special interest groups' requirements also needs to be identified and channelled to the relevant bodies in order to provide justice to the needs of members of such groups.

19.6   References

1 LEE, A.: 'A user's perspective on quality of service activities in world telecommunications', 1997, *Telektronnikk*, **93**(1), pp. 50–5

2 'The consumer in the information society', report by Ovum, for CEU/DG XXIV, June 1996, supplementary volumes

3 MACPHERSON, A.: 'International telecommunication standards organisations', (Artech, 1990)

4   'Comparable Performance Indicators' – Business customers, published quarterly by Oftel of UK

Web sites

1 INTUG (International Telecommunications User Group)
   http://www.intug.net (also contains coordinates of around 25 user groups from different parts of the world who are affiliated to INTUG)
2 National Consumer Council
   http://www.ncc.org.uk
3 Public Utilities Access Forum
   http://www.puaf.org.uk

Exercises

1   If your country does not have a user group draft a report on how service providers and users will benefit by the formation of a user group for the telecommunications industry. The report should include an analysis of the user population, identifying areas where user representation could be beneficial. Include in the report how to implement the recommendations.
2   If user groups exist in your country, identify areas if any, where representation could be improved. Mention the benefit of any additional legislation and or representation at national, European, regional level or at any other level.

Annex

Notes on user groups

1   INTUG

Full title: International Telecommunications Users Group.
Founded: 1974.
Objectives:
INTUG concerns itself with four major issues:

1   monopoly authority and the rights of users;
2   free access to telecommunications networks;
3   freedom in user choice of equipment and services;
4   constructive cooperation between public authorities and users;

Size: Total spending power of members on telecommunications estimated to be several millions of pounds. Membership of user groups throughout the world and large organisations.

2   CMA

Full title: Communication Management Association (previously known as Telecommunication Manager's Association)

Founded: 1966
Main objectives:

1   education and information exchange;
2   liaison with government and other official bodies such as Oftel, Parliamentary Information Technology Committee, ETSI, OECD and ITU;
3   dialogue with suppliers;
4   support for other Interest and Special Focus Groups.

Participated in Oftel led Comparable Performance Indicators Industry Forum, various joint BT and Mercury Quality Improvement Teams, membership surveys, coordinating consumer input to various official quality activities and is member of the British Quality Association.

Size: Over 1200 members with an annual telecommunications budget (for equipment and services) of the order of a few billions. Membership is for individuals working in large organisations.

3   TUA

Full title: Telecommunications User's Association
Founded: 1965
Objectives and principal activities:

To represent its member's telecommunications interests and requirements in various bodies such as Oftel, EU, suppliers and other relevant bodies.

Size: Membership is in excess of 1000. Estimated telecommunications expenditure is in excess of a billion a year. Membership is for organisations and not for individuals. Many multinationals are members.

4   ANEC (Association Europeene pour la Coordination de la Representation des Consommateurs pur la Normalisation)

ANEC (The European Association for the coordination of Consumer representation in standardisation) was established in 1995. It has forged strong relations with the European standards bodies and other related organisations. ANEC is directly represented on more than 55 Technical Committees of the European standards bodies (CEN, CENELEC AND ETSI) and coordinates input into the work of some 160 national consumer representatives. ANEC is also represented directly on the Board of the European Organisation for Testing and Certification (EOTC). At international level, ANEC has close links with International Organisation for Standardisation (ISO) and International Electro-technical Commission (IEC), both through Consumers International (an associate member of Consumers International) and directly as an observer on the Consumer Policy of Committee of ISO.

Consumers and standardisation: The objective of ANEC is to ensure that the interests of the European consumer are adequately addressed by the standardisation process and by any relevant initiatives dealing with standards directly or indirectly affecting consumers. International standardisation, following the GATT agreement and the cooperation agreements that have been concluded between European and international standardisation bodies, is taking more importance nowadays. ANEC will have a role to play to ensure even closer coordination with Consumers International representatives and consumer representatives on national delegations to ensure that consumer interests are adequately addressed in the case of European projects being drafted at an international level, for example, in ISO or IEC.

ANEC also plays a more dynamic role in transatlantic issues by participating in the Transatlantic Consumer Dialogue set up at the end of 1998.

Developments in standardisation: The process of standardisation is changing; formal standards-setting bodies have increasingly gained importance. In particular, in the area of the information society or services, it has to be recognised that a considerable amount of specifications which are of importance to the European consumer, are being developed outside the formal standards bodies such as CEN/CENCLEC/ETSI. With the increasing importance of the information society and services for the consumer, it is then paramount that the European consumer's interests are adequately addressed in bodies developing such de facto standards. ANEC is, therefore, in principle, open to address such bodies outside the formal standardisation process.

Chapter 20

Role of regulation

20.1   Introduction

Regulators in the telecommunications industry exist in many of the OECD countries and some countries outside it. The regulator's role is principally, to interpret the government's charter for the telecommunications industry. A regulator normally has responsibility for one industry sector in one country. The only exception is the regulatory role of the European Commission where the regulatory aspects apply to any one industry sector in all member countries of the European Union. The role of regulation and its involvement in Quality of Service (QoS) is becoming more complex.

The parties affected by regulation are the users, service providers and the regulators themselves. The management of quality takes on an additional dimension: that of managing the implications of regulation. This chapter looks at some regulatory considerations (some theoretical) of quality, the evolution of quality regulation in the United Kingdom, some of the current issues on quality of service in the USA and a profile of the ideal regulator. Discussions on regulation, in this chapter, have been confined to the relevance to QoS.

20.2   Regulatory considerations

20.2.1   Basis for regulation

For a theoretical analysis of regulation and quality the reader is recommended to the work of Bowdery [1]. The following is a summary of part of his work.

The basic case for the regulation of monopoly is that monopoly power leads to the production of a lower level of output than the optimum. By optimum level is meant the cost of production is minimum and therefore the price to the customer can also be minimum, subject to a standard mark up in the profit. An allocatively efficient level of output is produced when the marginal cost of the output equals its marginal benefit. Applied to quality, the quality is allocatively efficient when its marginal cost equals its marginal benefit. In this chapter and elsewhere in this book the terms allocatively efficient and optimal are used synonymously. In an unregulated market the monopolistic organisation has the following options:

- The restriction of output below the optimum permits the firm to push up the effective price to the customers. This can also lead to higher dividends to the shareholders.
- There is incentive to reduce costs and enhance profits by allowing quality to deteriorate provided that the demand for the firm's product is sufficiently quality inelastic that any loss of revenue following the quality deterioration is outweighed by the cost savings which accrue from it.

The above arguments can also be applied to quality. Optimum economic level of quality or the allocatively efficient level of quality can be offered to customers only when the cost of quality is minimum commensurate with demand and the resources at the disposal of the service provider. Normally the regulator is much more concerned with competitive aspects of telecommunications industry and the proper use of powers of the dominant company in the country. However, quality is increasingly becoming one of the key issues the regulator is concerned with. In the determination of optimum economic level of quality the following difficulties have to be addressed:

1   Under normal circumstances the benefit from the production of an additional unit of a homogeneous product (the marginal benefit), for example, a car, is only received by the marginal consumer (i.e. the consumer who buys the additional unit). However, an improvement in quality can usually be assumed to impact upon all units of output. Thus the benefit from an enhancement in quality (the marginal benefit) of a car includes the valuations of the quality enhancement by existing as well as those of any additional customers who may be persuaded to buy the product because of its improved quality. Therefore, the simplistic approach of a homogenous product, such as a car, cannot be applied when quality is concerned. More work is needed to establish the relationship between quality, price and customer behaviour.
2   Changing tastes and technology make the optimal level of quality a moving target, even for a theoretical product with uni-dimensional quality for which there is undifferentiated demand. Changing technology impacts both upon the nature of quality enhancements available to consumers and upon the marginal cost of achieving a given quality enhancement.
3   Telecommunications quality is not uni-dimensional. It is multi-dimensional and many of the economic models apply to uni-dimensional quality. This complicates analysis.

Despite the above difficulties the concept of optimal level of quality is important and it must be asked how far the various schemes available to the regulator will help to address these difficulties. There are basically three options available to the regulator.

These are

- publication of information upon quality performance;
- service standards and compensation schemes;
- quality sensitive price regulation.

These are examined in the following subsections.

20.2.2 *Publication of information on QoS*

This is the most basic of regulatory instruments in the monitoring of quality from a service provider. It consists of a set of performance parameters on which the provider is expected to publish delivered quality. No targets of performance are specified for any parameter. In theory, the published levels of performance should indicate to the reader the performance attained by the provider. This form of quality data is cheap to publish and therefore easy to enforce by the regulator.

Some of the issues to be addressed in this form of monitoring quality are as follows:

(a) Should the delivered quality data be audited by the regulator?
(b) There is no incentive for the secure and dominant service provider to improve the performance.
(c) The provider can, if necessary concentrate on the published parameters to give a good performance to the detriment of other parameters.
(d) Data could be selectively published to give a misleading picture. For instance, the percentage of calls answered by the directory services within 'x' seconds may be extremely high but not give any indication of the number of calls not answered due to network congestion or due to engaged tone (i.e. not enough lines provided for this service).
(e) Publication of delivered data need not have any direct relationship with the allocatively efficient level of quality.
(f) No attempt is made to compensate aggrieved customers for poor quality.

20.2.3 *Service standards and compensation schemes*

In this form of regulation of quality, standards of performance are set for specified performance parameters and, if not met, providers are liable to pay compensation to customers. The compensation is aimed to address the balance of price/quality trade-off, which had originally been offered to the customer. Another rationale is that the compensation scheme is expected to be an incentive for the provider to meet the individual standards set.

The following issues have to be addressed by the regulator in the formulation of service standards and compensations:

(a) In the assessment of trade-offs, the regulator has to make a judgement of the price–quality trade-offs. With the knowledge available in this aspect of economic analysis the consensus is that such a trade-off cannot be done accurately.

(b) In the setting of standards, the allocatively efficient level of service has to be determined. Any standards set without this consideration would not optimise the efficiency of the company.
(c) Setting of compensation should consider realistic disbenefits to the customer.
(d) Compensation levels should act as an incentive to the provider to meet levels of service promised.
(e) Choice of standards and parameters should reflect customer specific needs.
(f) Exclusion clauses in compensation schemes should reflect local business culture. For example, exclusion of weather conditions on performance could be argued both for and against inclusion in compensation.
(g) Universal service obligations have to be considered.

20.2.4   *Regulator oriented quality sensitive price cap*

In this form of regulation the provider's prices are subject to levels of QoS delivered and other factors affecting the cost of the service to the service provider. The regulation takes the form of limiting permitted price increases to RPI $-x$, where, RPI is the retail price index and x is a factor determined by the regulator. In choosing the x-factor the regulators take into consideration the capital costs, operating costs, and an allowable rate of return on capital. The x-factor is reviewed at intervals and the efficiency gains 'revealed' by the provider can be built into the x-factor for the next period. The net effect is to improve the efficiency of the provider. Ideally, successive x-factor regulation would drive the provider to provide a QoS that is the allocatively efficient level.

Bowdery goes on to state that in the absence of the ability to identify allocatively efficient quality levels, it might be assumed that quality levels are currently suboptimal. He offers the following arguments to support it:

(a) In conversation, a number of regulators and companies have stated that they suspect that current quality levels are sub-optimal, although the degree of confidence with which this is suggested carries across both industries and quality dimensions.
(b) Technological developments and the dynamics of customer expectations might be expected to push optimal levels upwards.
(c) Rovizzi and Thompson [2] examined the belief that publicly owned utilities in the UK had 'gold-plated' and set quality above the allocatively efficient level. They looked at former state monopolies that were subsequently liberalised and concluded that it is difficult to believe that the results are consistent with the systematic overprovision of quality under traditional public ownership. Such findings provide some comfort to the view that even some years after transfer to the private sector quality levels still remain sub-optimal.

Some of the issues to be addressed by the regulator in the study and treatment of x-factor are:

• How to close or curb the loophole whereby the provider can 'beat' the price-cap by carrying out certain improvements only after the next review has been carried out. This would enable the provider to just meet the price-cap requirement of the previous period and not carry out further improvement until the next phase of *x*-factor has been announced.

- Identification of the allocatively efficient level has been found to be fraught with difficulties and the identification of the optimal level of performance of the provider has been almost neglected. This is an area that requires further research, perhaps to be commissioned by the regulator.
- How does the regulator deal with the situation whereby demand is quality insensitive? The provider has no incentive to improve quality.
- The RPI−*x* formula rewards the provider for past performance of quality. This does not deal with current performance levels and does not satisfactorily address the issue of future quality requirements. There is some evidence that regulators are beginning to address this issue.

20.2.5   Customer oriented quality sensitive regulation

As the telecommunications industry becomes more sophisticated, and with the increase of competition both nationally and internationally, service providers will be subject to more pressures. Under these pressures a spectrum of reactions could take place, varying from the emergence of world class suppliers to those that cut corners and offer the poorest service the market will stand. Such poor service may even extend to areas within a nation where there is little or no competition and, therefore, without incentive to provide a higher level of service. It will, therefore, be in the customers' interests for the regulator to ensure that poor QoS is penalised, and provision of good quality is rewarded. It would be necessary for the regulator to establish a mechanism to identify good and poor service in an equitable manner.

One such method has been proposed by Lynch, Buzos and Berg [3]. Lynch *et al.* develop a method of determining a formula for calculating the overall QoS of a telephone company and suggest a method for the reward or punitive measures based on the delivered quality relative to the overall quality target. The authors propose a methodology for the determination of a representative index for the achieved or delivered quality of a service provider. Using techniques associated with the information integration paradigm [4–7] it is possible to determine a mathematical function relating the numerical value of technical measures to perceive overall quality:

$$Q^{\wedge} = f(x_1, x_2, \ldots, x_n)$$

where '$Q^{\wedge}$' is the index of quality, 'f' a function, 'x' the parameter and 'n' the number of parameters.

The authors claim that this system of estimation of quality index has four main advantages. These are as follows:

(1) The estimated model leads to greater consistency in the evaluation of quality, capturing reliable portions of their judgement policies and eliminates the unreliable elements as found by Bowman [8].
(2) Both political considerations and the psychological consequences of information overload are avoided. Thus the mathematical model can be a more valid tool for assessing overall quality than the unaided judgement of the regulator.

(3) This method leads to discussion where everyone can focus on dimensions of quality which are more or less important, rather than on the particular company being evaluated. This conclusion is supported by work by Edwards [9].

(4) It communicates clear and appropriate incentives to the regulated companies that should allow them to raise quality while reducing costs.

A detailed analysis of this work is not undertaken in this book. Readers are recommended to read the work of Lynch *et al.* [3]. According to NRRI report [10] this type of weighted index has two limitations:

(1) The measurement requires that weights be assigned to each service attribute by the regulator. Lack of information on costs and customer preferences, which is necessary for establishing appropriate weights on each attribute, may lead to inefficiencies. In the work of Lynch *et al.* a method to minimise this limitation is by using the expert opinion of the regulators.

(2) The approach did not necessarily encompass all service attributes valued by customers but simply overlooked those unmeasurable by regulators. These excluded quality dimensions provide another potential source for inefficiency. Therefore, the measurement proposal remains susceptible to some of the same informational problems and economic biases characterising the traditional forms of direct quality regulation.

Despite the apparent complexity of this method of determining a 'performance index' to ascertain a fair assessment of a service provider's delivered quality the concept is appealing and further studies may develop this to a more feasible formula for application in local situations in any given country.

20.2.6   *Different price/quality combinations*

Demand for telecommunications services could exhibit differentiated quality demands. In other words, different segments of the customer population could require different levels of quality, to suit their particular needs. This means the service provider has to offer different price/quality combinations. The regulators favour these offerings. However, in order for the regulator to monitor whether the offerings are optimal the following issues have to be addressed:

- Are the different offerings having an unfavourable effect on the standard offerings or the customers who require standard offerings? How are the possible resource misallocations to be identified?
- How does the regulator set standards for the different quality segments?
- How does the differentiation take place? Where does each different segment's boundaries lie?
- How does the regulator set compensation schemes for differentiated quality offerings?
- Differentiated segments could also include the disabled with special quality requirements. The regulator has to identify the needs of such special segments of the customers and address their quality needs to ensure that the provider does meet their quality requirements.

Notwithstanding the above issues to be addressed, this is an area, where, according to Bowdery [1], the regulators are enthusiastic for the providers of regulated industries to provide differentiated price/quality offerings.

20.3 Regulation and QoS in the UK

An industry workshop forum was set up around October 1993, with Oftel acting as a facilitator to arrive at a set of performance indicators in which service providers could report delivered performance. These parameters were measured so that different suppliers could produce comparable data. The first of these data were published by Oftel in spring 1996, for the October–December 1995 quarter and continue to be published for every quarter. Comparable Performance Indicators (CPI), the delivered quality, is published on a six monthly interval with information on quarterly performance for objective measurements, and on a six monthly basis for subjective measures, that is, customer satisfaction measures (see also Chapter 22 and Appendix 7).

Evolution of quality regulation in the UK shows that most, if not all, steps taken have been reactive. There is little evidence of foreseeing problems associated with quality delivered to the customer and attempting to address these.

Regulatory aspects have taken on increasingly more sophistication. Peter Walker [11] discusses the role of regulation in the UK and the proposed OFCOM (Office of Communications) designed to replace five current regulatory bodies including Oftel in the next few years. A bill in the UK Parliament is scheduled for 2003 and the new OFCOM may come into existence perhaps in 2004. It is too early to predict what the nature of the telecommunications part of this body would be, but it is thought that the principal functions of the present Oftel would perhaps be encased in the new organisation. This could result in the present policy on QoS being continued or perhaps becoming under increasing influence of the European Commission in the form of EU Directives.

20.4 QoS issues and regulation in the USA

20.4.1 Background

Both federal and state authorities oversee telecommunications service quality in the United States. The Federal Communications Commission (FCC) has adopted a very much 'hands off' stance on regulating service quality, choosing to emphasise publication of the performance of large companies on key quality indicators. The FCC requires major Incumbent Local Exchange Carriers (ILECs) to report data on switch outages and downtime, trunk blockages, installation and repair intervals, customer complaints and customer satisfaction. The FCC maintains the information in the Automated Reporting Management Information System (ARMIS). Most of the 50 states, on the other hand, extend their reach beyond requiring publication of results [12]. At last count, at least 47 state public utility commissions set standards for regulated companies. Ordinarily the states may impose penalties if a company fails to meet the standards. The National Association of Regulatory Utility Commissioners issued a handbook in 1992 that provides suggested standards, but there is no national requirement that each state adhere to the NARUC standards. Standards set by state commissions do vary, while often staying at or close to the NARUC objectives. Commission standards are in place for most states for installation, operator-handled calls, noise and transmission, call completion, trouble reports, disconnection, billing and outages. A number of states also set standards for payphone operation and for emergency services. Four states (Delaware, Louisiana, Maine and Rhode Island) have customer satisfaction standards.

Before passage of the Telecommunications Act of 1996, which opened telecommunications markets to competition, the Florida Public Service Commission briefly applied an elaborate service quality index to regulated telephone companies. The index comprised 38 separate, weighted quality indicators. The index was touted as a means of providing a consistent single measure that could be used to compare the performance of incumbent local exchange companies in the state or to track trends in company performance [13]. At one time, the Commission was considering requiring the companies to meet the index or face the possibility of enforcement. The index continued to be used for informational purposes but was never tied to performance incentives, whether penalties or rewards. The Florida PSC informally suspended use of the service quality index because of weighting problems when one or more tests were not performed. The Commission was revising service quality rules in 2002 and expected to eliminate the index.[1]

20.4.2   *Current issues – declining levels of local service quality*

US regulators were concerned about decline in the quality of local telephone service in many areas of the country through the late 1990s. A report by the US General Accounting Office (GAO) in 2000 reviewed data on customer complaints and customer satisfaction, finding increases in the former and decreases in the latter [14]. The GAO is the 'watchdog' of the US Congress, charged with investigating federal expenditures and evaluating federal programs.

The GAO study found that complaint levels per 1000 access lines increased between 1996 and 1999 for five of the eight major ILECs included in the analysis. Within ILECs, there was no difference in the complaint level for urban and rural customers, but residential customers reported a higher level of complaints than business customers. For customer satisfaction, the GAO study found variation from company to company, type of customer (residential, small business and large business) and type of service. The GAO did not discern an overall trend for the 1996–99 period but said the data do suggest customers were more dissatisfied with telephone service in 1999 than in 1998.

---

[1] Rick Moses, Bureau Chief, Bureau of Service Quality, Florida Public Service Commission, e-mail communication, April 16, 2002.

20.4.2.1   Analysis of declining local service quality – distinguishing between 'people' quality and 'equipment' quality

Clements, a senior analyst at GAO who participated in the study of decline in quality, proposes a framework for analysing quality in the local telephone industry that distinguishes between two categories of service [15]. 'Equipment and system oriented' service quality improved in the US between 1991 and 1999, concludes Clements, while 'people and process oriented' service, measured by customer complaints, customer satisfaction and other ARMIS measures, declined. Equipment and system oriented service quality consists of activities directly related to the telephone network. People and process oriented service quality consists of activities that support and augment services provided over the telephone network. The paragraphs below are excerpted from his article.

According to Clements, although the frameworks for analysis of service quality proposed by Oodan *et al.* (1997) [16] and Richters and Dvorak (1988) [17] improve upon earlier, less comprehensive definitions of quality, they are inadequate for theoretical or empirical analysis. First, they do not consider the different incentives that motivate a firm's provision of different types of service quality; second, they do not allow for a distinction between objective and subjective quality.

Equipment and system oriented service quality is directly associated with the telephone network. To provide equipment and system oriented service quality, the local telephone company must invest in network assets. Clements classifies these investments into two categories: (1) capacity and reliability and (2) advanced services. Investment in capacity and reliability is necessary to meet existing and future demand for current services. These investments can include redundant systems to facilitate reliability, sufficient switching capacity to meet anticipated demand, and fibre optic cable for capacity and clarity. Investment in advanced services provides consumers with access to emerging services. Digital subscriber line (DSL) technology is an example of investment in advanced services. Thus, equipment and system oriented quality is associated with reliable signal transmission and the availability of advanced services.

People and process oriented service quality involves activities that support and augment services provided over the telephone network. To provide people and process oriented service quality, the local telephone company must invest in staff, both in quantity and training, and internal processes. Clements classifies these functions and services into three categories: (1) pre-service activities, (2) service provision activities, and (3) post-service activities. The pre-service activities include functions and services provided by the local telephone company prior to the consumer using the network. These functions and services can include new consumer information and education, assignment of telephone numbers, and installation of drop lines where necessary. The service provision activities include functions and services provided by the local telephone company that support activities over the network. These functions and services can include education that helps consumers efficiently and effectively use the network and repair of the network to keep it operating. Finally, the post-service activities include functions and services provided by the local telephone company after the consumer has used the network. These activities include functions and services related to consumer account maintenance. For each category of activity, the local telephone company must maintain a sufficient number of trained engineers, operations (e.g. operator, directory assistance), and support (e.g. billing) staff as well as well-designed internal processes.

The distinction between equipment and system oriented service quality and people and process oriented service quality is important because the economic incentives to provide quality are different for each. The equipment and system oriented service quality requires a long-term investment in physical assets. At the same time, people and process oriented service quality requires a shorter term investment in people and internal process. Since the nature of the investment is different, the company's response to regulation or competitive entry through the provision of quality could be different. In the economics literature, Averch and Johnson [18] have shown that rate-of-return regulation can induce an over investment in long-term physical assets. Also, investment in long-term physical assets can have strategic purposes for blockading, deterring or accommodating entry (Tirole 1988) [19]. When the investments are short-term and the assets can leave without the company's control, the company's incentives can be different. While a complete definition of quality is important, theoretical and empirical analysis requires an understanding of the differences. These differences can provide insights on the mixed performance on quality.

The second component of Clements' framework is the nature of the evaluation of the QoS. The evaluation of the QoS can be either objective or subjective. The evaluation is objective when informed parties can agree on the measure and the relative performance while the evaluation is subjective when informed parties can disagree on the measure and the relative performance. The distinction between objective and subjective evaluation is important because existing paradigms provide no clear guidance. The paradigms of economics and operations research suggest that quality is objective, while the marketing-based paradigm suggests that quality is subjective. By considering both the objective and subjective aspects of quality, Clements' framework incorporates aspects from each of the major paradigms.

Table 20.1 illustrates the framework. In the first row, the equipment and system oriented service quality can be either objective or subjective. Examples of equipment and system oriented service quality measures include the percentage of access lines served by digital switches (objective) and consumers' satisfaction with voice clarity over the network (subjective). Similarly, the people and process oriented service

Table 20.1   Framework for local telephone quality-of-service

Nature of quality	Objective evaluation	Subjective evaluation
Equipment and system oriented	e.g. percent of access lines served by digital switches	e.g. consumers' satisfaction with voice clarity
People and process oriented	e.g. average installation interval	e.g. consumers' satisfaction with operator response time

quality can be either objective or subjective. Examples of people and process oriented service quality measures include the average installation interval (objective) and consumers' satisfaction with operator response times (subjective).

There are two advantages to this framework. First, the framework is not narrow. It incorporates the many aspects of local telephone service and definitions of quality. This is consistent with the expanding notion of quality in the telecommunication literature, as developed by Richters and Dvorak [17] and Oodan *et al.* [16]. Second, the framework recognises the different incentives to provide QoS in the local telephone environment. To understand QoS in the local telephone industry, a framework should incorporate the different incentives to invest in long-term physical assets versus short-term mobile assets. Just because a company invests heavily in network facilities and improves equipment and system oriented service quality does not imply that the company will invest in staff and training. Alternatively, a company with poor equipment and system oriented service quality may invest heavily in staff and training. In general, investment in one type of quality does not imply investment in the other. Thus, the framework can provide insights into empirical trends in local telephone QoS.

There are two related questions that motivate the empirical analysis. First, are there any empirical differences between equipment and system oriented quality measures and people and process oriented quality measures? Second, are there any empirical consistencies within these two categories of quality? These two questions will shed light on the applicability of the framework discussed in the previous section. To address these two questions, Clements (1) examines US trends in local telephone QoS for the period 1991–2000 and (2) examines the relationship between equipment and system oriented quality measures, and people and process oriented quality measures.

For the empirical analysis, Clements employs ARMIS data for the ILEC in each state, excluding Alaska. For the QoS and infrastructure investment reports in ARMIS, the unit of analysis is the local telephone company's operation in a state. Local telephone companies that meet ARMIS reporting requirements (revenues exceeding $114 million per year or price cap regulation) file annual QoS and infrastructure investment.

Clements found that the trends in local telephone QoS were consistent with his framework. All five equipment and system oriented quality measures considered are consistent with improved quality. At the same time, all seven people and process oriented quality measures are consistent with declining quality. These trends indicate that the framework may be applicable for theoretical and empirical analysis of the local telephone industry. First, there are consistent empirical differences between company performance on equipment and system oriented quality measures and people and process oriented quality measures. Second, there are empirical consistencies within the equipment and system oriented quality category and people and process oriented quality category.

ARMIS QoS and infrastructure investment data for the period 1991–2000 are consistent with improved equipment and system oriented quality. These data are consistent with telephone companies investing in networks, both for capacity and reliability and for advanced services. Thus, equipment and system oriented quality is improving.

ARMIS QoS data for the period 1991–2000, on the other hand, are consistent with a general decline in people and process oriented quality. Of the seven measures of people and process oriented quality, all are consistent with lower quality in 2000 than in either 1991 or 1996.

The relationship between infrastructure investment and people and process oriented quality measures and equipment and system oriented quality measures provides additional support for the framework. Clements compares performance on three people and process oriented quality measures and one equipment and system oriented quality measure at different quartiles of infrastructure investment. The results indicate that higher levels of infrastructure investment are consistently associated with lower levels of people and process oriented quality. This implies that there is an empirical difference between equipment and system oriented quality and people and process oriented quality. In addition, the results indicate that higher levels of infrastructure investment are generally associated with higher levels of equipment and system oriented quality. This implies that there is an empirical consistency within the equipment and system oriented quality category.

What is the current status of QoS in the US local telephone industry? This is an increasingly difficult question to answer with the evolving and expanding notion of quality. The answer most likely is 'it depends'. Some measures of quality show improvement while others show deterioration. Clements' framework provides insights on this inconsistent pattern. Based on the empirical results from 1991 to 2000, equipment and system oriented quality is improving while people and process oriented quality is deteriorating.

20.4.3   *Current issues – quality of DSL service*

A keystone of federal policy on the Internet has been to let it alone, allowing competition and use to develop with the least possible government intervention. Access to the Internet is provided by local exchange carriers, cable companies, wireless providers and satellite providers, all traditionally subject to different regulatory regimes. The incumbent local exchange carriers, in particular, which are the providers of DSL service, are subject to state regulation of local telephone rates as well as service quality. Both Congress and the FCC in 2002 took steps that would remove the states from regulation of DSL with the goal of making it easier for companies to deploy broadband capabilities. In Congress, a bill to bar both the FCC and the states from regulating any high-speed data, Internet access or Internet backbone service passed the House of Representatives but was not expected to pass the Senate and become law in 2002. The FCC, however, was moving ahead with a rulemaking proceeding to classify ILEC high-speed services as information services rather than telecommunications services, which could take them out of state regulation on both the wholesale and retail sides [20]. The policy is consistent with the historic policy of light regulation of the underpinnings of the Internet. Most states do not impose much regulation on retail advanced telecommunications services. They do have responsibility under the Telecommunications Act of 1996 for encouraging competition through interconnection agreements between incumbents and entrants. Classification of high-speed

Internet access as an information service rather than a telecommunications service could mean the incumbents are no longer subject to the unbundling, interconnection and collocation requirements of the Act.

Many DSL customers are not satisfied with the service provided by the incumbent telephone companies. In a survey of Internet service quality in the fall of 2001 that resulted in over 14,000 responses, the NRRI and BIGresearch found that 47.0 per cent of the 1138 DSL users had complained to the company providing them with Internet service about the quality of their service. Of those, 63.3 per cent complained about outages or service interruptions; 31.2 per cent about data rate speed; 40.9 per cent about technical support, 27.5 per cent about installation and 28.8 per cent about billing [21]. Thirty per cent of the respondents said they were dissatisfied or very dissatisfied with the company response to the problem. Twenty-nine per cent reported that they had complained to the state agency that regulates telephone service.

Most US states regulate DSL lightly. In a survey of 38 states by the National Regulatory Research Institute in 2001, 21 state regulatory commissions reported that they tariff advanced services but only seven reported that they set rates. Fourteen state commissions reported that they have regulatory authority over service quality issues involving DSL as part of their general authority over service quality of local exchange carriers. Faced with customer complaints that come to them because of their existing regulation of the incumbents, at least two states, California and Florida, were proceeding down a road towards active regulation of service quality for DSL providers. In April 2002, the California Public Utilities Commission found that it has jurisdiction over the quality of DSL service provided by SBC, the Bell company incumbent serving California [22]. The Commission noted the FCC notice of proposed rulemaking that would classify wireline broadband Internet access services as information services, but said the NPRM and the FCC adoption of tentative conclusions on DSL services did not affect the PUC ruling.

Florida has explicit statutory authority over quality for advanced telecommunications services. The Florida Public Service Commission has been considering adoption of a tariff for DSL service largely because of customer complaints on installation and repair intervals [23].

DSL in the US has competition, in the form of cable modem service which is diffusing to close to the same demographic and geographic population at about the same pace. Cable modem service, according to the NRRI/BIGresearch survey, has a different profile of user concerns than DSL. Of the 2511 respondents to the survey using cable modems to reach the Internet, 50.7 per cent had complained to the company providing them Internet service. Of those, 73.4 per cent reported problems with outages and service interruptions, more than for DSL; many fewer than for DSL complained about installation and billing.

Given adequate information, customers will choose the service that provides the quality they want. Yet if the FCC, by classifying high-speed Internet services as information services rather than telecommunications services, frees incumbent DSL providers from requirements on the wholesale side that enable Internet service providers not affiliated with the ILEC to interconnect with the ILEC networks, consumers will have less choice of DSL providers. Thus, within the DSL market, they will not have choices of either quality or price. Removal of regulatory barriers in the near term, by either Congressional or FCC action, could inhibit development of a robust, wide array of service providers in the long term.

20.5 QoS issues and regulation in European Union member countries

The main elements of the philosophy of the European Commission, the body responsible for issuing directives for the member countries are:

- harmonised QoS parameters and measurement methods (but not harmonised targets) for services covered by universal service obligations;
- encouragement for users and suppliers to work together (self regulation coregulation) to improve performance, with strong reserve powers for regulators if suppliers do not deliver the performance users need.

In 2002, the European Parliament and the Council adopted five directives that extends the regulatory framework for telecommunications to cover all forms of electronic communications infrastructure, including cable networks, satellite networks, networks used for broadcast transmission, IP networks, power line communications systems, as well as the traditional fixed and mobile networks used for voice or data. This adaptation of the regulatory framework was made in response to the convergence of technology, which increasingly makes it possible for all forms of content to be delivered over all types of networks.

The five directives are

- Directive 2002/21/EC of the European Parliament and of the Council on a common regulatory framework for electronic communications networks and services (Framework Directive);
- Directive 2002/20/EC of the European Parliament and of the Council on the authorisation of electronic communications networks and services (Authorisation Directive);
- Directive 2002/19/EC of the European Parliament and of the Council on access to, and interconnection of, electronic communications networks and associated facilities (Access Directive);
- Directive 2002/22/EC of the European Parliament and of the Council on universal service and users' rights relating to electronic communications networks and services (Universal Service Directive);
- Directive [. . .] of the European Parliament and of the Council concerning the processing of personal data and the protection of privacy in the electronic communications sector *(to be adopted – Directive number will be assigned at the time of adoption)*.

The new framework becomes effective from mid-2003. At that time, the new directives replace the previous directives in this area, namely the leased lines directive 92/44/EEC, the voice telephony directive 98/10/EC, the authorisation directive

97/13/EC, the interconnection directive 97/33/EC and the telecoms data protection directive 97/66/EC.

Formally there are two recommendations, one on ISDN and one on Packet Switched Data services (as described in the first edition of this book) that remain on the statute books, but they have little relevance today.

A recommendation is not legally binding to the member countries of the EU, whereas a directive is.

For QoS issues, the Framework Directive and the Universal Service Directive are most relevant.

20.5.1   Framework Directive

The Framework Directive sets out among other things the objectives for national regulatory authorities. The three top level objectives are to:

(1) promote competition in the provision of networks and services;
(2) contribute to the development of the internal European market;
(3) promote the interests of European citizens.

These top-level objectives contain a set of more specific objectives, which touch on consumer quality aspects, including:

- ensuring that users, including disabled users, derive maximum benefit in terms of choice, price and quality;
- ensuring a high level of protection for consumers in their dealings with suppliers, in particular by ensuring the availability of simple and inexpensive dispute resolution procedures carried out by a body that is independent of the parties involved;
- contributing to ensuring a high level of protection of personal data and privacy;
- promoting the provision of clear information, in particular requiring transparency of tariffs and conditions for using publicly available electronic communications services;
- addressing the needs of specific social groups, in particular disabled users.

20.5.2   Universal Service and User's Rights Directive

Universal service provision

The aim of this directive is to 'to ensure the availability throughout the Community of good quality publicly available services through effective competition and choice and to deal with circumstances in which the needs of end-users are not satisfactorily met by the market.' The Directive defines the minimum set of services of specified quality to which all end-users have access. This Directive also sets out obligations with regard to the provision of certain mandatory services such as the retail provision of leased lines (Article 1).

Internet access

The universal service obligation covers the provision of a connection to the network 'capable of allowing end-users to make and receive local, national and international telephone calls, facsimile communications and data communications, at data rates that are sufficient to permit functional Internet access, taking into account prevailing technologies used by the majority of subscribers and technological feasibility' (Article 4). Functional Internet access is explained in the associated recital as implying data rates up to 56 kbit/s.

QoS obligations for operators with universal service obligations

See Section 22.4.2 for the European Union requirements for QoS reporting.

20.6   Future of regulation

20.6.1   Global regulation

Early telecommunication regulation was mainly confined to domestic boundaries because service providers provided international services on a correspondent basis with bilateral agreements on delivery charges between countries. However, the recent emergence of regional regulators, the prime example being the EU, are increasingly influencing the domestic scene. Another example is the emerging pan American free trade area, commencing with the US–Canada–Mexico agreement. An integrated approach to telecommunications is starting with the Acapulco Declaration, agreed in 1992, which empowers the Organisation of American States to promote regional telecommunications development and integration, harmonisation on the region's standards and a greater uniformity of regulatory policies and pricing principles. The evolving Asia Pacific region is also one where economic relationships are intensifying and telecommunications cooperation amongst the ASEAN countries (Thailand, Indonesia, Singapore, Malaysia and the Philippines) is likely.

Additionally, any service provider who wishes to compete globally, whether to increase its revenue or protect its revenue from its multinational customers by meeting their needs on a worldwide basis will need to understand the regulatory framework in those countries in which it wishes to operate. Also, the impact of competition and domestic regulation is eroding the symmetry of international tariffs and this is being exploited by new third-party entrants who offer cheap calls by call-back from the country with the lowest tariff. This is obviously distortion of the traffic balance between countries.

20.6.2   World Trade Organisation

With the rapid developments in applications of technology and the global nature of telecommunications there is a need for a global regulator of some sort. The ITU-T is a standardising body and has no powers of regulation. In this context it is perhaps worth considering the role of the World Trade Organisation (previously known as

GATT) and its possible involvement in the future on the regulation of global services, for example, that of Internet.

GATT negotiations commenced in 1947 and made a significant contribution to the rapid economic growth in the 1950s and 1960s. Since the UK became a member of the EU it no longer negotiates in the GATT on its own behalf; in matters of foreign trade, member countries have pooled their sovereignty, and the Commission of the European Community negotiates on behalf of all member countries under a mandate from the Council of Ministers. This requires the Community line to be negotiated first but when a common position is reached, the EU is, with the USA and Japan, one of the most influential parties.

The GATT negotiations, called the Uruguay Round, were launched in Punda del Este in 1986 and have on a number of occasions seemed near collapse. Significantly, trade in telecommunications service was included in the Uruguay Round (the Group of Negotiations on Services (GNS) telecommunications annex) as an extension of the key GATT principle of 'Most Favoured Nation (MFN) Status'. Under this concept, a country concluding a bilateral agreement with a favoured nation, giving access to the domestic market, agrees to allow any GATT member access on the same terms. Led by the USA, the more liberalised countries blocked this initiative. This caused the FCC to defer a decision on C&W's application for licence to operate international services from the USA to the UK, on the grounds that it would weaken the USA's negotiating position in GATT.

The April 1996 round of negotiations failed to reach consensus when a group of countries lead by the USA insisted on keeping mobile satellite systems outside the scope of the agreement on liberalisation. However, in February of 1997, a new basic services liberalisation pact was agreed upon, where about 70 Governments have, in principle, consented to liberalise facilities-based access to their basic services markets, including foreign-company equity participation of up to 100 per cent. The USA, Europe and Japan (accounting for more than 70 per cent of global revenues) will open their markets in January 1998. Other countries will follow in the period up to 2003. However, a number of countries are seeking exemptions to the MFN principle, for example, the USA for satellite transmission of DBS television services; Brazil to the distribution of radio or television programming direct to consumers; Argentina for fixed satellite services; Bangladesh, India Pakistan, Sri Lanka and Turkey for differential measure, for example, accounting rates, in bilateral agreements with other operators or countries. The agreement will also result in the establishment of independent regulatory bodies, in signatory countries, to foster fair competition.

20.7   Profile of an ideal regulator's role on matters of quality

Based on earlier discussions one can derive the role profile of an ideal regulator. A regulator should fulfil the best interests of the customers, suppliers, equipment manufacturers, the government and other relevant bodies, national and international. As the telecommunications industry is still in the growth stage of the 'product life cycle', the profile of an ideal regulator in a few years' time could be vastly different from that of today. However, some of the general requirements and their obligations towards QoS can be confidently postulated. Such a profile would include the following characteristics:

- The regulator would be a facilitator in the telecommunications industry in relation to the following:
  - awareness of the quality implications of technological advances and new services;
  - liaison with standardising bodies for the development of standards for services at the right time, that is, influencing the prioritising of standards particularly for the customers' benefit;
  - management of assessment of service providers' delivered quality data;
  - development of an overall strategy for matters on quality;
  - developing and applying a formula for reward or punitive action of service providers based on supply of good and poor QoS.
- The regulators would adopt best practice from other regulators not only in telecommunications from other countries, but also from counterparts in other industries both at home and abroad.

A successful regulator would keep up to date on technical advances in telecommunications not only in its home country but in the rest of the world. It would also keep itself up to date on the implications on services. It would assess its implications on the telecommunications industry and, in particular, to the home country. It would try to forecast the implications on QoS and initiate early studies to assess if any action needs to be taken.

In the standards fora it would attempt to have an input based on the country's needs. It would encourage development or right standards at the right time. It would have to liaise with other regulators if it is to have any control at all, otherwise the control will be in the hands of the service providers, who form the majority of the membership of the standards bodies.

The regulator would play a proactive role in the selection of parameters on which delivered quality is to be published by the providers. The regulator could also assist in their definitions so that those who work with these understand the parameters unambiguously. The principal initiators of these parameters should normally be the user groups, as discussed in Chapter 14. However in certain cases, for example, residential and special interest customers, regulators may have to act on their behalf, as there may not always be user groups to represent their interests.

The regulator has to ensure that the performance assessments of the service providers are specified and audited. Performance assessment systems may include publication of results of delivered quality, compensation schemes, quality-sensitive price caps and differentiated price/quality offerings.

The overall approach of the regulator on matters of quality would be to attain maximum improvement in the quality of life for the people through the application of telecommunications. Regulators could consider commissioning studies into topics such as quality-sensitive price caps and identification of the allocative efficient level of quality. The regulator should take the responsibility to ensure that in this highly competitive industry, customers do not get the quality that represents the 'lowest common denominator' the market will withstand, but the best value technology can offer to customers.

An ideal regulator would liaise with other telecommunications regulators in other parts of the world. It would also liaise with regulators from other industries both from home and abroad. It will pick out the best practices and translate these to relevant applications in the home country. If any of these practices cannot be applied within the existing framework it would put these forward to the government (which is assumed to be the body responsible for the 'citizens charter') for consideration at the next review of the regulatory framework.

20.8   Summary

Regulators in the telecommunications industry can play a useful role in ensuring that QoS delivered by the service providers are optimal in the interests of the users and customers. Various methods can be used, for example, enforced publication of quality delivered, compensation schemes, carrot and stick approach etc. In some countries there is the tendency for the regulator to back off, leaving the market forces to ensure equilibrium is reached. However, it is debatable if abandoning to the market forces will ensure optimal quality–price relationship to the customer. This, together with the fact that telecommunications is very dynamic and rapidly changing, requires the role of the regulator to be reviewed regularly. Meeting this challenge is also the responsibility of professionals working in the telecommunications industry, perhaps by influencing the role and effectiveness of the regulator.

20.9   References

1   BOWDERY, J.: 'Quality regulation and the regulated industries', 1994, Discussion paper, Centre for the study of regulated industries (CRI), Research centre of the Chartered Institute of Public Finance and Accountancy (CIPFA)
2   ROVIZZI, L. and THOMPSON, D.: 'Price-cap regulated public utilities and quality regulation in the UK', London Business School, Centre for Business Strategy Working Paper, No. 111, November 1991; 'The regulation of product quality in the public utilities and the citizen's charter', *Fiscal Studies*, 1992, **13** (3), pp. 84–5
3   LYNCH, G. J. Jr, BUZAS, T. E. and BERG, S. V.: 'Regulatory measurement and evaluation of telephone service quality', *Management Science*, 1994, **40**, pp. 169–94
4   ANDERSON, N. H.: 'Methods of information integration theory' (Academic Press, New York, 1982)
5   LOUVIERE, J. J.: 'Hierarchical information integration: A new method for the design and the analysis of complex multiattribute judgement problems,' in T. Kinear, (Ed.): 'Advances in consumer research' (Association for Consumer Research, Provo, UT, 1984, vol. 11)

6   LOUVIERE, J. J. and GAETH G. J.: 'Decomposing the determinants of retail facility choice using the method of hierarchical information integration: A supermarket illustration', *Journal of Retailing*, 1987, **63**, pp. 25–8.

7   LYNCH, J. G., Jr.: 'Uniqueness issues in the decompositional modelling of multiattribute overall evaluations: An information integration perspective', *Journal of Marketing Research*, 1985, **22**, pp. 1–19

8   BOWMAN, E. H.: 'Consistency and optimality in managerial decision making', *Management Science*, 1963, **9** (2), pp. 310–21

9   EDWARDS, E.: 'How to use multiattribute utility measurement for social decision making', *IEEE Trans. on Systems, Man. and Cybernetics*, 1977, **7**, pp. 326–40

10 Telecommunications service quality, The National Regulatory Research Institute, Report NRRI 96-11, The Ohio State University, Columbus, Ohio, USA

11 WALKER, P.: 'www.regulation: the why, what and whither of communication regulation', *Electronics and Communication Engineering Journal*, December 2001

12 A compilation of state standards prepared by Lilia Perez-Chavolla of the National Regulatory Research Institute is available on the NRRI web site www.nrri.ohio-state.edu/publications

13 BERG, S. V.: 'A new index of telephone service quality: academic and regulatory review', in W. Lehr, (Ed.), 'Quality and reliability of telecommunications infrastructure' (Mahwah, NJ: Lawrence Erlbaum Associates, 1995)

14 US General Accounting Office, 'Telecommunications: issues related to local telephone service' (Washington DC: GAO, 2000)

15 CLEMENTS, M. E.: 'Local telephone quality of service: a framework and empirical results', Paper presented to the NARUC Committee on Consumer Affairs, 17, July 2001

16 OODAN, A. P., WARD, K. E. and MULLEE, A. W.: 'Quality of Service in Telecommunications', IEE, 1997

17 RICHTERS, J. S. and DVORAK, C. A.: 'A framework for defining the quality of communications services', IEEE Communications Magazine, October 1988

18 AVERCH, H. and JOHNSON, L.: 'Behaviour of the Firm under Regulatory Constraint', *American Economic Review*, 1962, **52**, pp. 219–28

19 TIROLE, J.: 'The theory of industrial organization' (Cambridge, MA: The MIT Press, 1988). GREENSTEIN, McMASTER, and SPILLER (1995) and Lee (1997) examine the incentives of local telephone companies to invest in modern infrastructure. By investing in modern infrastructure, the local telephone company can be interpreted as improving the equipment and system oriented quality. GREENSTEIN, Shane, Susan McMASTER, and Pablo SPILLER. 'The Effects of Incentive Regulation on Infrastructure Modernization: Local Exchange Companies' Deployment of Digital Technology', *Journal of Economics and Management Strategy*, 1995, **4**, 187–236. LEE, SANGJIN. 'The Impact of Alternative Regulation of Telecommunications Infrastructure Deployment: The Case of Price Cap Regulation.' Ph.D. diss., The Ohio State University, 1997

20   Federal Communications Commission, 'Appropriate Framework for Broadband Access to the Internet over Wireline Facilities', Notice of Proposed Rulemaking, CC Docket 02-33, CC Dockets 95-20 and 98-10, 15 February 2002

21   Sample results from the survey are available at www.nrri.ohio-state.edu/whatsnew/ The survey methodology controlled for representation of the US population by age and sex. Since respondents were self-selected, a disproportionate number may have had complaints

22   California Public Utilities Commission, 'Assigned Commissioner's and Administrative Law Judge's Ruling Denying Defendants' Motion to Dismiss', *California ISP Association* v. *Pacific Bell Telephone Company, SBC Advanced Solutions and Does 1-20*, case 01-07027, 28 March, 2002

23   Florida Public Service Commission, 'Intrastate Tariffing of xDSL Service by BellSouth Telecom, Verizon Florida and Sprint Florida', Docket 001332-TL, Agenda Conference, 6 February 2001. Comments of Commission Chair Leon Jacobs

Further reading

BARNES, F.: 'Quality regulation: the UK experience of regulating BT', *Consumer Policy Review*, 1992, **2**

HOGBEN, D.: 'Telecommunications regulation in the UK', Structured Information Program 16.1, British Telecommunications Engineering, *Journal of the Institution of British Telecommunications Engineers*

SPENCE, M. A.: 'Monopoly, quality and regulation', *Bell Journal of Economics*, 1975, **6**, pp. 417–29

VICKERS, J. and YARROW, G.: 'Privatisation: An economic analysis' (MIT, 1988)

Chapter 21

Role of standards

21.1 Introduction

Internationally accepted standards enable parties involved in the provision and maintenance of telecommunication networks and services such as the service and network providers and other pertinent bodies in the telecommunications industry, that is, manufacturers, users, regulators and research organisations, to benefit. Specifically the benefits related to Quality of Service (QoS) management are

- providers can use definitions of performance to declare to the public levels of performance that are planned and delivered;
- these could facilitate specification of desired levels of performance where these are considered necessary;
- form a basis for users specifying unambiguously service level agreements where quality and performance levels may be specified;
- specification of interconnect performance at the interfaces between different networks (these may be between competing operators or between different service interfaces in the same business);
- facilitates manufacturers to produce telecommunications equipment that can be used worldwide, thereby reducing manufacturing costs and facilitating delivery of services.

Due to the global nature of telecommunications (indeed this is the only piece of engineering which has a physical link around the world – cable) it is necessary for internationally accepted standards to be produced and applied by all nations. In this chapter the principal beneficiaries are identified, a review of the standards bodies is carried out, their contributions and shortcomings identified, a methodology is proposed for systematic study of QoS issues followed by an illustration of the use of such a methodology for the legacy POTS service. A case for an architectural framework for the study and management of QoS is proposed. This should enable a more coherent study of QoS issues, something that is lacking at present.

21.2   Beneficiaries of standards

21.2.1   General

The benefits of standards in relation to QoS in telecommunication may be discussed under the following headings:

- network providers;
- service providers;
- manufacturers;
- users/customers;
- regulators.

21.2.2   Standards and network providers

Network providers benefit from national and international standards on network technical performance. Standards could assist network providers to deploy systems faster and also free them from being locked into a single manufacturer. It is important that the standards for a new service or application or technology are published at the right time. Publish too soon and it is subject to costly amendments. Publish too late and it becomes ineffective.

Many major network providers have interfaces with networks of foreign service providers or network providers' administrations for the provision of international services. In order to specify the performances at interfaces and the projected end-to-end performance, it is necessary to agree on the performance provided by each national portion of the connection plus the international part. This task will be made much easier with internationally-agreed performance standards.

Standards are necessary for the interworking among network providers within a country. In the UK there are, at present, two major network providers, BT and Cable and Wireless Communications (CWC) and many smaller providers. Also in the UK an increasing number of cable TV operators and cellular radio operators have appeared on the scene. Standards could become a necessary means to ensure satisfactory end-to-end network performance for inter-networked services.

Standards will be useful in specifying SLAs with major customers. Targets could then be negotiated for improvement by mutual agreement between the network provider and the customer.

21.2.3   Standards and service providers

In cases where the network provider is not the service provider, nationally or internationally recognised standards will enable both parties to form the basis of contract for performance levels. If a network provider provides the network and a service provider has responsibility for the provision of mobile phone service, agreed performance parameter definitions and standards will provide an easier and more manageable service agreement between network provider and service provider.

21.2.4   Standards and manufacturers

Recognised international standards enable manufacturers to develop equipment which can be sold throughout the world and will interwork with the equipment of other manufacturers. Well known examples of these are, Signalling System No. 7, and GSM for mobile telephony. In the absence of standards, these two areas of telecommunications would not have penetrated and developed to the current extent. Manufacturers can assist in the identification of areas where such standards are to be developed and the practicality of embodying them in equipment design. Perhaps they could contribute by leading discussions in this area in the international standards forum.

21.2.5   Standards and users

Multinational companies may require standard definitions and values of performance to enable them to compare offerings from various service providers. A multinational company with private network services to its sites in different parts of the world will benefit from standards in their estimation of end-to-end performance of QoS. Large customers may also wish to connect their private networks with public networks and specifications of interface performances would facilitate easier SLA agreements with network providers. On a national level, standard definitions for performances will enable users to compare performance offerings of various service providers.

Service providers who do not cooperate in the provision of performance data to internationally agreed parameters, or who obstruct the creation of such parameters will be seen to be inward looking and protectionist and not really interested in catering for the benefits of the customer. Such attitudes are to be discouraged, not only from the customers' viewpoint but also for the good of the service provider, as customer confidence and loyalty will decline in the long run.

21.2.6   Standards and regulators

Regulators usually have a mandate to look after the customers' interests. In order to enable them to carry out this task it is necessary for standards bodies to define clearly performance parameters for the principal services on a service-by-service basis and any performance standards that may apply. Regulators also have an important role to play in stimulating the creation of standards for the principal services and in ensuring that the service providers publish achieved performance results according to these defined standards. The role of regulators in the management of QoS is dealt with in more detail in Chapter 20.

21.3   Review of standards bodies

21.3.1   General

There are three layers of standards bodies: international, regional and national. The only international body responsible for the telecommunications standards is the ITU-T. Examples of regional bodies are European Telecommunications Standards

Institute (ETSI), American National Standards Institute (ANSI) and others. The national bodies are usually affiliated to the ISO. In the UK it is known as the British Standards Institution. For a list of standards bodies see Macpherson [1].

21.3.2   *International standards bodies*

The International Standards Organisation (ISO) was founded in 1947 by 25 national standards bodies in a conference held in 1946 in London. Its objective is to encourage standards, which will facilitate exchange of goods and services throughout the world. It works through the national standardising bodies such as the BSI in the UK. In areas close to telecommunications, such as information technology, they liaise with the International Eletrotechnical Commission (IEC), which also has a similar interest, in the development of standards to avoid duplication of effort. Such standards are published by ITU-T as one of its recommendations with the logo of ISO and IEC to indicate their participation. The ISO publishes the same standard in their own standards numbering scheme. The IEC was formed in 1904, to facilitate the coordination of standards in eletrotechnology. Today their spheres of activities cover telecommunications, electronics and electrical and nuclear energy. They liaise with the ISO in the development of standards in telecommunications.

Without the output from ITU-T, international telecommunications would not be possible. However, an examination of the recommendations and the work in hand (see questions under various study groups in Tables A21.1–A21.2) result in the following observations:

(i)   Most of the QoS related standards have been developed primarily to meet the technical performance needs in the planning of a network and for interworking with other networks. However, some recommendations (notably the P series) were developed considering customers' responses to transmission quality; this includes a subjective element.

(ii)   QoS studies have been mostly stand-alone and do not form part of an architectural framework. An examination of the QoS related recommendations indicates that many of the studies have a specific application and are not related to other recommendations. While these studies have been useful in their own right, the maximum benefit that might have accrued with the use of an architectural framework is missing. Absence of an architectural framework for the study of QoS has resulted in ad hoc and individual studies.

(iii)   There is a lack of agreed definitions on user related QoS parameters, which would be of benefit to both customers and service providers. For example, in the case of basic telephony there is an absence of performance parameter definitions. This makes it impossible for the performances of service providers of one country to be compared with those of another country (see Chapter 22). A selected number of end-to-end QoS parameters, for the principal services, will not only benefit customers' requirements but will also be of benefit to the service providers.

(iv)   There is little or no mention of the cost of quality, nor any attempt to study in detail the concept of optimal level of quality. Optimal level of quality is governed by the customers' needs, service providers' infrastructure, cost of quality components, and the price customers are willing to pay.

(v) There is no separate section (or series) for recommendations on QoS. Those relating to QoS are distributed in various recommendation series.

The classification of recommendations has evolved over the years. Recommendations series are sometimes purely functional (e.g. D Series on tariffs), sometimes part functional and part service (e.g. R Series on telegraph transmission) and sometimes certain activities are grouped together (L Series on construction, installation and protection of cable and other elements of outside plant). A series dedicated to QoS would, in the future, assist in focusing the issues related to quality and therefore should be considered by the ITU-T.

Perhaps ITU-T could adapt to the increasing pressures and demands of the telecommunications industry by taking note of the above findings. Consultations with user groups, regulators and leading service providers would greatly contribute to a better understanding of the industry requirements.

21.3.3   Regional standards bodies

Regional bodies are concerned with an acceptable level of quality and recommend standards to achieve this to a particular geographic area. The ETSI is concerned with the development of standards for the geographical Europe. They liaise with the ITU-T where such standards may be of global interest. Where ITU-T are ahead in the development of recommendations, ETSI manicures these, where necessary, to suit the needs of Europe. The ETSI publish a plethora of categories of documents as standards, reports, technical guides etc. They claim to have a wider coverage of applications than the ITU-T.

Like the ITU-T the choice of standards is largely influenced by the network and service providers and less influenced by the users. The tendency to consider telecommunications services as a feat of engineering and less as a commodity for use by the users to meet specific needs is obvious. However, there are signs of user inputs being taken a little more seriously. Like the ITU-T there is no architectural framework for the study and management of QoS except one document, which is very often quoted in other QoS related documents.

21.3.4   National bodies

Many countries have their own national standards bodies, such as the British Standards Institute in the UK. In the USA, due to decentralisation, there are very many standards bodies. The ANSI has the de facto status of a national standard body. The national standards bodies are usually responsible for the introduction of standards for a range of topics from agriculture to telecommunications to X-rays. These bodies usually tailor standards from the 'parent' organisation the ISO, to suit local needs, where necessary, or publish the ISO version where no change is considered necessary. The national bodies may also produce specific national standards where necessary.

In countries where there is no liberalisation in the provision of terminal equipment the end-to-end service provision and maintenance is probably under the responsibility of one dominant service provider who is also the network provider. In such cases, the service provider ensures the terminal equipment interworks with the network for optimum end-to-end quality.

In liberalised environments, where there are multiple suppliers of CPEs, it is necessary to have common standards of performance of terminal equipment and the interface at the access to the network. The specification for terminal equipment needs to address issues such as

- electromagnetic compatibility;
- effective use of the radio frequency spectrum, where appropriate;
- protection of the network from harm by induced voltages from the terminal equipment;
- interworking aspects with the network;
- performance of the terminal equipment for its principal function (e.g. voice quality in the case of a telephone).

In the UK the terminal equipment for telecommunication services are approved by the British Approvals Board for Telecommunications (BABT). The majority of approved terminal equipment work satisfactorily when connected to the network in the UK. However, with the progress in harmonisation within the European Union, the European Commission now requires all member countries to comply with the requirements in one of their directives, No. 1999/5/EC [2]. This deals with the connection of terminal equipment to the network. The requirements have become less stringent and there is no guarantee that all terminal equipment, which meet all requirements in the directive, will function satisfactorily. The EC is currently developing a new directive 'Conformity assessment of connected telecommunication equipment in Europe' which will give manufacturers the opportunity to state the compliance with EC requirements. This would enable national approvals bodies to relinquish some of their responsibilities in favour of the manufacturer.

The principal point to be noted is that an effective standards or approvals body is necessary to ensure optimum end-to-end QoS. To ensure optimum QoS, the CPE quality should reflect the capabilities of the network and this in turn requires a recognised and authoritative body to lay down minimum standards of performance. In addition to approvals the BABT are also involved in the Comparative Performance Indicators and Revenue Assurance schemes.

21.3.5 Shortcomings of standards bodies

The principal shortcomings of standards bodies may be grouped under the following:

(i) the QoS documents have been developed under the influence of the network and service providers and the interests of the user and customer is less obvious;
(ii) most of the QoS documents have been developed without an internationally agreed architectural framework for the study on an end-to-end basis.

What framework exists is based on technical performance and does not entertain user relevant QoS criteria;

(iii) a long lead time exists between the development of a new service or application and the development of QoS related standards.

To counteract the above shortcomings it is proposed that an internationally agreed architectural framework incorporating user relevant end-to-end QoS be developed. Standards based on this framework will stand a better chance of meeting the needs of the user. This in turn will bring benefits to the network and service providers in terms of better asset utilisation factor and result in greater profit per unit asset. Apart from the economic benefits to both the user and the provider the quality of telecommunications services is also greatly improved. A proposal for an architectural framework is put forward in Chapter 27. A start has been made in the ITU-T with the introduction in year 2001 of a user related document ITU-T Rec G 1000 [3]. Further documents in the series would ensure this objective would be met at some future date.

21.4 A methodology for the development of standards

21.4.1 General

The methodology offered for the development of standards has two parts:

- identification of areas requiring standards; and
- development of standards.

Accurate identification of the areas requiring standards could eliminate unnecessary or overlapping standards. Success in such an identification would be assisted by two factors, first, starting from the end-to-end user experienced performance and second, mapping the user related performance criteria against current standards and identifying areas requiring new standards.

21.4.2 Identification of areas requiring standards

The following steps could ensure the determination of the areas requiring standards:

Step 1: Select service: Performance criteria that are of relevance to the users are best specified on a service-by-service basis. This is followed by the service specification, listing the service features and identifying its borders of capabilities.

Step 2: Select the relevant elements from the QoS framework offered in Section 2.7. By an iterative process of comparing current knowledge of performance parameters with what is desirable a set of user related performance criteria could be arrived at.

Step 3: Map against current standards: The requirements are compared against current standards for either the existing service or similar service/s. Gaps requiring further standards or amendments to existing standards are identified.

Step 4: Evaluation: The areas requiring standards are evaluated for the following considerations:

* priority of resources;
* possibility of merging performance standards with other standards without losing the benefit of standardisation.

A set of work proposals could then be produced and an action plan prepared for the timely production of standards.

21.4.3   *Development of standards*

The principal linkages in the relationship between standards and QoS are illustrated in Figure 21.1.

Standards may be classified into two categories:

* parameters defined but without target values;
* parameters defined but with target values.

An example of the first category is bids per circuit per hour (BCH). BCH is an indication of the average number of bids per circuit, in a specified time interval. It will therefore identify the demand and, when measured at each end of a both-way operated circuit group, identify the direction of greater demand. It has no target but can be monitored to give an indication of network congestion, if a prescribed threshold (based on the size of the traffic route) is exceeded.

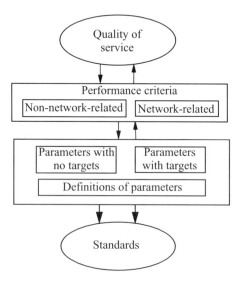

Figure 21.1   Schematic illustration of relationship between quality of service and standards

An example of the second category is the propagation delay. ITU-T recommendation G.114 defines not only the propagation delay but also recommends international and national limits. If these are exceeded standards are not met.

The process of developing standards are well established in ITU-T. The principal suggestion for improvement would be to speed the process for the development and issue of standards.

21.5   Mapping and review of ITU-T Recommendations on QoS for a legacy service – basic telephony over PSTN

An example of mapping of user related QoS criteria for legacy POTS obtained from the QoS matrix of the framework against current ITU-T Recommendations is given in Figure 21.2.

Time for provision
No recommendations exist specifically dealing with the time for provision. This parameter is considered an essential QoS parameter by most service providers, regulators and users. At present there is no internationally recognised definition for this parameter. Variables in the definition of this parameter are

- calendar time or working time;
- if calendar time, consideration of holidays;

Service function \ Service quality criteria		Speed 1	Accuracy 2	Availability 3	Reliability 4	Security 5	Simplicity 6	Flexibility 7
Sales	1	1	1	1	1	1	1	1
Service management	Provision 2	2	1	1	1	1	1	1
	Alteration 3	1	1	1	1	1	1	1
	Service support 4	3	1	1	1	1	1	1
	Repair 5	2	1	1	1	1	1	1
	Cessation 6	1	1	1	1	1	1	1
Call technical quality	Connection establishment 7	1	1	1	1	1	3	1
	Information transfer 8	1	3	1	1	1	1	1
	Connection release 9	1	4	1	1	1	1	1
Billing	10	1	3	1	1	1	1	1
Network/service management by customer	11	1	1	1	1	1	1	1

Figure 21.2   *Matrix indicating status of customer related QoS Recommendations*
   1 – No recommendations exist and none considered necessary;
   2 – No recommendations exist, but some considered desirable;
   3 – Recommendations exist but refinements considered necessary;
   4 – Recommendations exist and considered adequate.

- definition of the start and finish times for the provision, that is, when the effective waiting time commences from contract and when the service is deemed provided to the customer;
- whether the definition should be on a service-by-service basis or on a basket of services.

Desirable: An unambiguous and commonly agreed definition, developed by the ITU-T will assist service providers and customers of different suppliers, nationally and internationally.

Time to resolve complaints

The principal Recommendation that exists for this parameter is E 420. However, this recommendation does not define the time for resolution of complaints. The issues to be addressed in the formulation of a definition are

- identification of a complaint from an enquiry, clarification and other forms of queries from customer;
- the issues on the measurement of time as discussed above for the 'time for provision'.

Desirable: A universally agreed definition for the time for resolution of complaints could benefit customers and possibly service providers.

Time for repair

No recommendations exist specifically dealing with time for repair. Arguments for the time for provision, discussed earlier, apply to this parameter. The possible variables in the definition of this parameter are as follows:

- The time when the fault is deemed to have taken place – is it when it was reported by the user, or when identified by the service provider?
- When is the repair considered to be completed?

Desirable: A universally agreed definition or set of definitions developed by the ITU-T will assist both service providers and customers.

Simplicity in connection establishment

Recommendations specifying ring tones (indication when called person has been reached) exist. It would help customers if a uniform set of tones were specified for the following:

- indication of called customer's terminal engaged;
- indication of equipment engaged/busy/not available (network congestion).

Desirable: At present there is a wide range of such indicators making it difficult for the customer calling an international number to identify which tone indicates which network state. Some standardisation could be useful.

Call quality (accuracy of information transfer e.g. speech)

A measure of call quality for various types of calls may be helpful to customers. However, it must be added that with digitalisation the transmission quality of basic speech is considered less of a problem. Nevertheless, this parameter should still be studied, as it would be of interest in some parts of the world for some more years to come.

The principal Recommendations to cover transmission quality are

- ITU-T E.432 [4]
- ITU-T E.855 Rev. 1 [5]
- ITU-T G.101 [6]
- ITU-T G.712 [7]
- ITU-T P.11 [8]

The following are the degrading parameters contributing to lack of call quality (covered in ITU-T Rec. P. 11):

- loudness;
- circuit noise;
- sidetone;
- room noise;
- attenuation distortion;
- group delay distortion;
- absolute delay;
- talker echo;
- listener echo;
- non-linear distortion;
- quantisation distortion;
- phase jitter;
- intelligible crosstalk.

A model to establish the relationship between customer opinion on a transmission path under various combinations of loudness, circuit noise, sidetone, room noise and attenuation distortion does exist. At present it is not possible to determine the combined effect of all parameters as it becomes unwieldy and cumbersome to combine the effects of more than a few parameters.

Desirable: An indication of the level of call quality (expressed as the percentage of customers likely to find the circuit good for call quality) which users can expect for basic telephony for local, national and international calls under various circuit conditions.

Billing accuracy

The principal Recommendation on billing accuracy is E. 433 [9]. This Recommendation quotes a maximum figure for the probability of under and over charging a customer. However, this is not considered adequate. For the sake of completeness the magnitude of the maximum error should also be specified. Therefore, an additional clause to this Recommendation stating the maximum permissible inaccuracy expressed in a suitable manner is considered desirable. For example, the maximum error should be expressed as not more than $x\%$ of the total bill or not more than $y\%$ of the correct charge to the call. The ITU-T could debate to determine the form of expression for the maximum permissible error.

Desirable: Refinement of the existing Recommendation to specify the magnitude of maximum error, in addition to the maximum frequency of error.

Examination of customer-related QoS Recommendations may be carried out on a service-by-service basis for their suitability. Refinements of existing Recommendations or addition of new ones should not introduce a conflict of interest with service providers. If such an exercise is carried out for all principal services, international comparisons of performance will be feasible. Additionally, customer appreciation of service quality will also be enhanced.

21.6   Standards for emerging services

The illustration of mapping basic telephony and identifying the desired areas for further recommendations in the previous subsection may be applied to any emerging service. The overall QoS framework described in section 2.7 could facilitate the identification of QoS parameters of concern to the users. Mapping of existing standards and identifying areas requiring further work may, hence, be deduced. A case for a standardised QoS framework is put forward in Chapter 27.

21.7   Future role for standards bodies

It is useful to paint a picture of the ideal of a standards body for the future. The activities, as far as QoS standards are concerned may be indicated as shown in Figure 21.3.

Service providers

The service providers have been largely influential in the past in the initiation of topics to be studied in standards bodies. This situation is likely to continue but they should not have the principal voice in initiating and finalising standards. ITU-T members will have to take into account competition and the resulting need for a different emphasis in standards. There is a need for standards on end-to-end performance parameters that will be of concern to customers. There could also be a need for more generic forms of standards, so that each service provider has the choice to provide a range of quality levels in a competitive environment. Standards need not be restrictive but should allow freedom to service providers to offer enhancement in quality offerings.

Network providers

The role of network providers in the standards bodies may remain unaltered in so far as their prime interest is concerned. Their primary role would be to ensure standards are developed for interworking with other networks, both national and international and to ensure compatibility of service usage. In the future, they will have to keep pace

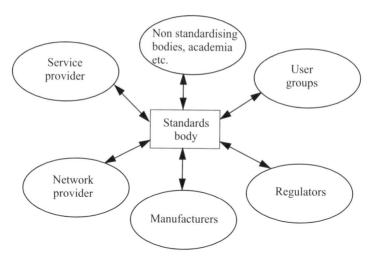

Figure 21.3   Membership of the standards body, ITU-T

with new services and endeavour to initiate development of standards for publication at the right time. Multimedia services are currently being developed and there is a need for standards on various issues. Network providers have a contribution to make in this area, not in isolation but in consultation with other bodies.

User groups

The user groups could influence the standards bodies in the choice and definitions of QoS parameters for

(1) reporting achieved performance by the service providers and network providers;
(2) specifying the offered QoS parameters.

The role of the user groups is discussed in more detail in Chapter 19.

Regulators

The regulator's role can be both reactive and proactive. Where quality has not been adequate it can set national standards or suggest performance parameters on which to publish achieved results. With the introduction of new services it could initiate development of new performance standards. The role of the regulator is discussed in more detail in Chapter 20. Regulators must carefully consider their role. One school of thought says that they should leave matters to the market forces and there ought to be minimum regulation. The other school of thought says that there ought to be near restrictive regulation. The determining factor ought to be not the lowest common denominator as far as QoS delivered to the customer, but the highest common denominator. The regulator ought to find means to achieve this in their country. No common formulae can be offered.

Research organisations and academia, and other organisations

Bodies such as the IETF, EURESCOM, European Commission, academic organisations and many other bodies carry out studies on specific QoS topics. Some of the findings could form the basis for development of standards. Further studies ought to be encouraged for an adequate and representative set of standards to be developed and issued.

Manufacturers can bring practical experience of the application of technology and its resulting impact on the definition of the standards. Standards are economically applied if they can be easily embodied in telecommunications systems development. In this respect manufacturers usually have a significant contribution to make. In the future, with more applications and technology being enhanced their contribution will be even more valuable.

The Standards bodies

Standards bodies have to address the following issues:

(i)   How do the ITU-T and other regional standards bodies decide when standards are to be developed? The computer industry is an example where lack of certain standards has helped development and expansion. Do standards inhibit innovation?

(ii) Are standards necessary in a competitive environment? Or should the market forces be allowed to find their own equilibrium? When should there be intervention?

(iii) With the convergence of telecommunications, information technology, information services, computers and broadcasting how should the different standardising bodies of these industries work together?

(iv) Should the standards be *de jure* or *de facto*? If there is room for both types of standards, what mechanism should be in place to channel the future standards into these categories? How much influence should market forces have in this aspect?

(v)   Should there be more regional standards? If so should there be standards for inter-operability?

(vi) Where standards are required, how can these be developed in time for maximum use? How can this be facilitated? What are the catalysts?

(vii) In the past ITU-T have concentrated mainly on technical performance of the network. In the future there will be additional requirements in the form of standards more oriented towards customers. How will the standards bodies adjust to these?

(viii) Since standards are agreed on by a majority of the delegates in an international forum, how are the needs of the less well developed countries and those with special needs catered for? Should these have a separate forum, part of the ITU-T? Who sets the level of standards? In other words how many standards and to what degree are standards required? Should this be left to market forces or should it be the prerogative of an educated few? Should this be technology led?

A conscious and reasoned effort must be made to choose a set of standards bodies to cater for most, if not all, known needs for telecommunications. The obvious choice is for one body to cover the whole world and regional bodies to specify requirements unique to a particular area, for example, the European Commission for Europe. When regional standards bodies are recognised and granted authority and autonomy, the scope of their activities and their relationships with other similar bodies should be clearly specified.

Where regional standards bodies exist their functions should, where possible, be synchronised with the international body. They should specify changes in performance requirements from the global standards clearly and state how they are relevant to the particular region. Such an explanation will enable the network providers and service providers to be able to reconcile, where necessary, the differences in performance requirements.

Future standards work would benefit if these are based on an architectural framework for the study of QoS. At present, there is no internationally agreed framework for studying all QoS parameters. Recently, a framework has been agreed upon in ITU-T Rec G 1000 [3]. This could form the basis for a comprehensive set of architectural framework for the study and management of QoS.

The standards should cater for the viewpoints of all the principal parties associated with the telecommunications industry. These parties are: the network provider, service provider, user/customer, regulator and manufacturer. The individual requirements are sometimes subtly different; if these differences are not identified and addressed in the standards, opportunity for maximum benefit from the standards activity is lost.

Standards, by their very nature, are time critical both in terms of time required for evolution and its timing of availability for maximum benefit. For this reason alone it is very desirable that the standards to be developed be prioritised according to some credible criteria. Prioritisation is necessary because of limited resources.

21.8 Summary

Standards bodies have played a useful part in the past. The changing requirements due to increased competition, formation of global alliances, convergence of telecommunications, broadcasting, computing and information industry will test the ingenuity of standards bodies. Their contribution in the future will also be tested severely by the increasing range of technical applications and the speed with which these applications are put into large-scale use. Correct timing of developing standards will be a major issue. Other bodies, such as regulators, user groups and research organisations could alleviate the difficulty of producing standards at the right time by making contributions to standards bodies. There is an overwhelming case for an internationally agreed architectural framework for the study and management of QoS. On this framework the current standards may be mapped and areas requiring further standards may be developed. This could lead to better QoS provided by the network and service providers.

21.9   References

1   MACPHERSON, A.: International Telecommunication Standards Organisations, 1990, Artech House
2   Directive 1999/5/EC: 'On radio equipment and telecommunications terminal equipment and the mutual recognition of their conformity'
3   ITU-T Recommendation G 1000: 'Communications Quality of Service: A Framework and definitions'
4   ITU-T E.432: B-ISDN User–Network Interface – Physical layer specification
5   ITU-T E.855 Rev. 1: Connection integrity objective for the international telephone service
6   ITU-T G.101: Transmission plan
7   ITU-T G.712: Transmission performance characteristics of pulse code modulation
8   ITU-T P.11: Effect of transmission impairments
9   ITU-T E.433: Billing Integrity

Web sites

1   ANSI   **http://www.ansi.org**
2   BABT   **http://www.babt.com**
3   BSI   **http://www.bsi.org.uk**
4   CEN   **http://www.cenorm.be**
5   CENELEC   **http://www.cenelec.be**
6   ETSI   **http://www.etsi.org**
7   IEC   **http://www.iec.ch**
8   IEEE   **http://www.standards.ieee.org**
9   ISO   **http://www.iso.ch**
10 ITU   **http://www.itu.ch**

Exercise

1   For a service of your choice carry out an analysis described in Section 21.3 of this chapter and identify areas where additional standards would be useful. For any one or more of these areas draft a standard that you would consider ITU-T ought to issue as a Recommendation.

Annex

Table A21.1 List of questions to be studied by ITU-T study group 12 during the 2001–04 study period

Question number	Title	Status
1/12	Evolution of the work programme	Continuation of Q.1/12
2/12	Speech transmission characteristics and measurement methods for Terminals and Gateways Interfacing Packet-Switched (IP) networks	Continuation of part of Q.9/12
3/12	Transmission characteristics of speech terminals both for fixed circuit-switched and mobile networks	Continuation of part of Q.9/12 and Q.12/12
4/12	Telephonometric methodologies for hands-free terminals and speech enhancement devices (including AEC and Noise Reduction)	Continuation of Q.6/12
5/12	Telephonometric methodologies for handset and headset terminals	Continuation of Q.8/12
6/12	Analysis methods using complex measurement signals	Continuation of Q.7/12
7/12	Methods, tools and test plans for the subjective assessment of speech and audio quality	Continuation of Qs. 14 and 22/12
8/12	Extension of the e-model	Continuation of Q.20/12
9/12	Objective measurement of speech quality under conditions of non-linear and time-variant processing	Continuation of Q.13/12
10/12	Transmission planning for voiceband, data and multimedia services	Continuation of Qs.16 and 17/12
11/12	Speech transmission planning for multiple interconnected networks (e.g. public, private, Internet)	Continuation of Q.18/12
12/12	Transmission performance considerations for voiceband services carried on networks that use Internet Protocol (IP)	Continuation of Q.23/12
13/12	Multimedia QoS/performance requirements	New
14/12	Effects of interworking between multiple IP domains on the transmission performance of VoIP and voiceband services	New
15/12	QoS and performance coordination	New
16/12	In-service non-intrusive assessment of voice transmission performance	New

Table A21.2   List of questions to be studied by ITU-T study group 13 during the 2001–04 study period

Question	Title
1/13	Principles, requirements, frameworks and architectures for an overall heterogeneous network environment
2/13	ATM layer and its adaptation
3/13	OAM and network management in IP-based and other networks
4/13	Broadband and IP related resource management
5/13	Network interworking including IP Multiservice networks
6/13	Performance of IP-based networks and the emerging Global Information Infrastructure
7/13	B-ISDN/ATM Cell transfer and availability performance
8/13	Transmission error and availability performance
9/13	Call processing performance
10/13	Core network architecture and interworking principles
11/13	Mechanisms to allow IP-based services to operate in public networks
12/13	Global coordination of network aspects
13/13	Interoperability of satellite and terrestrial networks
14/13	Access architecture principles and features at the lower layers for IP-based and other systems
15/13	General network terminology including IP aspects
16/13	Telecommunication architecture for an evolving environment

Section VI

Management of QoS

Management of QoS has many strands of associated activities and some of these are distributed in various parts of this book. In this section four topics of specific interest have been focused upon; comparisons of performance, economics of QoS, security and fraud.

Chapter 22

Comparisons of performance

22.1   Introduction

Comparison of QoS has become necessary in the competitive environment in which telecommunications service provider companies operate. Comparisons are important for users, regulators, service providers and network providers. The user needs to compare delivered performances of various providers within the country and perhaps to a lesser extent that of the international providers. The business user sometimes needs to compare the offered quality both nationally and internationally. Service providers find it necessary to compare their offered and delivered performance against the competitors'. They could also benefit from comparing customer perceptions of their services with those of competitors'. This chapter gives a brief review of some of the existing comparisons, identifies the limitations and proposes solutions to make meaningful comparisons possible.

22.2   Categories of comparisons

22.2.1   Comparisons within a country

Comparisons of performance within a country may be carried out on each of the four viewpoints of QoS. Though the four viewpoints arose from the Quality Cycle model principally designed for the legacy services, the principles may be applied to the emerging services such as Internet supported services and mobile communication services.

22.2.1.1   Offered QoS

Comparison of offered QoS would benefit customers, regulators, service providers and the network providers. Customers can assess if their telecommunications needs can be met by the service quality offerings of the various providers. The regulator can assess if any further action is to be taken for the improvement of quality even though such decisions, in the past, have tended to be based on the delivered quality. The service and network providers could benefit from the comparisons from a competitive angle. Such comparisons would enable providers to review, if necessary, their offerings.

It is necessary for the service providers (who could be independent of the network providers) to be aware of the network access performance at the interface with the network providers. The national regulator would, in certain cases, stipulate guidelines on access and the interface specifications. Quality would be one of the principal considerations in such specifications.

22.2.1.2   The delivered QoS

Users and regulators can compare how well individual providers have performed against each other. If the comparison is sufficiently detailed, a variety of users and user groups, such as business users, special interest groups and residential users could study the performances from their viewpoints and determine which provider is best in terms of quality. Caution must be exercised in comparing achieved performance. For example the 'average' can mask a bad 'tail' or a 'black spot'. Where there is reason to suspect the average does not reflect the true state of affairs, data must be provided for the spread of results. Suitable explanations should identify the black spots or areas of poor performance. The achieved QoS must be provided either by the service provider with audit by an independent body, or by an independent body which has scrutinised the collection of performance data (see Section 6.3.5).

For the provider, delivered-quality data would enable comparisons to be made against their competitors. Such comparisons would highlight the need to review their quality improvement strategies.

22.2.1.3   Perceived QoS

The principal beneficiaries of customer perception ratings are regulators and providers. These bodies can see if customer perceptions match the delivered performance of providers and, additionally, if there is a difference between customer perceptions between different providers. If two providers deliver similar quality, customer ratings should be similar. However, since perception is subjective (see section 3.2.4), there could be variations in customers' perceptions of providers. Should such variations exist, these comparisons will enable their identification and any necessary investigation or study may be initiated.

The fourth viewpoint, the customer's QoS requirements, is not a strong candidate for national comparisons. The prime interest of this viewpoint would be to the service provider with commercial considerations.

22.2.2   *International comparisons*

Comparisons of performance among different countries may be carried out on each of the four viewpoints of QoS.

22.2.2.1   Customers' QoS requirements

Comparisons of performance requirements of customers could be of benefit to service and network providers, as these parties will be associated with their counterparts in another country for the provision of international services. Comparisons of quality requirements of the users in the respective countries could be an invaluable input to the process of establishing a common and agreed level of performance, between the two countries.

22.2.2.2   Offered QoS

The principal, additional beneficiaries from national comparisons are multinational customers who have private or leased networks in different parts of the world. These companies would benefit by being able to compare what is available in the market from different providers.

Network providers and service providers could also benefit from comparisons. International performance comparisons of networks could enable service providers to choose, on performance criteria, with which carrier companies they wish to associate for the provision of international services.

22.2.2.3   Delivered QoS

In international comparisons of QoS this is the principal viewpoint that is being compared. The delivered quality is considered the most meaningful for the users in terms of what is achievable and therefore used as the benchmark for the less performing countries. However, as discussed in Section 22.5.4 there is a case for an internationally agreed set of definitions on QoS parameters for the principal services to enable such comparisons to be made.

22.2.2.4   Customer perception of QoS

Though international comparisons of customer perception would be desirable, data to carry this out is very limited. There is very little obligation on the part of service providers to publish such figures. What data are available are the result of comparisons carried out by independent studies. Such studies have to be examined in close detail for their relevance to international comparisons of service providers. This is outside the scope of this book.

22.2.3   *Benchmarking*

Benchmarking has been dealt with in some detail in Section 5.4.1.2. Comparison of performance of competitors within a country and with counterparts in another country, with a view to establishing best performance, is another benefit arising from comparisons of quality.

The important criteria for benchmarking are performance parameters defined and accepted internationally. In their absence, one is left to infer meaning from incompatible data. The case for a common set of definitions is made in Chapter 27.

22.3 National comparisons

22.3.1 General

The onset of liberalisation and competition in many countries has resulted in a national requirement for the publication of delivered performance data. The publicly available comparative QoS information is reviewed here for three countries, the UK, Australia and the USA.

22.3.2 The UK

In the UK the regulatory body, Oftel (Office of Telecommunications), has worked with the service providers to arrive at a set of QoS performance parameters on which achieved performance is to be reported [1]. These parameters, which are reported on, are listed below. The publication is available every six months and includes objective data on a quarterly basis and subjective data (customer satisfaction ratings) on a six monthly basis. These are listed in Appendix 7.

Very recently, limited mobile performance data were made available by four of the major operators on the Oftel website. The following parameters are reported on, on a quarterly basis:

- calls connected and completed successfully;
- successful set-ups;
- successfully-held calls.

The reporting of parameters is under continual review, though changes are slow to materialise. The driving forces are the market (principally user's lobby), service provider's motivation, Oftel's interest and the flavour of the Directives from the European Commission.

22.3.3 Australia

In Australia continuous evolution of formal reporting of service providers' delivered performance has resulted in revision of parameters since the first set of parameters were published in 1994. As of spring 2002, between the Australian Communications Authority (ACA) (regulator) and Australian Bureau of Statistics, performance data are provided for three principal services:

- basic telephony;
- internet; and
- mobile communications.

The principal features of the reporting mechanisms are given in Appendix 8.

The key features of the Australian method of reporting QoS are as follows:

1 QoS statistics are reported quarterly.
2 In the report there is usually a commentary by the ACA on the performance achieved by the various providers. This is not present in the UK reports.

3 Performance targets are laid down in the form of 'Customer Satisfaction Guarantees' (CSG) for some key parameters. If the standards are not met the provider is expected to fulfil an economic benefit, as laid out by the ACA.
4 The statistics are divided to reflect the local needs, that is, regional divisions are urban, major rural, minor rural and remote. In addition there are residential and business divisions where appropriate.
5 Different providers supply performance data on certain parameters. This contrasts with the UK practice of going for the least number of parameters that all providers can deal with.
6 Australia has QoS parameters for basic telephony, mobile communications and Internet unlike the UK.

The parameters and the CSG are under review and the reader is advised to check the web site of the ACA and ABS (listed in the references) for the latest information.

22.3.4 The USA

In the USA there are two sets of 'national' QoS statistics to be published: the first is the requirement from the Federal Communications Bureau (FCC) and the second is the local state requirement. The move towards publication of the QoS of local operating companies in the USA started with the divestiture of AT&T of its local services. Over the years, improvements were made to meet improved operational requirements and currently a quarterly return is required from each of the principal local operating companies. The parameters on which the FCC requires regular publication of QoS statistics and a list of parameters required by Colorado are given in Appendix 9.

The rather detailed reporting of availability of access and of outages probably reflects the needs of the users. Every country tends to publish performance parameters to reflect its local needs and the definition of parameters would, therefore, be different in various countries and will not reflect any international comparisons requirement.

22.4   International comparisons

No formal arrangement exists for publication by any international body of performance statistics of all the countries of the world. However, there are some regional bodies, the principal ones being the Organisation for Economic Co-operation and Development (OECD) and the European Commission.

22.4.1 Member countries of the OECD

The aim of the performance indicator group within the OECD is to publish comparable performance indicators from the member countries. Performance indicators are published in 'OECD Communications Outlook' [2], which is published every two years. This publication reviews the communications scenario among the thirty member countries on topics such as telecommunications growth, market statistics and outlook for the future. The QoS parameters, on which the OECD has published achieved performance, in 1999, but for a previous year, are as follows:

1   network access: waiting time for new connection;
2   network access: outstanding applications for connection;
3   number of payphones;
4   payphones per 100 inhabitants;
5   percentage of payphones that are card phones;
6   percentage of payphones in working order;
7   fault incidence;
8   fault repair time;
9   answer seizure ratios;
10   availability of itemised billing;
11   cost of itemised billing;
12   director assistance charges.

The fault rate per 100 lines per annum and the fault repair rates (as reported for year 1999) are reproduced in Appendix 10 to illustrate what was achieved.

The principal shortcomings of the OECD statistics on quality are as follows:

1   For some parameters comparative data are not to commonly agreed definitions.
2   Comparative data are not on a service-by-service basis except in a few instances.
3   Not all member countries report all the performance figures.
4   The OECD publication is made once in 2 years: therefore, many of the results are dated. The publication of year 2001 contains mostly performance statistics as of 1999.

22.4.2   *European Union member countries*

Operators with universal service obligations have to publish (and supply to the national regulatory authority) adequate and up-to-date information concerning their performance in the provision of universal service, based on a set of QoS parameters, definitions and measurement methods which are set out in Annex III of the Directive (Article 11). Annex III of the directive is reproduced here. Alongside the QoS parameters in Annex III, national regulatory authorities may specify additional QoS standards relevant to disabled end-users and disabled consumers.

If necessary, national regulatory authorities can set performance targets for operators with universal service obligations, in consultation with interested parties including consumers. Persistent failure by an operator to meet performance targets could result in a fine.

Annex III   Quality of Service parameters

Supply-time and QoS parameters, definitions and measurement methods

Parameter (Note 1)	Definition	Measurement method
Supply time for initial connection	ETSI EG 201 769-1	ETSI EG 201 769-1
Fault rate per access line	ETSI EG 201 769-1	ETSI EG 201 769-1
Fault repair time	ETSI EG 201 769-1	ETSI EG 201 769-1
Unsuccessful call ratio (Note 2)	ETSI EG 201 769-1	ETSI EG 201 769-1
Call set up time (Note 2)	ETSI EG 201 769-1	ETSI EG 201 769-1
Response times for operator services	ETSI EG 201 769-1	ETSI EG 201 769-1
Response times for directory enquiry services	ETSI EG 201 769-1	ETSI EG 201 769-1
Proportion of coin and card operated public pay-telephones in working order	ETSI EG 201 769-1	ETSI EG 201 769-1
Bill correctness complaints	ETSI EG 201 769-1	ETSI EG 201 769-1

Version number of ETSI EG 201 769-1 is 1.1.1 (April 2000)

Note 1 Parameters should allow for performance to be analysed at a regional level (i.e. no less than level 2 in the Nomenclature of Territorial Units for Statistics (NUTS) established by Eurostat).

Note 2 Member States may decide not to require that up-to-date information concerning the performance of these two parameters be kept, if evidence is available to show that performance in these two areas is satisfactory.

National regulatory authorities may specify the content, form and manner of information to be published, in order to ensure that end-users and consumers have access to comprehensive, comparable and user-friendly information. They can require independent audits or reviews of an operator's performance data, paid for by the operator, in order to ensure the accuracy and comparability of the data.

Public payphones

National regulatory authorities can impose obligations on operators in order to ensure that public payphones are provided to meet the reasonable needs of end-users in terms of the geographical coverage, the number of telephones, the accessibility of such telephones to disabled users and the QoS (Article 6).

Consumer contracts

Consumers have a right to a contract with a company providing connection and or access to the public telephone network. The contract must specify the services provided, the service quality levels offered (as well as the time for the initial connection), together with any compensation and the refund arrangements that apply if contracted service quality levels are not met (Article 20).

Improving QoS

Interested parties may develop mechanisms to improve the quality of electronic communications services. Such mechanisms could include, for example, codes of conduct and operating standards agreed on by consumers, user groups and service providers, under the guidance of national regulatory authority (Article 33).

However, national regulatory authorities are also able to take stronger measures, such as forcing operators and service providers to publish comparable, adequate and up-to-date information for end-users on the quality of their services. National regulatory authorities may specify the QoS parameters to be measured, and the content, form and manner of information to be published. Where appropriate, the parameters, definitions and measurement methods given in Annex III can be used (Article 22).

Leased lines

National regulatory authorities have to ensure the ongoing provision of the minimum set of leased lines set out in Directive 92/44/EC, until such time as they decide that there is effective competition in the relevant leased lines market.

The minimum set of leased lines is shown in the table here (reproduced from Annex II of Directive 92/44/EC as amended).

Definition of a minimum set of leased lines with harmonised technical characteristics

Leased line type	Technical characteristics	
	Interface presentation specifications	Connection characteristics and performance specifications
Ordinary quality voice bandwidth analogue	2 wire (1) (2) – ETS 300 448 (3) or 4 wire (2) – ETS 300 451 (4)	2 wire – ETS 300 448 (3) 4 wire – ETS 300 451 (4)
Special quality voice bandwidth analogue	2 wire (1) ETS 300 449 (5) or 4 wire (2) ETS 300 452 (6)	2 wire – ETS 300 449 (5) 4 wire – ETS 300 452 (6)
64 kbit/s digital (7)	ETS 300 288 ETS 300 288/A1 (8)	ETS 300 289
2048 kbit/s digital unstructured (9)	ETS 300 418	ETS 300 247 ETS 300 247/A1
2048 kbit/s digital structured (10)	ETS 300 418 (11)	ETS 300 419 (12)

(1)   The attachment requirements for terminal equipment to be connected to these leased lines are described in Common Technical Regulation 15 (CTR 15).

(2)   The attachment requirements for terminal equipment to be connected to these leased lines are described in Common Technical Regulation 17 (CTR 17).

(3) Previously provided in accordance with CCITT Recommendation M.1040 (1988 version) instead of ETS 300 448.

(4) Previously provided in accordance with CCITT Recommendation M.1040 (1988 version) instead of ETS 300 451.

(5) Previously provided in accordance with CCITT Recommendations M.1020/ M.1025 (1988 version) instead of ETS 300 449.

(6) Previously provided in accordance with CCITT Recommendations M.1020/ M.1025 (1988 version) instead of ETS 300 452.

(7) The attachment requirements for terminal equipment to be connected to these leased lines are described in Common Technical Regulation 14 (CTR 14).

(8) For an interim period extending beyond 31 December 1997, these leased lines may be provided using other interfaces, based on X.21 or X.21 bis, instead of ETS 300 288.

(9) The attachment requirements for terminal equipment to be connected to these leased lines are described in Common Technical Regulation 12 (CTR 12).

(10) The attachment requirements for terminal equipment to be connected to these leased lines are described in Common Technical Regulation 13 (CTR 13).

(11) Previously provided in accordance with CCITT Recommendations G.703, G.704 (excluding section 5) and G.706 (cyclic redundancy checking) (1988 version) instead of ETS 300 418.

(12) Previously provided in accordance with relevant G.800 series CCITT Recommendations (1988 version) instead of ETS 300 419.

For the types of leased lines listed above, the specifications referred to also define the Network Termination Points (NTPs), in accordance with the definition given in Article 2 of Directive 90/387/EEC.

National regulatory authorities have to ensure publication, in an easily accessible form, of certain information on the minimum set of leased lines, including

- the typical delivery period, which is the period, counted from the date when the user has made a firm request for a leased line, in which 95% of all leased lines of the same type have been put through to the customers. This period will be established on the basis of the actual delivery periods of leased lines during a recent time interval of reasonable duration. The calculation must not include cases where late delivery periods were requested by users;
- the contractual period, which includes the period which is in general laid down in the contract and the minimum contractual period which the user is obliged to accept;
- the typical repair time, which is the period, counted from the time when a failure message has been given to the responsible unit within the undertaking identified as having significant market power pursuant to Article 18(1) up to the moment in which 80% of all leased lines of the same type have been re-established and in appropriate cases notified back in operation to the users. Where different classes of quality of repair are offered for the same type of leased lines, the different typical repair times shall be published;
- any refund procedure.

In addition, where a Member State considers that the achieved performance for the provision of the minimum set of leased lines does not meet users' needs, it may define appropriate targets for the supply conditions listed above.

22.5   Management of comparisons of QoS

22.5.1   Structured approach

In this section a structured approach for an effective management of comparisons of performance is briefly explained. A methodology for identifying and addressing the key issues is illustrated in Figure 22.1.

The key issues to be addressed are

- whether the comparisons are to be carried out on a national or international level, and for what viewpoints;
- whether data for comparisons are to be on a service-specific basis or based on a basket of services;
- development of definitions of parameters;
- establishment of the logistics to produce performance data;
- establishment of an audit process;
- publication arrangements.

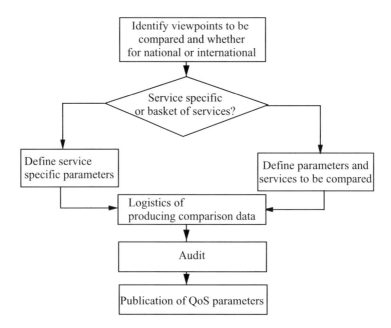

Figure 22.1   Schematic diagram for the management of comparisons of QoS

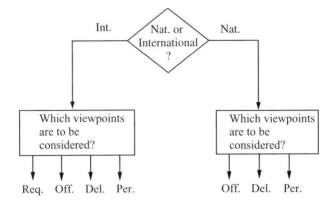

Figure 22.2   Logic diagram to determine the viewpoints for comparisons

22.5.2   Determination of viewpoints

The first task, is to identify whether comparisons are to be carried out at a national or international level. At a national level, comparisons of customers' QoS requirements are of little value. All other comparisons shown in Figure 22.2 may be considered.

22.5.3   Service specific or basket of services?

For simple comparisons of overall performance, a basket of services may be chosen. For example, the basic telephone service with payphone availability may be considered adequate for the performance comparison. More complex baskets may be produced for the higher end of the market. At the most sophisticated level the comparisons should be on a service-by-service basis.

22.5.4   Development of definitions

Performance parameters must be defined if these are to be compared on a service specific basis. Such definitions should be carried out at a national level for use within a country and at an international level for international comparisons. It is not necessary for the national definitions to be agreed on at an international level. However, it may be useful for parameters to be defined at an international level, as this would enable service providers to provide data, which can be equally used within the country, and for international comparisons without additional effort.

Definitions of parameters must be unambiguous and clear. For example, parameters to illustrate time for provision of a service should clearly state whether the time is in clock hours/days or working hours/days. There are many pitfalls in arriving at unambiguous definitions and these should be identified by the parties who will provide and use the comparative data.

22.5.5   Logistics

The logistics of producing performance data consists of specifying monitoring systems, collection of measured performance values and the measurement of end-to-end performance. These have been dealt with in Section 6.3.3.

22.5.6   Establishment of audit process

Performance data produced for purposes of comparisons require an audit process to add credibility. Audit processes are described in Section 6.3.5.

22.5.7   Publication arrangements

The publication of national comparisons would normally be carried out by the regulator. International comparisons would be published by a body such as the OECD. The principal issues relating to publication of delivered performance have been dealt in Section 6.3.4.

22.6   Proposed parameters for basic telephony

As a way of illustration, a set of parameters is indicated, in Table 22.1, which may be used to indicate the achieved or delivered performance of basic telephony. These parameters may be used within a country for purposes of comparison. A subset of these may be used for international comparisons. All parameters may not be relevant to all countries. For countries where the network is fully digital national call set-up time is of little interest. Call set-up time for international calls may be of more interest. Choice of parameters within a country would normally reflect local needs. Comparisons of international performances would depend upon the collective requirements

Table 22.1   Suggested parameters for international comparison of achieved quality of basic telephony service

1	Time for provision of service
2	Time for resolution of complaints
3	Time for repair
4	Repeat faults
5	Time for connection set-up
6	Measure of misrouted calls
7	Availability of service
8	Call quality
10	Cut off during conversation phase
11	Measure of overhearing
12	Billing accuracy

of member countries and the parameters would be chosen by the body responsible for such comparisons, for example, the OECD or the World Trade Organisation. A standardising body such as ITU-T would define the parameters.

22.7 Parameters for other services

Determination of parameters considered most relevant for users, service providers and others for the emerging services such as 3G mobile and Internet supported services may be arrived at after thorough research among a representative sample of the population. Studies have been carried out within the European Union for Internet Access performance parameters but these have not been validated against a representative user base. Such validation for the emerging services is expected in the future. Perhaps a selection of these will form the performance criteria for service providers to report delivered quality on a regular basis in the future. The same principle ought to apply for 3G mobile services.

22.8 Summary

Current performance comparisons systems, within a country, appear to have evolved to suit local needs. This is particularly true in countries where competition has been in place for some time, for example, in the USA. In other countries where competition has been in existence for a shorter period, for example, the UK and Australia, performance parameters are still being developed by the regulators for use by service providers. It will take some time before an optimum set of parameters has evolved on which delivered performances are required to be published.

The principal requirement for effective comparisons of performance – agreed definitions of parameters – is lacking. Within a country it is up to the local regulator to ensure such definitions are agreed upon. However, in the international scene an agreed set of performance definitions are required for effective comparisons of performance. The onus is upon the user groups throughout the world and the service providers to address this issue, perhaps through the international forum ITU-T.

In the international arena there has been little cause for anybody to influence countries and their service providers either to develop a set of performance parameters commonly defined or to publish achieved results. With the emergence of new services such as 3G mobile and Internet supported services, definitions for meaningful parameters for providers to report delivered performance would be highly desirable. This should be arrived at after validation of such parameters with a representative user base.

22.9 References

1 Telecommunications Companies: Comparable Performance Indicators, published half yearly, by Unitech on behalf of the Industry Forum of the UK.
2 'Communications Outlook', OECD, 2001.

Web sites

1   Australian Communications Authority: **http://www.aca.gov.au**
2   Quarterly bulletin on QoS performance in Australia:
    http://www.aca.gov.au/publications/performance/index.htm
3   'Customer Satisfaction Guarantee' on QoS in Australia:
    http://www.aca.gov.au/consumer/csg_index.htm
4   Australian Bureau of Statistics For Internet statistics: **http://www.abs.gov.au**
5   UK Comparative Performance Indicators **http://www.cpi.org.uk.**
6   USA – Federal Communication Commission **http://www.fcc.gov**

Exercises

1   Identify the service/s considered important by users of your country (broken down
    into residential and business user categories) for regular publication of statistics
    on delivered QoS by service providers.
2   Develop a set of QoS parameters for the service/s identified in Exercise 1. Outline
    an implementation guide for adoption by the service provider and the auditor.

Chapter 23

Economics of quality of service

23.1 Introduction

There are both costs and benefits associated with the provision of an acceptable QoS, but little work has been done to quantify the economic dimensions of QoS and its treatment is usually subjective and fragmented. It is not possible to accurately quantify the economic dimensions, which are often soft and speculative in nature. Nevertheless, it is necessary to, at least, carry out a qualitative cost–benefit analysis when introducing QoS initiatives to gain an understanding of the possible economic consequences. This requires an understanding of the economic drivers and their linkages, which is the subject of this chapter. The costs and benefits can be progressively expanded as shown in Figure 23.1.

The main categories of cost are those associated with the failure to achieve a satisfactory level of QoS, that is, failure costs, appraisal costs which relate to activities concerned with the assessment of QoS and those necessary to ensure that a desired level of QOS is obtained (i.e. prevention costs). The costs can be further decomposed into those directly affected by the network and others. Benefits usually accrue in terms of increased revenue. In a competitive environment this is often the result of minimising loss of market share to competing operators and service providers. This is particularly important to the incumbent operator who finds it difficult to compete on price because of the burden of the 'Universal Service Obligation' to serve all customers at a fair price, together with higher overhead costs. Such operators must therefore differentiate themselves from their competitors by the QoS they achieve.

23.2 Cost of quality

The traditional curves associated with failure and prevention costs are illustrated in Figure 23.2 [1]. The summation of the two gives the total quality costs, which would appear to indicate a minimum cost which provides optimum quality and differs from the zero-defects level. However, this ignores the benefits aspects of QoS and may not be the most profitable solution.

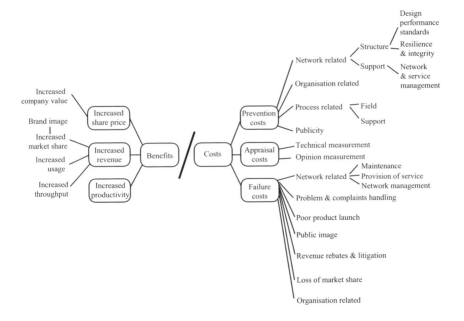

Figure 23.1   Quality of service costs and benefits

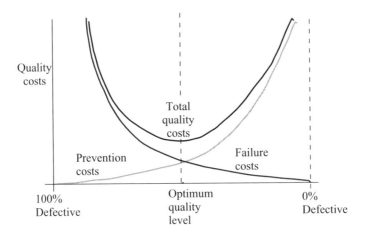

Figure 23.2   Cost of quality

23.3 Prevention costs

23.3.1 Definition

Prevention costs are those which are necessary to ensure that the products and services delivered to the customer, that is, the offered quality, meet an acceptable level of quality. They include costs related to the structure of the network as well as the costs of all activities required to prevent production and delivery of sub-standard service offerings.

23.3.2   Network prevention costs

Prevention costs associated with the network are primarily related to its design and management. The three main parameters that affect the design and dimensioning of the network are cost, performance and throughput. Generally, the design of networks is optimised for minimum cost for a prescribed throughput and performance level, although any of the three variables could be optimised with the other two fixed. Chapter 9 showed how many of the performance parameters can be mapped onto the network-related QoS elements, and also that the end-to-end performance for each parameter, and hence the offered QoS, was broadly the summation of the individual performance of each link and node making up the connection, as illustrated in Figure 23.3. For many performance parameters there is a trade-off between the cost and quality for each link and node. Hence, it is possible to arrange for the apportionment of the target end-to-end performance between nodes and links in such a way as to minimise the cost of achieving it. It is generally most costly to obtain good performance in the local access network (local loop) because of its low-grade technology, hostile environment and distributed nature. Therefore, the most stringent performance

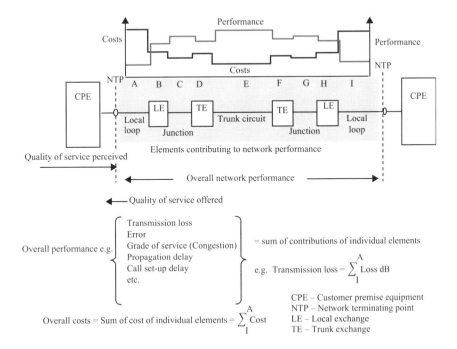

Figure 23.3   Apportionment of cost and quality

targets are allocated to the nodes and links in the core of the network where traffic is most concentrated and performance most easily controlled. For each node and link, costs and performance tend to be inversely proportional, as shown in Figure 23.3.

The use of equipment with less stringent Mean Time Between Failures (MTBFs) may reduce capital cost requirements. However, in order to meet customer's required availability targets, more maintenance effort (and hence more operating costs) may be needed. Thus, there is a trade-off between capital and operating costs to meet a given availability target and the optimum should be to minimise whole life costs, that is, the combination of capital costs (represented by their equivalent annual charges) and operating costs. In a similar fashion, the mean time to repair (MTTR) is a critical QoS parameter and is the aggregate time of the processes from fault reception to solution. It is a queuing system [2] where fault reports are inserted into a store at a variable rate according to reporting behaviour, weather conditions etc., and drawn out according to the availability of technicians. The dominant element of MTTR is queuing, travelling and repair, as illustrated in Figure 23.4. Queuing can dominate at low resource levels

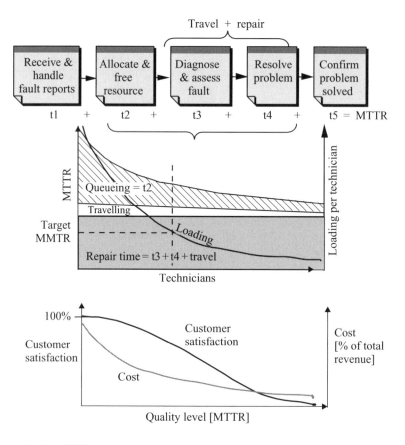

Figure 23.4   MTTR drivers

where the probability of a job arriving where there are no available resources is high; loading is also high in terms of faults per technician and travelling time increases because the area covered by each technician increases. As the manpower resource increases their effect becomes progressively less by the Pareto effect. It is therefore necessary to balance the needs of productivity cost savings and quality (MTTR). The latter is a particularly critical parameter in a competitive environment when incumbent Telcos may be forced to introduce contractually binding, service level agreements (SLAs) that give revenue rebates to customers when problems are not cleared within a prescribed timescale.

Such an analysis of process costs could provide a method of calculating the overall cost of improving QoS. Initially, it would be necessary to determine, by customer survey, the main QoS parameters and their relative importance to customers. An example of such a survey could be

1 – speed of provision	13.7%
2 – time to repair (MTTR)	12.8%
3 – accuracy of billing	12.4%
4 – connection availability	10.8%
5 – call quality	10.3%
6 – connection accuracy	10.1%
7 – service availability	9.9%
8 – time to resolve complaints	7.9%
9 – call set-up time	6.5%
10 – time of pre-contract activities	5.6%
	100%

For each parameter, an analysis of the process costs related to QoS would be carried out to provide cost and customer satisfaction graphs similar to that shown in Figure 23.4. The graphs, weighted for their relative importance could then be summated, as shown in Figure 23.5 to provide an indication of the additional costs necessary to improve QoS or the likely customer satisfaction for a given cost. In order to take account

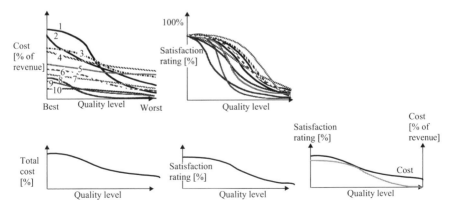

Figure 23.5   QoS relationships

of differing sizes of Telcos, costs are plotted as a percentage of revenue. There are obvious limitations to this method due to the difficulties of quantifying the softer aspects of QoS.

There is also a trade-off between capital (prevention) costs and operating (failure) costs for maintenance and provision of service in the local loop. Traditionally, capacity is provided in increments to meet growth for a period of time (the design period). However, due to the difficulty in forecasting growth at this distributed periphery of the network, provision of service often requires manual intervention at the flexibility points, that is, Primary Cross-connect Points (PCPs) and Distribution Points (DPs), to utilise available pairs. In addition to incurring failure costs in service provision, this intervention and disruption of the network creates faults, thus increasing the failure costs of repair.

In an extreme case, stringent QoS targets for time to provide service can lead to expedient provision, such as pair diversion, due to the non-availability of a pair when a new customer requires service. This is not only expensive, but it diverts effort from normal provision of plant to meet the growth forecast, lengthens the provision-of-service time and worsens the fault rate. Such reactive planning to provide service reduces effort available to provide plant to meet forecast requirements hence increasing the shortage and reducing the probability of pairs being available when and where required. Such a cumulative situation can rapidly run down network capacity and is exacerbated by tight staffing levels and short provision-of-service targets which may influence manager's targets.

This problem may be overcome by providing sufficient capacity to provide for all unserved properties, thus making the network independent of forecasts. Such a 'stabilisation' of the local loop can incur high (prevention) capital costs; however, provision of service no longer requires physical intervention, thus reducing failure costs. Prior and Chaplin [3] demonstrated that stabilising the local loop in a particular local exchange area (St. Albans in the UK) had a significant impact on reducing the fault rate. They argued that the whole-life fault liability of local loop plant follows the traditional 'bath tub' curve shown in Figure 23.6(a) and that when a cable joint or cross connect point is opened and physically altered to provide service the whole of

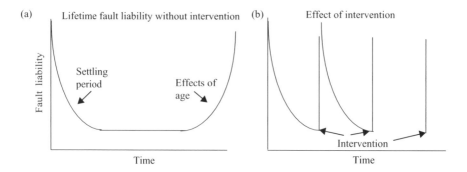

Figure 23.6   Effect of intervention on fault liability

the particular element (not just the pair) could be moved back to the start of the 'bath tub' giving the 'saw tooth' reliability curve shown in Figure 23.6(b).

Additionally, if the unserved properties are pre-connected to the local exchange and given a class of service (soft dial tone) which only allows calls to emergency services and the network operator's sales office and bars incoming calls, service can be immediately provided by the customer plugging in a telephone and calling the sales office. This dramatically improves the QoS (i.e. provision time) for service provision.

23.3.3   Network resilience

In order to minimise disruption to service due to equipment failure or traffic overload, resilience is often built into networks. This is because an efficient network design reduces spare capacity and hence increases susceptibility to traffic overload. The effect of network overload decreases with the use of common control exchanges and Automatic Alternative Routing (AAR) or Dynamic Alternative Routing (DAR) which can divert calls to less congested routings. Although AAR is valuable in absorbing small overloads it can, in itself, spread congestion through the network under severe overload conditions unless carefully controlled.

The main causes of overload are the reduction in network availability (i.e. the probability of free paths through the network) due to equipment breakdown, the effects of which become more pronounced due to the increasing trend for higher capacity transmission systems and larger more complex exchanges. Also, traffic surges due to the increasing sophisticated usage of the telephone by subscribers, for example, phone-in radio and TV programmes etc.

A selection of resilience measures is shown in Figure 23.7. It can be seen that the measures can be grouped into those provided at the physical layer (layer 1), which offer protection against transmission failure at the transmission systems and/or optical bearer layers. Those at the functional layers (layers 2 and 3) are concerned with the routing of traffic around failed or congested nodes and links, and those at the upper, service layers protect against problems with service application software. All these measures involve expenditure and it is therefore necessary to select a combination of measures that meets the needs of the network operator at an acceptable cost. It is not possible to evaluate costs and benefits, but Figure 23.7 illustrates the likely relationship between restoration effectiveness and cost.

Even though the Internet was originally designed as a 'bomb-proof' network with in-built resilience, the combination of rapid growth, commercial utilisation and very high capacity routers and core transmission links requires added resilience measures for the evolving IP networks of tomorrow. Veitch [4] argues that a combination of physical and functional layer measures are necessary to protect future IP networks against failures and overloads.

23.3.4   Network integrity

The increasing threats to network integrity due to the interconnection of networks and their increasing complexity and the provision of sophisticated services is described

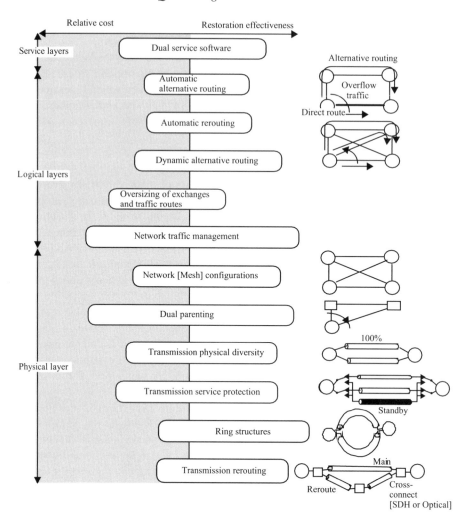

Figure 23.7   Network resilience measures

in Chapter 9. The economics of integrity can be grouped into failure and prevention. There are also commercial implications on customers and service providers of denial of interconnect.

The cost of failure of network integrity needs to be evaluated for the various failure modes, which could be as follows:

(a) Catastrophic failure, where large outages occur such as failure of a switching node or nodes, can be very costly in terms of lost revenue. This cost can, in theory, be evaluated if the number of affected customers is known together with their average traffic (and average charge per call) together with the duration of the outage. However, this does not take account of repeat calls after restoration of service, or contractually-binding rebates and litigation. Neither does it take account of loss of customer goodwill and confidence which may result in the loss of customers to other operators.

(b) Failure of interconnect, where all calls passing between the networks are prevented from completing; again, theoretical lost revenue can be broadly calculated and would affect both the operator in whose network the calls originated and the one in whose network the call terminates.

(c) Failure of a particular service (or services) in one or other network, or just those that require interoperability between networks; again theoretical revenue loss can be calculated.

(d) Degradation of performance that does not affect call completion, for example, high error rate; in this case, performance acceptability from a customer viewpoint often depends on the customer application, for example, data transmission is much less tolerant of error than voice. In this instance, it is almost impossible to determine loss of revenue. Even if loss of revenue can be established, there remains the problem of apportionment of blame and recompense.

It would be reasonable to expect that all operators would have a minimum degree of integrity protection in their networks as a result of normal network resilience measures. However, additional protection will need to be provided at the points of interconnect. These could, for example, be as follows:

(a) Detailed and unambiguous interconnect and feature standards; however, ensuring compatibility of implementation of standards requires close cooperation between operators. The cost of creating standards by, for example ETSI, and its funding are well established. However, the costs of implementation and compatibility assessment can be very high and are bound to give rise to debate about how such costs should be funded.

(b) Mediation and policing devices; their location (i.e. in one or both networks) and cost apportionment between operators will be a major issue.

(c) Conformance testing; this may require expensive captive models as well as internetwork trials and it raises issues regarding when such testing should take place. This could be when new services are interconnected or even when major software upgrades take place in either network. Major issues include responsibility for conformance approval, self certification, funding of test models and requirements for testing.

The commercial implications of an operator restricting access to its network in order to protect its integrity include the following:

(a) Potential loss of revenue to the service provider or network operator denied access – this is difficult to independently quantify.

(b) The costs to customers who might be denied cheaper services – this is impossible to quantify.

(c) The reducion in market growth, innovation and development of new services resulting from the reduction of competition and lack of creative service providers – this is analagous to the creation of new applications for PCs. There would be an economic impact on the customer community.
(d) Conflict with the regulator whose policy is to open networks for interconnection to competing networks, for example, the Open Network Provision (ONP) policy of the European Union and Open Network Architecture (ONA) policy in the USA.

23.3.5   Network and service management

Network management is the use of computer support systems [5] to improve both the efficiency of maintenance and service provision processes, that is, to reduce failure costs and to improve the performance of the network. Such systems are most effective when they interact with the network, each other and the operatives in a coherent fashion [6]. Hence much effort is being undertaken by the standards creation bodies (ISO, ITU-T, ETSI, ETC) to develop appropriate interworking standards under an architecture known as the Telecommunications Management Network (TMN). This has standard protocols and interfaces to allow maximum flexibility in interconnecting support systems and network elements in a multi-vendor environment. A well structured set of network management support systems will markedly improve QoS in terms of time to provide service and repair. Additionally, it should detect problems before they affect customers.

No matter how good network management is, the impact on QoS can be reduced by a poor interface to the customer. Whilst network management ensures that the operation of the network is optimised to the desired performance levels, service management should optimise services to the best customer QoS. Service management can be defined as the coordinated management of a portfolio of services that gives customers an integrated interface to the group of services that they use, and gives the network provider a single integrated view of its customers. It includes customer reception, to present a user-friendly single point for customers to interface with the network provider, hence making a significant contribution to the QoS perceived by customers. Whilst the development of network and service management support systems is very expensive, they make a significant reduction in prevention and failure costs as well as improving QoS.

23.3.6   Organisational costs

The way that the field force is organised and directed can have a significant impact on both prevention and failure costs. A modern network of processor-controlled digital exchanges and transmission systems with remote monitoring and control features requires centralised control of operations staff via a minimum number of operations centres [7]. Such an organisation could contain a small number of Network Operations Units (NOUs), the major control centres which initiate and control all activities. Also, a Central Operations Unit (COU), which provides top level support, is the centre at which single tasks for the whole network are carried out and generally includes the National Network Management Centre which provides overall network oversight for network management, major network rearrangements and overall performance management. Network Field Units (NFUs) control field staff who carry out work scheduled by the NOU. Close links are provided to the 'front office' Customer Service Units.

The development of work management support systems can do much to reduce prevention and failure costs by increasing the productivity of those engaged in volume activities of provision and repair; primarily for Customer Premises Equipment (CPE) and in the local loop. Savings can be expected to accrue from more efficient day fill, skill matching, reduction in travelling time and distance, better communications, reduced control requirements, avoidance of work duplication, better jeopardy management and reduced effort in collecting and analysing statistics. Additionally, it can be expected that improvements to QoS would result from reductions in time to provide and repair, together with better control of appointments and quicker reaction to changing circumstances. The broad objectives of work management are to get the right person to the right place at the right time with the right stores and information; and know where operations people are, what they are doing, when they will finish and what they will do next.

23.3.7　Publicity and public relations

Good advertising and public relations often provides an effective way to improve customers' perception of QoS, which may not be directly linked to the service offered. However, national publicity is expensive, particularly if television is used.

23.4　Failure costs

23.4.1　Definition

Failure costs are those costs incurred as a result of unacceptable QoS as perceived by customers and the cost of rectifying failures in the network.

23.4.2　Process related failure costs

A process can be defined as the logical organisation of people, materials and procedures into activities designed to produce a specific result. Individual activities within a process are termed components and are normally linked together by exchanges of information or materials. QoS processes generally begin with a customer request and should end with a satisfied customer. However, for telecommunications, such processes are often complex, with a large number of components for a range of services and product lines modified by customer type and spanning a number of business organisational units. The opportunities for failures that affect QoS are therefore many. Process management is the control and systematic analysis and, if necessary, redesign of processes in order to maximise their efficiency and effectiveness [8].

Processes may be analysed in terms of the cost of each component, the time taken to execute it and, if possible, its quality. Where a process spans a number of business units, SLAs are often negotiated between component owners at either side of the interfaces to record the QoS required across the interface to give the end-to-end target for the overall process.

BT have estimated that process failures account for $37 \pm 10\%$ of the total cost of serving customers [9]. They have built a failure model with processes and failure/success volumes, together with activity-based costs reconciled with budget allocations. Root causes of failure were then established through consultation with people who were affected by failure as part of their work experience; this revealed five service issues, namely,

- fault information quality and dialogue;
- self-generated faults and visits;
- order quality and dialogue;
- provision visits; and
- early-life failures.

The root causes for a number of the above areas were found to be common, for example, management of records, call-handling time constraints and poor dialogue with customers. Other root causes were no access, line plant, systems and practices, experience, training/knowledge, culture and resourcing. Although some of these root causes could not be quantified the failure model (Figure 23.8) presents a useful picture to drive down failure costs.

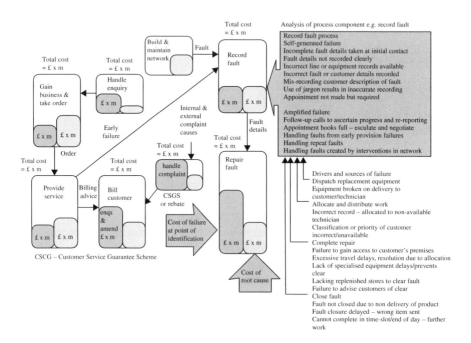

Figure 23.8   Failure/root cause model

23.4.3   Network-related failure costs

Network-related failure costs include a large proportion of the cost of maintenance. In a modern network, a high proportion of these costs fall within the local loop and CPE sectors. The minimisation of failure costs requires an understanding of the cost elements and their drivers, that is, those aspects that influence them. Although the manpower element for operating the network is diminishing with the penetration of modern equipment and the introduction of computer support systems, pay is still a substantial element of operating costs. Time can be classified as effective and ineffective. The former can be measured and analysed to determine areas for productivity improvement. But the latter, covering such things as sick and annual leave, training and travelling are 'softer' and difficult to understand and analyse. They are often influenced by organisational structure, leadership and morale.

Many of the cost drivers are themselves dependent on other influences, for example, QoS targets drive capital and operating costs and are significantly affected by the regulator; they also determine market share which affects network growth and hence investment. There are also interactions between other network costs and drivers, for example, capital costs increase assets which incur additional depreciation charges and so drive up operating costs. However, if the investment is in modernisation then maintenance costs are reduced. Therefore, the full picture of costs, drivers and the interactions is extremely difficult to comprehend but it is an essential element of managing a telecommunications business. An example of the interactive nature of a cost driver is shown in Figure 23.9.

Failure costs can be minimised by a systematic analysis of the costs to identify the drivers and thence determine the means of reducing costs. An example is given in Figure 23.10, where analysis of the operating costs has shown the local loop to be a high cost area. Examination of the loop costs indicates that maintenance (failure) costs dominate so these are analysed to identify areas for improvement. The results indicate that, although PCPs are relatively few, compared to cables and DPs, they have a high fault rate. Further investigation reveals that failure costs for PCPs are driven by their design, poor workmanship when rearrangements are carried out and the number of times they are visited. Potential means of reducing the high fault rate are to refurbish the PCPs, replace them with a better design or to lock them and strictly control access.

23.4.4   Network traffic management

A major cause of poor QoS from networks is call congestion caused by increasingly volatile traffic. This derives from the increasing usage of telephones, for example, on Christmas Day, Mother's Day etc. and media-driven events that cause short-duration, focused surges of traffic to, for instance, TV and radio phone-in programmes and advertisements. Also, failure of increasingly larger modules of capacity, with the increase in size of exchanges and transmission systems has a great impact on traffic. Such heavy congestion cannot be overcome by traffic routing strategies such as AAR and the resulting call re-attempts exacerbate the problem. This type of congestion may be classified as general overloads affecting the entire network, local overloads

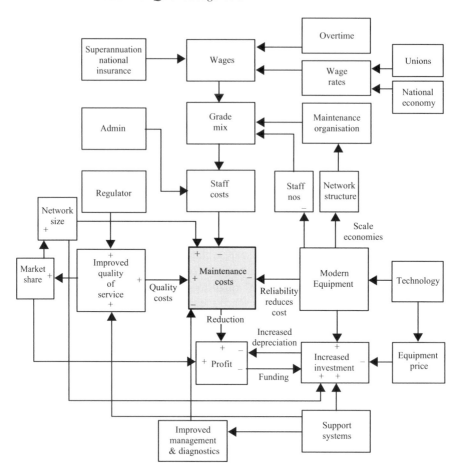

Figure 23.9   Maintenance cost drivers

affecting part of the network, and focused overloads where the calling rate to a specific destination is particularly high.

The effect of these overloads can be reduced by network traffic management which monitors certain network performance parameters from each exchange at regular, say five minute, intervals. When they reach a prescribed threshold, build up of congestion is indicated. In such circumstances, traffic is managed by the use of expansive controls and protective controls. The former typically re-route traffic over parts of the network where spare capacity exists, to avoid congested areas. Protective controls are designed to prevent otherwise-ineffective calls to focused destinations from entering the network in order to allow other calls to succeed. The usual protective control is call gapping at the originating source – only 1 in *n* calls or 1 in *T* seconds is allowed – to mitigate the effect of focused overloads. It does not improve QoS to the focused destination but improves it to other destinations. The cost of network traffic

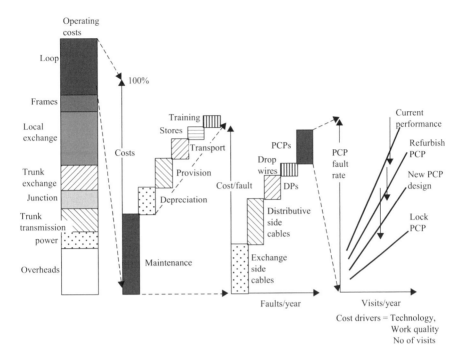

Figure 23.10   Investigation of failure costs

management is high. However, it not only improves QoS but also the throughput of calls, thereby increasing revenues.

23.4.5   Revenue rebates

In a competitive environment, where the customer has a choice of supplier, network operators are often obliged to offer contractually binding service level agreements (SLAs) that guarantee specific levels of service. For example, a standard contract for residential customers may specify a five-day time to provide and same-day repair; business customers may be offered superior QoS objectives. For any such contract, there is usually a penalty in terms of a revenue rebate for non-compliance; poor QoS can therefore result in a substantial loss of revenue. For example, consider a private circuit 'Reduced Charges Scheme' which guarantees that customers are rebated up to 100% of rental charges if six faults during a year are not cleared within five hours. If the achieved MTBF for the local loop is 15 years and if the target MTTR is no more than 1 day, then inaccessibility quota = MTTR/MTBF = 24/15 = 1.6 h per year. In practice, the Pareto principle applies and downtime is not evenly distributed over a customer population so a proportion receive a disproportionate number of failures, for example, 80% faults experienced by 20% of the customers. Using the same principle, a proportion of the failures will not meet the MTTR requirements. So, for a total population of, say, 15,000 private circuits giving an average of 1.6 faults per circuit per annum a calculation might be:

$$
\begin{aligned}
1.6 \text{ faults} \times 15,000 &= 24,000 \text{ faults} \\
80\% \text{ faults} &= 19,200 \text{ faults} \\
20\% \text{ circuits} &= 3000 \text{ circuits, get } 19,200 \text{ faults} \\
\text{faults per circuit per annum} &= 6.4 \text{ faults pa}
\end{aligned}
$$

100% rebate if 6 faults pa not cleared within 5 hours

$$
\begin{aligned}
\text{average rental} &= \pounds 8000 \text{ pa}
\end{aligned}
$$

assume 10% faults not cleared within 5 hours, then:

$$
\text{Annual rebate} \quad = \quad 8000 \times 1/10 \times 3000 = \pounds 2.4\text{m}
$$

In many countries, litigation is becoming increasingly commonplace and privatised network providers no longer have the protection of public ownership. Business customers who increasingly rely on good communications may well successfully sue for substantial damages if communications outages cause loss of income.

In an environment with contractual service targets against which failure penalties are paid, work management processes and support systems must closely monitor tasks to ensure timely completion. Imminent and missed service targets must be automatically selected and referred to field managers for corrective action; this is known as jeopardy management.

23.4.6   Product launch

In a competitive environment, getting new products and services to market early is important in securing a high market share. However, poor product-launch processes result in teething troubles and poor QoS to early customers and can cause loss of confidence in the service and reduced market share. Failure costs are therefore high, both in loss of revenue and retrospective action to correct problems.

The launch of a new telecommunications product or service is a complex exercise which can involve a substantial multi-discipline team working to tight target dates. The core product team will often consist of a Product Manager who is responsible for overall product profitability, a Marketing Manager who represents customer requirements and a Finance Manager who produces the financial data associated with the product and who provides budgetary support. Expert support is provided by R&D, network planning and operations, customer service operations, training, equipment procurement, logistics, commercial contracts, billing systems, tariffing etc. Good project management is therefore a key ingredient of the product launch process. Such a project plan often uses critical path techniques to sequence the activities, timescales, interactions, responsibilities and key milestones.

A feature-rich service realisation will provide good differentiation from competitors' services and could be more attractive to customers. This gives a high market penetration and share, hence substantial revenue. However, such an option would require a high-functionality network and the complexity would incur risks in delivering to time and on budget with high failure risk and cost. High-functionality services launched late may well lose market share to an earlier but simpler offering from

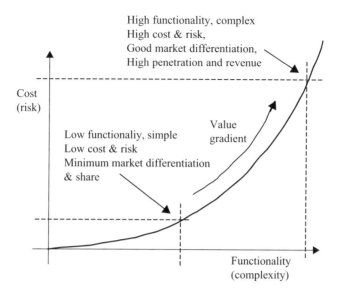

Figure 23.11   Risk assessment of network options

a competitor. A low-functionality option will generally be of lower risk but less competitive in the market. A compromise may be to launch early with a simple solution to capture the market and then migrate the customer up the value gradient as shown in Figure 23.11. However, this requires a flexible network realisation to allow this to be done in a cost effective manner.

For advanced services, for example, intelligent network (IN) services, the network control complexity can give rise to a feature interaction, whereby new services interact with existing services in an undesirable and unpredictable manner. This can require high prevention costs when developing new services, to identify such problems by modelling using formal methods such as Specification and Description Language (SDL) [10].

23.5   Appraisal costs

Appraisal costs represent the cost of measuring QoS and network performance. The measurements range from, for example, customer satisfaction surveys to the sampling of 1 in n live calls by exchange software and the processing of the samples to provide meaningful network performance statistics [11].

23.6   Benefits

23.6.1   General

In a similar fashion to costs, the benefits of QoS are difficult to quantify. In a competitive environment, it is difficult for the incumbent network operator to compete on

price for POTS, its main source of revenue, because of the 'universal service obligation' to serve all customers, even those who are unprofitable, at a fair price. For other
services, the regulator will often insist that tariffs are cost based to prevent unfair
competition by predatory pricing as a result of cross-subsidisation. Such operators
often carry high overhead costs as a result of their historical monopoly culture. Therefore, in order to retain market share, the network operator is forced to differentiate
itself from its competitors by offering a superior QoS and more extensive portfolio
of services.

23.6.2   Network modernisation

Good QoS and new services are difficult to achieve when a network operator is mod-
ernising its network because of the legacy burden of obsolete plant. Its competitors
can build from scratch with modern plant. It is therefore necessary for the network
operator to schedule carefully the order in which exchanges are modernised. Early
priority must obviously be given to large towns with high business penetration. These
are most at risk in a competitive environment, but they are also likely to generate
most revenue from new services. It becomes more difficult to prioritise the residual
majority of exchanges to ensure that the supply of digital exchanges is deployed in
the most cost-effective manner, particularly when replacing exchanges.

The financial worth of modernisation can be assessed by trading off the capital
required to modernise each exchange against the savings in operating costs and the
incremental revenue from new services, usage stimulation from improved QoS and
the loss of revenue protected from competitors. There is evidence that improved QoS
can stimulate usage. For example, modernisation by replacing analogue exchanges by
digital reduces call establishment time and the probability of congestion whilst also
improving call clarity due to the reduction of transmission loss and noise. This tends
to encourage customers, particularly in the residential market sector, to make more
calls. Conversely, in a partially modernised network, the variability of performance
due to the mix of analogue and digital routings can give rise to increased complaints
about QoS.

Clearly, the reduction in congestion due to, say, automatic alternative routing
and dynamic alternative routing strategies together with network traffic management
increases traffic throughput and hence revenue. Some calls lost due to congestion
are eventually successfully repeated. The 'repeat call' button on most modern telephones assists in this, although it can exacerbate the problems of heavy congestion
by exponentially increasing call attempts.

23.6.3   Productivity

It is generally acknowledged that failure costs in telecommunications outweigh prevention costs by many times. It therefore follows that increasing prevention costs to
reduce failure costs will result in overall productivity gains.

23.6.4 Share price

When network operators are privatised, their worth, at any point in time, is determined by the price of their shares which is, in turn, determined by their perceived value by the stock market. The prime purpose of the Stock (or Equity) Market is to provide a structured environment within which investment capital is raised. Hence, all equities of an equivalent risk class are priced to offer the same expected return. Shares are purchased by investors who require either high dividends, that is, 'income shares', or capital appreciation of the share price, that is, 'growth shares'. There is no fundamental difference between the two approaches, each delivers an equal return to the investor.

It is therefore possible to express 'Shareholder Value' as the sum of the dividends (D) and stock appreciation (SA), that is, Shareholder Value $= D + \text{SA}$. Hence the most attractive investments are those with high dividends and stock appreciation. To maintain a high and stable share price and hence company worth, a network provider needs to generate high profits from which to maintain good dividends or to be perceived by the market analysts and institutional investors as having good growth prospects and thus share appreciation.

Market share and company image in terms of the QoS delivered to customers are important influences on share price. A study by Aaker [12] found a strong relationship between changes in consumers' perceived quality and the stock performance of a corporation. It is suggested that major investors are directly influenced by changes in consumer goodwill for key brands, and they are sensitive to initiatives that have influence on, or may influence consumer quality perceptions. Maintenance of shareholder value can be a significant driver of business policy of privatised network operators. If they do not exhibit growth potential then, to maintain share price, they may need to trade-off against poor stock appreciation by paying high dividends which reduces the proportion of profit available for network investment.

23.6.5 Market share

It is generally accepted that customers make purchases on the basis of 'Value For Money' (VFM), where VFM is a function of perceived quality and perceived cost. Furthermore, as long ago as 1944, Scitovszky [13] argued that people judge quality by price on the basis that the forces of supply and demand would lead to a natural ordering of competing products on a price scale with a strong relationship between price and quality. Also, assuming that good quality costs more to produce, there would be a relationship between cost and quality and hence between cost and price if sellers set price in terms of cost plus profit margin. Much work has been done in an attempt to justify this but generally in the context of the retail market.

Hill [14] argues that as the level of quality required by customers increases, the value of a product also increases, as does the price. But the rate of increase in value decreases with the increase in quality level, whilst the rate of increase of cost, and hence price, will accelerate in the same circumstances. Hence, the best implied quality level for the customer is where the value/cost relationship has maximum gap, as illustrated in Figure 23.12.

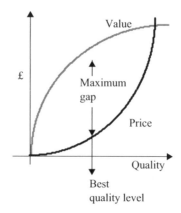

Figure 23.12   Price/value relationship

It is not clear what the price/quality relationship is in the relatively-immature competitive telecommunications market. Generally, the previous monopoly supplier would be perceived as overcharging with poor QoS and such an image is difficult to dispel. Furthermore, telecommunications pricing is complex in relation to normal purchases made by customers, that is, a basic call charge structure based on distance, duration, time of day or day of week, with possible discounts and price promotions plus rental and other services charges. It is therefore not surprising that customers' perception of price is normally poor and that they generally grossly overestimate call prices. This not only affects market share but also market size as witnessed by the poor utilisation of the local loop, that is, of the order of four to eight minutes per day on average for residential customers. This has led to advertising campaigns that are aimed at increasing the awareness of call prices, for example, by comparing them with common retail products. Simplicity of pricing may well improve VFM perception.

Price elasticity and quality elasticity do exist but they tend to vary according to market segment[1] as illustrated in Figure 23.13.

It is likely that the market segment that is most price sensitive is low-calling-rate residential customers; this is normally the loss-making segment that the universal service obligation forces network operators to serve. This segment is, however, least sensitive to quality. On the other hand, businesses which are dependent on good communications, for example, the finance segment, tend to be highly sensitive to quality but less concerned with price. There are, of course, businesses that have a large communications spend that are also very price sensitive.

23.6.6   Brand

In a competitive market, brand is an important differentiator. This is particularly important for the established telecommunications operator who finds it difficult to

---

[1] Segmentation is the process of identifying groups of customers with sufficient common characteristics to make it possible to determine the market needs that each group requires.

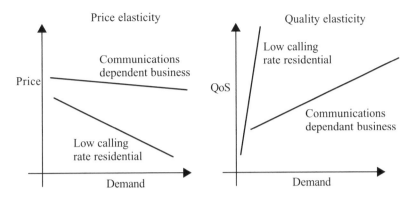

Figure 23.13   Price and quality elasticity

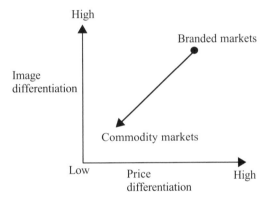

Figure 23.14   Brand behaviour

compete on price because of its large overhead-cost legacy from its monopoly era. According to Majaro [15], a strong brand brings reassurance to customers by providing a perception of permanence and quality. Not only does brand have a major influence on market share but any damage to brand image causes a slide towards the commodity end of the market with an inevitable reduction in price and hence revenue, as shown in Figure 23.14.

The perception of brand quality has a major impact on sales according to 'Total Research' who have been running the EquiTrend consumer survey since 1990. The central measure of EquiTrend is consumer perception of brand quality on a 0–10 scale. It is estimated that one point gain on the perceived quality scale can result in a 30% increase in sales [16].

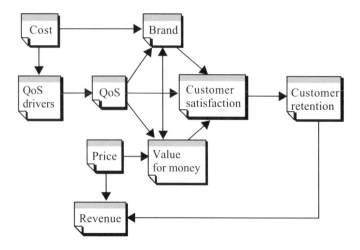

Figure 23.15   Customer satisfaction model

23.7   QoS model

A possible but subjective approach to evaluating the economics of QoS is illustrated in Figure 23.15. It assumes that the three main drivers of customer satisfaction are QoS, value for money and brand image. Thus, if two are held constant and the third varied, it would, theoretically, be possible to measure, by a survey questionnaire, the change in customer satisfaction. If the thresholds of customer satisfaction at which various proportions of customers might defect were known, then it would be possible to calculate the loss of revenue.

If it was assumed that perceived QoS was the same as that achieved, then it might be possible to change a QoS parameter, for example, time to repair, measure the cost and determine the likely retention of customers and hence revenue. Likewise, the effect of a price change and an advertising campaign could, theoretically, be evaluated. The effect of changing the variable could, perhaps, be measured by suitable questionnaires. However, there are many aspects that influence customer satisfaction, which vary according to the market segment, and isolating the three key drivers could be most difficult.

23.8   Summary

This chapter has dealt with the economics of QoS in a qualitative manner, since it is not possible to establish direct relationships between all of the cost and benefit elements. Nevertheless, these elements are large in financial terms and can have a major impact on network operator profitability. It is therefore important to understand the economic dimensions of quality and their drivers in order to have an informed view of the financial consequences of QoS initiatives or opportunities.

23.9 References

1 PENCE, J. L.: 'Is zero defects economical'. Proceedings IEEE Globecom, 1985, Paper 5.3.1

2 BELL, P.: UCL MSc Dissertation

3 PRIOR, J. and CHAPLIN, K.: 'The St. Albans study', *British Telecommunications Engineering*, 1995, **14** (1), pp. 46–9

4 VEITCH, P.: 'Resilience for IP-over-DWDM backbone networks', *IEE Electronics and Communication Engineering Journal*, 2002, **14** (1)

5 FURLEY, N.: 'The BT operational support systems architecture framework', *British Telecommunications Engineering*, 1996, **15** (2), pp. 114–21

6 Network Management Forum, 1994 – http://netman.cit.buffalo.edu/WGs/ByBds/NMF/ovw.html

7 MILWAY, N. and WRIGHT, B.: 'NAIP – The realisation of a network vision', *British Telecommunications Engineering*, 1995, **13** (4), pp. 268–73

8 FINEMAN, L.: 'Process re-engineering: measures and analysis in BT', *British Telecommunications Engineering*, 1996, **15** (1), pp. 4–12

9 JONES, S., KORYCKI, A. and ROBERTS, D: 'The truth is out there – Analysing and addressing the cost of failure in customer services processes', *The Journal of the Institute of British Telecommunications Engineers*, 2001, **2** (1)

10 WOOLLARD, K.: 'What's IN a model? Modelling IN services using formal methods', *BT Technology Journal*, 1995, **13** (2)

11 HAND, D. and ROGERS, D.: 'Network measurement and performance', *British Telecommunications Engineering*, 1995, **14** (1), pp. 5–11

12 AAKER, D.: 'The financial information content of perceived quality', *Journal of Marketing Research*, May 1994

13 SCITOVSZKY, T.: 'Some consequences of the habit of judging quality by price', *Review of Economic Studies*, Winter, 1944, pp. 100–5

14 HILL, T.: 'Production/operations management' (Prentice Hall, 1983) pp. 275

15 MAJARO, S.: 'The essence of marketing' (Prentice Hall, 1993) pp. 87–9

16 FOX, H.L.: 'As you like it', Marketing Focus, 1995

Chapter 24

Telecommunications security

24.1 Introduction

Now, more than ever, is the time for telecommunications operators (Telcos) and their suppliers to refocus attention on the issue of security in networks and support systems. Many operators are burdened with debt and suppliers are facing a downturn in fortunes as their customers take a long hard look at future volumes. In this climate, attention is naturally focused upon eliminating avoidable costs, and this should also encompass losses due to security failure and fraud.

This new focus ought, in an ideal world, to build upon already established good governance. Often, however, security is viewed as a barrier to progress. The opposite view may also be taken and a creative Telco can use security as an enabler or value proposition. On a global basis losses due to security failure and fraud have been estimated by the International Forum for Irregular Network Access (FIINA) to be as high as $40 Bn per annum in the telecommunications sector. This is based upon their estimate of a loss of up to 6% of turnover. The telecommunications sector cannot afford to ignore this opportunity.

The convergence of voice and data applications and services is upon us. Therefore, this chapter will move freely between those domains as it looks at security definitions (based upon the CIA-A model explained later), principles, relevant threats, vulnerabilities, countermeasures and current developments. The impact of security failure on Quality of Service (QoS) will be felt through degradation of confidentiality, service integrity and service availability.

24.2 Protect, detect, react and deter

Figure 24.1 illustrates an overarching model, which can be applied to security management in all its dimensions. This model, a closed loop system, has the benefits of simplicity and ease of application.

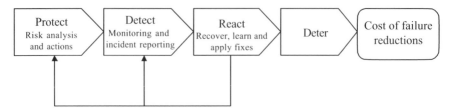

Figure 24.1   Protect, detect, react and deter model

Protect

When considering one's approach to a security issue the logical attitude to take is 'how do I protect myself against this situation?' The answer lies in risk management. It is simply not economical to protect against every eventuality. Risks are the product of threats, vulnerabilities and likelihoods which give rise to business impacts. Based upon risk assessment of vulnerabilities of assets against predicted threats and their likelihood, it follows that protection against highly unlikely events may well be uneconomical. As an example, for most operators it would be prohibitively expensive to locate all central office or trunk switches in underground bunkers as protection against physical threats. A logical, perimeter protection mechanism such as a firewall might be considered. Firewalling of narrowband telecommunications networks is difficult and potentially expensive whereas in the data domain this form of protection is commonly seen and is written about extensively [1].

Detect

In situations where protection is not an option, the next consideration must be to detect the occurrence of an infrastructure threatening event. In other words, some form of Intrusion Detection System is required. This approach is common in the data domain where the debate over the relative merits of host-based or network-based intrusion detection reigns.

React

Protect and *Detect* are of little value if some form of *Reaction* posture is not present. This will usually consist of processes and may well reuse established business continuity measures and recovery is paramount among these. *React* is arguably the most crucial component as it is here that organisational learning takes place by the application of countermeasures thereby ensuring that situations either cannot recur or that they may be contained.

Deter

There are two outcomes to the application of the Protect–Detect–React–Deter (P–D–R–D) 'mantra'. In most cases protective measures or reactive repulsion measures will result in a deterrent effect and an attacker will seek out a softer target. There will always be a small number of individuals that will view protect and detect measures as a challenge. All the more reason to make these measures rigorous. The second outcome should be a reduction in losses due to security failure with a direct effect upon the profitability of the business.

24.3   Confidentiality, integrity and availability

Perhaps the most common model of security, frequently used in the government and military arena, is that of Confidentiality, Integrity and Availability (CIA). Many definitions of these characteristics have been offered. Here is one interpretation slanted towards telecommunications transmission.

- *Confidentiality* – Assurance that information transmitted over a medium has not been accessed by an unauthorised individual.
- *Integrity* – Assurance that information transmitted over a medium reaches its destination unaltered.
- *Availability* – Assurance that contracted facilities are available to meet customer needs.
  There is one other dimension which can be added and which has significance in the telecommunications world.
- *Authentication* – Assurance that the individual who sent a message is who they say they are.

The acronym thus becomes CIA–A.

Failure of any of these attributes will result in degradation of QoS which may be reflected in mainstream QoS measures, for example, circuit availability, or it may be more insidious, for example, loss of customer confidence from which it can be difficult to recover.

24.4   Trust

Trust derives from the CIA–A model above. In the past, telecommunications companies have implicitly trusted one another across interconnects, for example, internationally. In the voice communications space, signalling protocols, such as ITU-T 7, which are essential to the interconnect, contain no mechanism for authentication of the parties to the interconnect. The physical existence of a signalling medium and traffic channels is regarded as authentication enough. Trust is therefore implicit between network operators. The potential consequences of this; the impact on network integrity and the countermeasures available to carriers, are discussed in Chapter 9 of this book. As more network operators demand interconnect the level of risk increases.

Similarly, in a local wireline access network, usually, no authentication is made of a customer terminal to the network. Compare this to the GSM mobile system where customer terminals are authenticated to the network by a strong cryptographic process. The consequence is that security of the copper access network can be compromised, for example by clip-on. The threat posed by an illicit carrier-to-carrier interconnect is much more serious.

24.5   Threats and vulnerabilities

Threats and vulnerabilities combine with likelihood to produce risk of business impacts, which are usually financial but can be QoS affecting. The threat and vulnerability landscape is changing rapidly in the telecommunications field as products, technology, interconnect and user expectations and skills develop.

24.5.1   Threats and vulnerabilities of Customer Premises Equipment (CPE)

For convenience we will take a domestic/commercial split.

Domestic CPE

In its simplest form, customer voice CPE is either hard wired or relies upon cordless technology. A hard wired telephone presents little inherent vulnerability to a carrier other than by physical disruption or by listening in. Internal wiring may be subject to interference, for example, in a multi-occupancy environment and such situations may lie outside the carrier's control. However, in a dispute situation, possibly billing related, the carrier will become involved and the inevitable outcome is the significant cost of dispute resolution. Early Cordless Technology (CT1) had very little security. Developments led to simple access coding techniques to authenticate handsets to base stations but the available codes were limited, typically to 65536 combinations. As the air path was not encrypted, overhearing was a real possibility with only eight discrete channels available. One offence or scam was to hijack calls after the handset to base station authentication by cruising a neighbourhood with a handset until a conversation was overheard. The offender would then wait until the call cleared. If the call was incoming to the targeted customer and the caller cleared first, dial tone would be offered after the usual guard period leaving the offender in control of the customer's line. In reality the chances of being hit by this scam were very small. Modern DECT technology has eliminated the issue for all practical purposes due to the difficulty, both technically and economically, that an offender would have in subverting or spoofing the DECT access protocol.

Commercial CPE

Commercial customers from small to medium up to major Businesses will often use Private Branch Exchanges (PBX). Modern PBX equipment varies enormously in capacity (and price) and in major corporations is often networked. As regulation over features which could be provided by PBXs has relaxed over the years, so the opportunities presented have been picked up, not only by customers but by fraudsters. The most commonly abused feature of PBX equipment today is Dial Through. This basic feature may be accessed by a number of services offered by the PBX and these are developed in Chapter 25 on Fraud. Essentially though, offenders are able to extend calls through the PBX and out onto public networks. The charges for those calls, often international, and therefore likely to be substantial, are rendered to the PBX owner. This invariably comes as a nasty surprise. While carriers and PBX vendors continue to work closely on this problem it remains one of the most significant security and fraud issues.

24.5.2   *Always on technology*

In the fixed network, carriers worldwide are now providing high speed data services over copper access pairs using Digital Subscriber Line (DSL) technology and its developments, generically xDSL. The end result from a customer perspective is that they may well present a long term IP address to the Internet in contrast to a limited lease IP address on a dial-up data service. It is not unusual for a customer with an xDSL access product to be port scanned by individuals many times an hour. These scans are usually looking for

- vulnerabilities in the local operating systems which might be exploited;
- running applications which have vulnerabilities which might be exploited;
- Trojan horses, such as Back Orifice or Sub Seven, which allow an intruder to gain access to and manipulate a target system.

Anyone operating an IT business with an Internet connection will, or should be, aware of this situation and will usually have some form of firewall or network address translation interface between their host systems and the Internet in the wild. Domestic users will usually not be so aware but the consequences to them of a breach of security can be just as painful.

A possible solution to this problem is the use of a personal firewall on the customer's PC. These are readily available, sometimes free of charge, and when backed up with virus protection will provide protection from the most likely threats. A word of warning though: even the simplest personal firewalls to operate require some IT competence and, therefore, in some cases this will result in a, hopefully low, residual risk. For ordinary consumers or small-to-medium enterprises, carriers may choose to offer a personal firewall bundled with their always on service. This choice must be weighed against the following risks:

- how to ensure the firewall is properly configured at all times;
- compatibility issues between the firewall software and its host machine;
- how to deliver software updates;
- the possibility that the firewall software itself may present vulnerabilities;
- claims from customers for data loss due to Internet borne attacks which the firewall cannot repel;
- false claims for damages.

In the mobile world, the rollout of General Packet Radio Service (GPRS) and Third Generation Mobile (3G), both always on technologies, will present similar problems. As mobile handset complexity grows and more functionality is added we uncover the possibility of them either being hacked directly or being used as a hopping off point to attack another system. Protection against the threat by good design is given, but the industry must be ready with a Detection and Reaction mechanism.

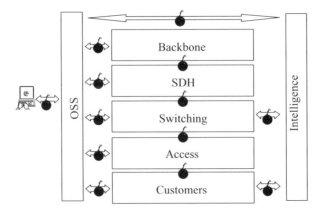

Figure 24.2   Layers and vulnerabilities

24.5.3   Threats to networks

Each component of a telecommunications network presents its own problems in terms of security. There is a danger that the growth in complexity of network components and their support functions is not matched by an understanding of the vulnerabilities.

Figure 24.2 is one way of looking at a generic network in a layered fashion. Vulnerabilities may exist both within the layers themselves and in the interfaces between layers by virtue of the way they interact, protocol vulnerabilities for example.

24.5.3.1   Access

Incumbent telecommunications carriers around the world have a massive investment in copper access networks. Its physical presence renders it vulnerable to malicious attack. The motivation for such an attack may be simple vandalism or as a means to an end, for example to disable auto-dialing burglar alarm systems in order to facilitate theft of property.

Figure 24.3 depicts a wireline access network with each component representing a potential vulnerability from exchange Main Distribution Frame – Exchange Manhole – Primary Cable – Primary Cross Connection Point – Secondary Cross Connection Point – Final Distribution Point – Final Drop to the Customer, including any joints in footway or carriageway boxes along the distribution.

Any policy for the protection of access network components must take into account

- the range of countermeasures available – lockable flexibility points, armoured cable, overboxes for distribution points, cable (de-)pressurisation alarms;
- decisions on levels of enhancement – base, medium or high – determined by risk analysis;
- decisions to enhance all the network or just hot spots – based upon risk analysis.

Figure 24.4 presents a possible policy and process structure for access network security.

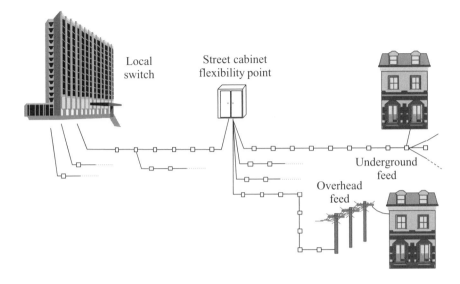

Figure 24.3   Typical wireline local access network

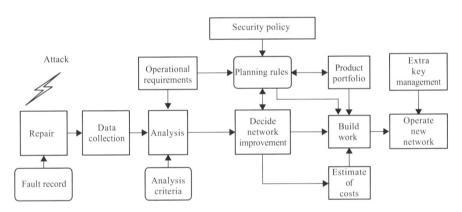

Figure 24.4   Overall security policy and process

To deliver against such a process requires a responsible entity. If such an entity exists then, armed with the inputs shown in Figure 24.5 it can deliver the benefits shown in the same figure.

The most obvious countermeasure to wireline access vulnerabilities is therefore physical and process oriented. However at least two carriers worldwide have experimented with authentication schemes. Figure 24.6 illustrates such a scheme. In this arrangement an off-hook condition anywhere on the access line is detected by the challenge/response module. A cryptographic challenge is sent back to the master socket using V.23 data protocol. If the calling loop originated inside the customer premises the master socket responds to the challenge based upon a burned in serial

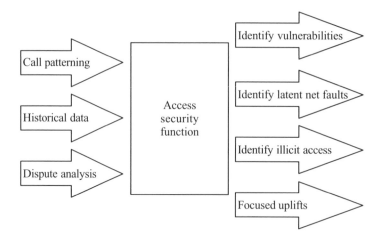

Figure 24.5   Inputs and outputs of an access security function

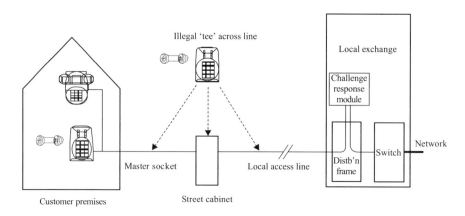

Figure 24.6   Wireline authentication scheme

number. If the response matches the challenge, dial tone is connected to the line. This process can take less than 0.3 s and is not much of an inconvenience to the customer. If an incorrect response or no response is returned to the challenge/response module this implies that the loop was produced by an illegal tee across the line. Two options are open. One could choose to deny dial tone and assume a protective stance. Alternatively, one could allow dial tone and log the digits dialled in the challenge/response module to support an investigation and eventual prosecution of the offender.

Such an arrangement is also attractive in the resolution of intractable billing complaints. If all calls can be proven to be correctly authenticated then the calls must have originated from inside the customer's premises. Practically, this tactical use is likely to be the only application for such technology until the per line costs of an authenticating master socket can be driven very low and until the challenge/response module functionality can be incorporated into the core network, possibly as an intelligence function.

24.5.3.2 Switch

Logical threats to switching network elements derive principally from their interfaces. For element management, vulnerability may arise from the access arrangements made for Operational Support Systems (OSSs). In the case of traditional TDM/ATM switches these arrangements might be via

- a command and control Wide Area Network (WAN)
- X.25
- dial-up remote access

The choice of access method will depend on a number of influences: geography, manpower availability, costs etc.

Whichever is chosen it is essential that adequate access controls are in place. If a WAN is the chosen mechanism then it should be part of a separate security domain from any corporate network. Consideration should be given to resilience of the WAN. For example if a WAN outage prevents the extraction of billing records, for how long is that tolerable and what is the likely restoration time for the failed WAN?

If X.25 is the chosen medium, Closed User Group working should be rigorously applied.

If dial-up access is used then at the very least a dial-up/dial back arrangement should be used. It is also preferable to provide further cryptographic authentication, perhaps token based.

For data networks where in-band element monitoring is attractive from a cost point of view, the risk of loss of control must be weighed against the provision of a separate overlay command and control network. No one solution will be mandated for every situation. So, for example, in the case of a DWDM backbone, given its criticality, element management should preferably be by overlay.

Switch elements may also be put under stress by attacks delivered over their interface to the wider network. In the narrowband TDM environment the most likely candidate is the almost ubiquitous ITU-T 7 signalling system. This is considered later in this chapter.

24.5.3.3 Transmission

The principal threat to transmission systems is physical disruption, either deliberately (e.g. to facilitate crime), or accidentally, perhaps as a result of road works. Networks will generally be designed to minimise disruption caused by single points of failure but not all transmission links will be provided with back-up routes as a result of economic and risk based decisions. In these cases, restoration and recovery mechanisms (Reactions) become important.

In the case of radio relay systems, disruption may be caused by electrical interference or perhaps by the interposition of an interfering structure. The reliance, or otherwise, on radio relay for prime service provision is a risk based decision.

24.5.3.4   Intelligence

Intelligence entities share vulnerabilities with switch elements in that they present interfaces for management and for network connectivity. The need to interface to ITU-T 7 networks presents certain vulnerabilities as does the element management infrastructure. There is a further issue and that is to do with customer access. It is attractive to be able to offer customers the ability to alter their virtual network by means of a simple, usually web based, thin client. This requires an interface to the carrier's intelligence platform. The PARLAY group working in this area has recognised the need to secure this interface.

24.5.3.5   OSS/network management

OSS and network management functionality is critical to the *continued* operation of a network. Loss of its functionality for a brief period of time would not cause loss of network availability directly but it would prevent the carrier from configuring his network around operational difficulties such as transmission failures or network overloads. Therefore fallback arrangements must be in place to replicate the important functions at a lower layer in the network management hierarchy. Full network management centre functionality would not be required at the fallback centres; it is possible to manage without some performance statistics for a period of time. Network re-routing capability is an essential fallback element. It should go without saying that the right skills must be present at the fallback centres and the fallback process should be regularly exercised as part of business continuity management.

24.5.3.6   Protocols

Telecommunications networks rely upon signalling protocols for their operation. The protocols used have developed over the years from

- in-band DC, for example, loop disconnect;
- in-band voice frequency, for example, DTMF, CCITT 5, CCITT R2;
- hybrid in-band/out-band, for example, CCITT R2(Digital);
- full out-band, for example, ITU-T 7;
- IP oriented, for example, BICC, SIP.

Each of these carries associated vulnerabilities to the operation of networks. There are differing degrees of exploitability. Beginning with humble loop disconnect we have seen the ability to bypass payment mechanisms in very early pre-payment call offices. In-band voice frequency systems like CCITT 5 have been exploited heavily in commission of fraud. This is developed in Chapter 25.

Most prevalent in the narrowband TDM world is ITU-T 7. This signalling system carries with it much power to disrupt. Not only is it responsible for call control but there are powerful network controls embedded into the protocol. Chapter 9 gives an insight into some network integrity issues. As the proliferation of interconnected carriers continues, the chances of misoperation of ITU-T 7 has increased. The recently introduced Licence Condition 20 for operators in the UK has ramifications for interconnected operators. The task group commissioned by the Network Interoperability

Consultative Committee (NICC) to produce best practice guidelines, has also drafted example criteria for disconnection of service. Under these criteria, the importance of ITU-T 7 signalling is recognised. If an interconnected operator can be proven to be sending disruptive signalling sequences into another carrier's network, then subject to certain procedures, the offending carrier's interconnect may be disconnected from a signalling point of view. The condition of proof must be met and carriers should consider the benefits of ITU-T 7 monitoring systems, particularly on network interconnects. Benefits are

- early warning of potentially damaging signalling events;
- assistance in fault finding;
- verification of interconnect billing;
- source of business information;
- data source for fraud management.

Chapter 9 gives us some clues as to countermeasures, such as gateway screening using policing masks. As interconnected operators move towards allowing STP working over the interconnect the usefulness of screening diminishes. Also, whilst ITU-T 7 policing measures are very useful in preventing propagation of damaging signalling events into a network, they are a limited resource in most switches. Their Protect function is not therefore universally available and this reinforces the argument for monitoring of network interconnect points.

24.5.3.7 Physical and People

Most operators will have well developed physical security arrangements for their switching nodes and in some cases, access network elements. Within this remit should also rest consideration of gas, fire and flood detection. Access to buildings and plant should be restricted to those personnel who need access for their work and this should be the cornerstone of physical security policy. There is a critical dependency for any operator on its people. Certain job functions such as data build and network control, are crucial to the day-to-day operation and survivability of a network. Consideration should therefore be given to commercially vetting staff who are in critical posts. The objective should be to eliminate the risk caused by individuals who may be manipulated into causing damage to the enterprise. In any case, the need for basic physical security health lies with individuals and this should be recognised by entries in job descriptions and reinforced by periodic briefings and web based training.

A new threat has arisen due to the drive to outsource effort, for example,

- switch maintenance by switch vendors;
- network order handling and provisioning by network vendors;
- third-party running the access network;
- buildings and facilities management.

Such cases should be managed at a minimum by Service Level Agreement with a further consideration to commercially vet suppliers in certain critical situations. Where access to the operator's internal systems (e.g. IT systems) is required this should be secured by Authorisation, Authentication and Accounting (AAA) techniques as a matter of policy.

The advent of Local Loop Unbundling has also brought with it the situation of other operator's staff having access to incumbent operators' sites. In such cases, physically securing the access arrangements and separating access within a building may be problematic. In such a case, where Protect fails, the Detect option of a Closed Circuit Television installation may be considered.

24.5.3.8   Internet Protocol (IP) and next generation security

The convergence of the IP world and traditional TDM, in the form of Voice over IP for example, means that issues and opportunities specific to IP networks are becoming significant for network operators. Using the P–D–R–D model we can examine some of the components.

Protect – firewalls

Firewalls are an essential component to any network or host wishing to interface with the Internet. More generally we can say that a firewall is mandated on an interface with an untrusted network or entity. In the telecommunications environment, firewalling of TDM networks has been problematic, largely due to the processing involved and the problems of the ITU-T 7 protocol which carries network management control intimately interwoven with call control. From this perspective the situation is better in the Next Generation world where firewall functionality may be built into IP network elements, at the hardware layer in addition to the provision of dedicated firewalls at strategic points. Also call control and network control can be carried by independent protocols, an improvement in risk over ITU-T 7.

Protect – encryption

This is a major enabler in protecting confidentiality and is in widespread use in data networks. Its use in telecommunications networks in general is therefore certain to develop but it is not without problems in some countries. There are also issues of key management and information recovery in the Legal Interception area. In the developing Voice over IP, setting issues over performance is problematic when considering encrypting voice data streams – latency, an inevitable by-product of encryption/decryption, must be kept to a minimum.

Protect – authentication

Cryptographic techniques in conjunction with IP based network elements provide the means for authentication of those elements to each other to form a trusted network. This potentially provides us with a feature hitherto denied in narrowband networks. The challenges, in a large network, are those of key and certificate management.

Detect – intrusion detection

This functionality is well developed in the IP world and provides a valuable tool to detect unwelcome events on hybrid networks. At least one manufacturer of ITU-T 7 monitoring probes has partnered with a customer to develop an intrusion detection function. Operators choosing IP based telephony products to replace traditional TDM methods must also factor in the robustness of the chosen product to traditional IP threats like Denial of Service attacks, maybe due to message flooding.

Some risks are shared between the old (TDM) and new (IP) technologies. It is possible for a denial of service attack to be launched on a TDM network simply by sending multiple call set-ups towards a destination which cannot answer all the calls. If no restrictive network controls are applied, and this is quite possible if the event is not foreseen, then not only will the destination customer be effectively isolated but there is a real danger that the processor in the destination switch will become overloaded and fail. Likewise, a rogue signalling end or relay point might send multiple messages to a Destination Point Code (DPC) which could be forced into failure. In IP networks, particularly the public Internet, denial of service attacks are fairly common events. The differentiator between the two technologies is that pretty well anyone can get an IP address but only a few can get an ITU-T 7 Originating Point Code (OPC). This leads to the inevitable conclusion that IP networks used for public telephony etc. should be running in separate domains and not accessible from the public internet.

Some countermeasures for IP based networks are

- intrusion detection systems
  - packet filtering for invalid source address
  - consideration of both host based and network based solutions;
- strong (i.e. cryptographic) authentication of network elements;
- inventory and version control;
- physical protection of network elements;
- logical and physical separation of resource and backup (e.g. call control servers);
- loading of only necessary software on elements.

24.6 Some drivers for change

24.6.1 Mobility

Since their inception, mobile networks have been a target, usually with the aim of fraudulently obtaining service. Way back when mobile services were operator connected and in the early direct dialing systems, lack of security mechanisms in the mobile equipment gave rise to fraud. This was often simple to commit by cloning a mobile using readily available hardware and cheap 2764 EPROMs or even fuse-link PROMs. Early cellular systems did not prove much of an impediment to offenders and cloning of TACS and AMPS systems was rife. GSM, CDMA and later developments have proven much more robust and the offenders have sought alternative means of leverage rather than overcoming the inbuilt security measures. That is not to say that some of the security measures used in modern cellular systems have proven to be beyond attack but they are arguably adequate in risk management terms. The impact of the mobility aspects of IP terminals and always on technologies has yet to hit us.

24.6.2   Electronic trading

Security of transactions is of paramount importance. Whilst the security of transport mechanisms can be reasonably assured, by means of Secure Sockets Layer (SSL) or Secure Electronic Transaction (SET) with appropriate key lengths, the major threat to e-trading still comes from fraudulent transactions, prevention of which is outside the capability of the technology. The lesson is not to ignore the low-tech threat and to ensure that the technology solutions are wrapped in robust processes.

24.6.3   Legal interception

As complexity of networks increases and convergence develops, the difficulties faced by law enforcement agencies are a matter of live debate. Changes in legislation often lag behind the introduction of new technologies and they are frequently challenged by civil liberties campaigners. There can be little doubt about the necessity for legal interception of traffic but its sensible management and application in a converged world needs careful attention.

24.6.4   Common equipment means shared vulnerabilities

Bespoke systems are becoming a rarity in the telecommunications world. IP based networking means that operators will tend to use similar equipment and share the vulnerabilities that go with it. This provides an opportunity for offenders who have knowledge of these vulnerabilities to launch exploits much further afield than before. This uncovers another debate which is running at present. Should vulnerabilities in network elements be public domain knowledge or should their existence be known to only the chosen few? Those with an academic background would probably choose the former approach as a fix would be subject to peer review and could be expected to be robust. Those, perhaps more commercially oriented, individuals may choose the latter approach in an attempt to protect their interests. On balance the concept of peer review leading to a robust solution seems the best approach whilst remaining cognizant of the risks.

The key to keeping up with vulnerabilities is a Protect function:

- Maintain an accurate inventory of network elements:
  - build level;
  - patch status;
  - owner (must be kept current).

- Monitor for vulnerabilities and patches:
  - contact with manufacturers;
  - Computer Emergency Response Team (CERT);
  - web based information, for example, bugtraq.
- Validate, test and apply patches promptly.

24.6.4.1   Malicious code – viruses, worms and Trojans

Whilst this book does not attempt to cover the subject of malicious code in detail, its impact on networks and OSS, including customer relationship management systems, is such that it is worthy of mention.

The US Institute for Telecommunication Sciences (www.its.bldrdoc.gov) defines malicious code as,

> Software or firmware capable of performing an unauthorized function on an information system.

Definitions of viruses, worms and Trojans are many and varied. Furthermore, the creators of malicious code, as they are not constrained by definitions, will 'design' in features of all three thereby frustrating lexicography of the subject. With this caveat in mind the following working definitions are offered.

Virus

A virus is a program which is usually unauthorised and which may insert itself into an 'innocent' pre-exisiting program, maybe even the host's operating system. It is also capable of replicating itself, usually by infecting other innocent files when it runs. If those files are shared with another host, the virus can propagate. A virus may or may not carry a payload which may or may not be destructive, for example, through deleting or altering stored data. Viruses may go to some lengths to avoid detection, for example, by mutating over time – one reason why anti-virus software must be kept up to date.

Worm

A worm is malicious code which is capable of self replicating, usually by means of network connectivity between hosts. It is usually differentiated from viruses by not attaching itself to existing programs. It will typically consume operating system and network resources which can give rise to QoS degradation both for the host and the connected network. Worms may also deliver payloads.

Trojan

Derived from the ancient tale of the Trojan Horse, this is a program which may have been delivered by a virus or a worm. A trojan might function as legitimate software, maybe a screen saver, whilst at the same time replacing existing functional-ity, for example, masquerading as a login prompt and capturing username/password information. It might remain dormant as a back door to facilitate easy compromise of a system by an aggressor.

Regardless of the purity of definition, all malicious code is objectionable and must be countered. Malicious code writing kits are available in the wild and the phenomenon can truly be said to be pandemic. Carriers must respond to the malicious code threat and the minimum response set should be

- policy led enterprise anti-virus protection with automated virus signature updates;
- running of minimum software by hosts in order to carry out their design function. For example many servers were unnecessarily hit during the fairly recent 'Code Red' worm incident because they were running Microsoft's IIS when it was not actually required;
- gateway screening for at-risk e-mail attachments, that is, all executables as a minimum;
- e-mail server screening for at risk e-mail attachments to suppress internal spread over an intranet;
- consider blocking intranet access to public web based e-mail providers.

24.7   Information Warfare

In our connected world, military Information Operations take their place alongside Land, Air, Sea, Space and Psychological Operations. This leads to the concept of Information Warfare (IW) and its synonym Cyberwarfare.

The more dependent national infrastructure becomes on new ways of trading or administration, through e-Business, e-Government etc., the more critical those entities become and the more attractive to a potential aggressor. Thus attacks upon the information infrastructure of a community have become a reality. Even intra-national cyber-attacks are expensive. For example, in 1999 the FBI and the US Chamber of Commerce announced that US companies were losing around $2 billion per month to corporate espionage, including those launched through computer systems (www.fas.org/irp/ops/ci/docs/fy99.htm). If a telecommunications operator has any doubts about the general level of cyber-activity out there they should ask their firewall administrator how many packets are rejected per day. Not all of these are attacks by any means, but should a weakness in the firewall be detected, the aggressor may escalate the impact of his attentions rapidly. The range of impacts of such an attack will span from simple web site defacement through to denial of service attacks leading to loss of business. Direct impact aside, there is no doubting the adverse effect on brand value of even the most 'benign' of attacks should the News media become aware.

In an effort to differentiate the response to IW from Business as Usual security, an IW attack is defined here as

A structured, unauthorised and targeted attack ... **On**; Critical information assets ... **To**; Derive intelligence, deny service, corrupt data or channel an attack elsewhere ... **By**; Entities who are able, motivated and probably funded.

24.7.1 Critical national infrastructure

Telecommunications operators provide a component of any nation's Critical National Infrastructure (CNI).

Key elements of the CNI are

Energy	Water
Finance	Government
Transport	Food
Communications	Sewerage

and, in some countries, Weather Services.

Telecommunications is arguably the most important contributor to the CNI. There are interdependencies within the elements of this table but by far the most 'interconnected' of those elements is (Tele)Communications.

24.7.2 Response to the IW threat

The following quotation from the website of the US National Communications System (www.ncs.gov) sets the scene in the US from 1990:

> In April 1990, the Chairman of the National Security Council (NSC) Policy Coordinating Committee for the National Security Telecommunications and Information Systems requested that the Manager, National Communications System (NCS), identify what actions should be taken on the part of Government and industry to protect critical national security telecommunications from the threat from computer intruders. Working together, the Manager and the President's National Security Telecommunications Advisory Committee (NSTAC) established a structure and a process for addressing network security issues.
>
> Central to this process are separate, but closely coordinated, Government and NSTAC Network Security Information Exchanges (NSIEs). Government member organizations include departments and agencies that are major telecommunications services users, represent law enforcement, or have information relating to the network security threat. Industry member organizations include telecommunications service providers, equipment vendors and major users. NSIE representatives are individuals who are engaged full time in the prevention, detection, and/or investigation of telecommunications network software penetrations, or who have security and investigative responsibilities as a secondary or collateral function. Both Government and NSTAC NSIE representatives are subject matter experts in their fields.
>
> The NSIE's primary focus is to exchange information on issues of unauthorized penetration or manipulation of the Public Network (PN) software and databases affecting National Security and Emergency Preparedness (NS/EP) telecommunications services. Periodically, the NSIEs conduct a risk assessment of the PN. Previous risk assessments were completed in 1993 and 1995. The current version, 'An Assessment of the Risk to the Security of the Public Network' was completed in April 1999.

In July 1996, John Deutch, then Director of the US Central Intelligence Agency, was quoted thus:

> The U.S. will face very, very large and uncomfortable incidents at the hands of cyber-terrorist ... The U.S. Government is not well organised to address the threat of foreign attacks on public switched networks and government systems ...

The UK Government responded in December 1999 with the formation of the National Infrastructure Security Co-ordination Centre (NISCC) with the specific goal of working with government departments and industry to coordinate and develop their work in defence of the Critical National Infrastructure against electronic attack. This was greeted with some enthusiasm within industry, in particular telecommunications, who had been working for some time on the subject whilst seeking a steer from the government. NISCC is also responsible for operating the Unified Incident Reporting and Alert Scheme (UNIRAS) which, amongst other things, is the UK Government's Computer Emergency Response Team (CERT) and is a member of the Forum of Incident Response and Security Teams (FIRST) organisation.

Information Assurance

The term Information Assurance (IA) puts a positive spin on the activities of CNI members in countering the IW threat, often termed Defensive Information Warfare or DIW. Academia and Industry members of the CNI community in the UK formed the Information Assurance Advisory Council (IAAC, www.iaac.org.uk), with strong links to the government and with representation from major CNI players. Its overarching objective is to promote knowledge and understanding of Information Assurance aspects. IAAC defines Information Assurance as '. . . operations undertaken to protect and defend information and information systems by ensuring their availability, integrity, authentication, confidentiality and non repudiation'.

24.7.3   *Telecommunications operators and IA*

In recognition of their contribution to the CNI, operators should consider whether their existing mechanisms for business continuity management, response to risk and threat etc. are appropriate to the IA scenario. With the current and developing multiplicity of interconnected networks, technologies and products, there is arguably only one network worldwide and any weak point presents a risk to all parties. Threats can arise from

- nation states;
- terrorists;
- commercial competitors;
- criminals;
- disaffected employees;
- technical delinquents.

The impact of failure to mitigate these threats, should they occur, can be difficult to quantify. But one need only to look at fairly recent events with widespread network outages in North America, Singapore and Japan, together with power outages in Canada and New Zealand to gain an appreciation of the potential impact.

Figure 24.7 positions IA activity as distinct from Business as Usual (BaU) security. In the real world, however, overlaps will exist. IA aggressors are postulated to possess relatively high capability and to concern themselves with high impact targets.

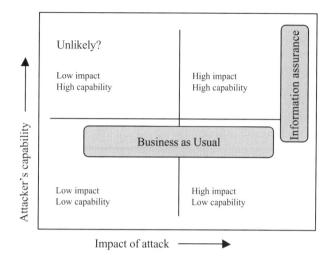

Figure 24.7 IA vs Business as Usual (BaU) security

This is helpful in directing risk analysis work for IA as distinct from BaU security risk analysis. The number of dimensions for analysis is reduced to those forming part of the CNI (or a dependency thereon) or enterprise threatening and which may be subject to attention by a skilled attacker. Formal risk analysis processes can then take over and produce a list of recommendations for corrective action which may be prioritised for action. A further refinement is to model the capabilities of the potential attacker in order to divine a likelihood score for each vulnerability against each potential attacker. This should enable realistic prioritisation for actions.

Risk analysis leading to the generation of countermeasures fulfils the *Protect* requirement of an IA response. This must be underpinned by a *Detect/React* posture bearing in mind that an IW attack may be

- preceded by low level probing, probably not serious enough to raise alarms;
- delivered on a wide front, not simply one system and not restricted to the cyber domain for delivery. There may be a kinetic element (physical disruption);
- arranged so as to frustrate recovery mechanisms;
- targeted elsewhere, using the carrier as a conduit.

Therefore, a 'joined up' approach across the carrier's business is mandated. This may build upon business as usual or it may take a more IA focused form.

24.7.4 Some generic best practice statements

- Remove system defaults.
- Only run the minimum acceptable set of services on your system.

- Audit firewall logs and have supporting response processes in place which support user's requirements and align with other access control mechanisms.
- Consider processing audit and accounting logs offline, using a so-called 'drop log'. This can offer some protection against certain firewall attacks and provides some chance of analysing an event.
- Watch the alert resources on the net, for example, bugtraq, for vulnerability warnings and apply patches in good time.
- Have a rehearsed recovery plan.
- Be prepared to disconnect from the Net in extremis.
- Ensure backups (especially legally binding documents) are current and accessible.
- Responding to a back-traced attacker with a retaliatory attack is not advised.

24.8   The role of policy

Policies are statements of company attitude or direction on particular subjects.

Figure 24.8 depicts how security policy can be driven by adoption of a security management system, in this case BS7799/ISO17799. In the right hand box, the elements of an enterprise security framework are listed.

Specific security policies should encompass the following areas:

- electronic information security
    - e-mail and Internet use and access,
    - working from home,
    - media handling and destruction,
    - data communications and networks,

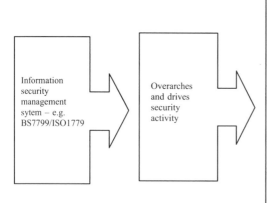

Figure 24.8   Relationship between security management and policy

- computer installations,
- logical access/passwords,
- Personal Computer users (including portables),
- viruses,
- disaster protection,
- development and support services;
- people security
  - recruitment,
  - leaving the company,
  - visitors,
  - senior managers travelling abroad,
  - staff protection;
- physical security
  - physical access control,
  - key management,
  - valuable items in transit,
  - local network security,
  - protection of the core network,
  - vehicles,
  - fire/bomb alerts and threats;
- security process policies
  - fraud,
  - investigation,
  - incidents and reporting,
  - product protection,
  - information leak investigations,
  - internal communications networks (including PBX);
- information security policies
  - data protection legislation,
  - security alert status,
  - information security,
  - government security,
  - suppliers,
  - partners.

24.9 Prosecution

The final element of the P–D–R–D model, deterrence, leads us to a brief discourse on the basic elements of supporting a courtroom prosecution of a telecommunications related offence. Although based upon English law the principles should hold good for all practical situations as they are based upon common sense. For a complete view of the P–D–R–D model the principles are covered here, though they are also applicable to Chapter 25 on Telecommunications Fraud.

24.9.1   Evidence preparation

Courtroom audiences are almost invariably not technically literate. One must therefore make allowances when describing complex systems and the way they operate. Even the basics of what happens when making a telephone call, when broken down to its components, can be a challenge to describe in non-technical terms. A broad knowledge of switching, billing (accuracy), data communications/networks and protocols (robustness) will stand the potential witness in good stead. Further points to note are:

- System time clocks may not be synchronised and they may drift but this can and must be explained and measured.
- In terms of billing accuracy, no system is perfect and certification to a standard (e.g. Oftel OTR003) can help.
- Cordless telephones can prove a nightmare in annoyance call cases as security can be weak.
- Know and explain the differences between analogue cellular and GSM systems.
- Your evidence must have good provenance.
- You should avoid hearsay unless specifically allowed.

24.9.2   Conduct in court

In general, juries, judges and magistrates will take into account your demeanour. Here are some commonsense suggestions:

- You might be aware of some weaknesses in your case, such as the apparent insecurity of an access network. If this is the case then a high risk strategy can be used to explicitly draw attention to its weaknesses. You would then build up a picture which establishes that the chances of a security failure are actually very slim. The essence here is candour.
- Do not be evasive.
- Do not lose your temper.
- Know when to say you do not know.
- Ask for clarification of questions – it can give you thinking time.

24.9.3   Some general principles

- Consider jurisdiction – where was the offence committed, detected, prosecuted?
- What is the burden of proof; for example, balance of probabilities or beyond reasonable doubt?
- Admissibility of computer produced evidence might be problematic if it cannot be proven that the system in question was working properly at all material times.
- Be sure that what you are prosecuting is actually an offence.
- Be certain of who the suspects are.
- Protect evidential provenance – no tampering.

- Do you prosecute a minnow or wait to get the shark?
- Be aware that the opposition might engage an expert witness.
- Is private prosecution an option?
- The media have in the past spun courtroom stories so that the network operator looks bad.

24.10   Summary

Telecommunications operators face mounting security issues as the drive towards convergence continues apace. As they move into the content provision sector, if they are not already there, the issues extend from protecting the infrastructure CIA-A to protection of delivered content. Cryptographic techniques provide the way forward but heed should be taken of the difficulties faced by cable and satellite TV operators in user authentication. The fact is that cryptosystems will be examined, probed and attacked. As networks evolve, the security issues facing data networks such as denial of service attacks, node-to-node authentication etc. will come into the province of the erstwhile narrowband operator.

The QoS effects of security failure derive from failure to attend to CIA-A. Most obvious is the direct effect on QoS of loss of availability caused, say, by a denial of service attack. If doubts exist over the likelihood of such an attack one need only to look back to the 1960s in the UK and the so-called 'MiniCab Wars'. Here rival minicab operators would tie up the lines of their competitors by dialing them from payphones and leaving the handset off-hook – not an attack on the network per-se but with definite QoS implications for customers. Packet flooding on hybrid or IP networks may also cause focused overloads giving rise to QoS degradation in bandwidth availability, jitter, delay and packet loss. At present no measures are specified for the QoS implications of security failure. Within the ITU, its Quality of Service Development Group (QSDG) is looking at the issue and so are the Communications Quality and Reliability forum of the IEEE. If the consequences of security failure impact upon QoS then the potential exists to use QoS degradation as a possible indicator of security failure.

There is much work to be done in building a language to describe QoS related security issues and associated metrics, one of the major barriers being the understandable reluctance of operators to disclose potentially share price impacting events. A radical solution to this may be for regulatory authorities to engage in the debate with the objective of engendering openness in reporting of events without resort to compulsion.

24.11   References

1   W.R. CHESWICK and S.M. BELLOVIN: 'Firewalls and Internet security, repelling the wily hacker', (Addison-Wesley), Reading, Mass USA, 1994

Web sites

1 IAAC   Information Assurance Advisory Council **www.iaac.org.uk**
2 IEEE CQR   Communications Quality and Reliability **www.comsoc.org/~cqr/**
3 Parlay   PARLAY group **www.parlay.org**
4 QSDG   Quality of Service Development Group **www.qsdg.com**
5 US ITS   US Institute for Telecommunication Sciences **www.its.bldrdoc.gov**
6 US NCS   US National Communications System **www.ncs.gov**

Chapter 25

Telecommunications fraud

25.1 Introduction

Fraud in the telecommunications context may be defined as

> The obtaining of telecommunications services or delivered content with intent to avoid payment in full or in part

This leads to some questions:

- About where an offence occurs physically, for example, on whose network in an interconnected world?
- Who is the owner of any content fraudulently obtained?
- Who is responsible for protection of content in transit?
- What is a reasonable level of physical and logical protection for products, services and content?

Network operating costs incurred by an operator will be recovered through billing and attempts to fraudulently subvert this process fall within the operator's responsibility. The situation is less clear cut where interconnection of networks, public or private, is involved and where the boundaries of responsibility for the physical network and its ephemeral content are not well defined.

Technology drivers for fraud
- mobility – can provide anonymity and present difficulties in detection;
- blurring of the network edge – new risks of vulnerabilities in networks being exploited;
- network complexity – may hinder an operator's ability to detect subversion as technical skills dilute over time;
- always on technology, mobile and fixed.

Commercial drivers for fraud

- High value services – maximise the benefit to a fraudster for a given exposure time.
- Multiplicity of products and services – each product or service carries its own vulnerabilities and they may be used in concert to facilitate fraud.
- Multiplicity of competing operators – fraudsters will move between operators to avoid detection. Where does the responsibility for fraud detection lie?

Fraud and security

- Fraud opportunities may arise from a lack of security – but lack of security is not necessarily a prerequisite for fraud.
- Security (failure) is not necessarily to do with money; fraud invariably is.

The crime triangle

The diagram in Figure 25.1 will be familiar to criminologists. The principle is that if any one side of the triangle is missing then a crime will not take place – in theory at least.

Figure 25.2 gives us some steer as to where countermeasures may be sought but it is not a foolproof approach. The psychopathic element in society will not necessarily conform to societal norms with respect to motivation.

When considering a telecommunications operation, the most difficult element of the crime triangle to affect is that of *capability*. Telecommunications networks are in the main built to open standards and from experience, those who are inclined towards committing fraud are prepared to spend inordinate amounts of time in preparation for their misdeeds. Reducing the argument in the extreme, the capability required to dial a telephone call is minimal but the potential fraud exposure if a fraudster uses another person's telephone line illegally may be great. Therefore, there is little that can practically be done to mitigate the capability element of the crime triangle. Fraud management policy should therefore concentrate on good design, detection, reaction and deterrence. Readers will identify the parallel with *Protect, Detect, React and Deter* from Chapter 24.

Figure 25.1   The crime triangle

Elements of the crime triangle	Countermeasure strategy	Potential for realistic countermeasures
Motivation	Increase the risk of detection and prosecution, especially where high value services are concerned.	High
Opportunity	Increase the effort required to commit a fraud, for example by good product design.	High
Capability	Protect intellectual property about design of a product or service.	Low

Figure 25.2   Elements of the crime triangle

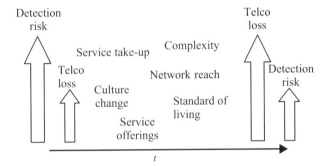

Figure 25.3   Detection risk and loss over time

History and change

Figure 25.3 presents a hypothesis from both an operator's and a fraudster's perspective. As time *t* has passed, operators have moved from a low potential loss to a high potential loss situation. This is due both to technology and commercial effects and also to lifestyle and standard of living changes over the years. Over time, the fraudster has moved from a high risk of detection to a low risk of detection situation – *in the absence of any direct countermeasures to his activity*. In simple small networks, perturbations from the norm due to fraud should be easy to spot using simple observational methods. In large complex networks involving interconnects the problem scales up by orders of magnitude and a more considered approach is necessary.

25.2   Cost of fraud

Opinions vary widely in the telecommunications industry of the exposure to fraud that the industry faces. There is one common conclusion though – that it is both significant and worth addressing.

The Forum for International Irregular Network Access (FIINA), which has been studying network abuse and fraud since before 1987, has estimated that the loss due to fraud for an operator may be as high as 6% of annual turnover. From experience, a well managed world-class operator can expect to reduce this exposure to around 0.2%. However, even at this level, which compares well with a UK retail sector average of about 1% (http://www.retailing.uk.com/index.html), the hit is on the bottom line and it is worth adopting a risk management approach to continuous fraud management improvement.

25.3   Scene setting

Telco fraud has traditionally been viewed, by perpetrators in particular, as a victimless crime. As we have seen, the reality is that losses due to fraud hit an organisation's bottom line and therefore will impact all its stakeholders; directors, shareholders, suppliers, employees and customers.

The nature of Telco fraud, or crime against Telcos in general, is that the offending acts are usually committed remotely. The offenders' perception, is therefore, that they are unlikely to get caught. As far as the offender is concerned this, is therefore, a positive motivational influence. Telcos worldwide are working hard to counter that mindset.

Telco fraud is often committed without the use of tools. Instead of a jemmy or a crowbar, many of these crimes involve the use of knowledge only; for example, a calling card or credit card number.

Telco products which are frequently part of a commodity market may be easily converted to cash; for example, in Call Selling or Money Laundering.

The dilemma facing Telcos is how to ensure that 'Products are user friendly and *abuser* unfriendly' and this must be addressed at product design – not after, or, least of all, post implementation. The fraud scenarios covered in this chapter are worldwide phenomena, no one Telco is immune from fraud exposure and those that claim to be cannot be trading.

25.4   The effect of products on fraud

The inherent attributes of a Telco product have a direct affect on its susceptibility or attractiveness to fraud. Figure 25.4 is a brief illustration of some generic attributes and some associated specific vulnerabilities.

25.5   Subscription fraud and call selling

All Telcos face the problem of bad debt and in all cases bad debt falls into two categories:

Can't Pay   Won't Pay

Generic	Specific
Access method	Wireline or mobile air interface, direct or indirect other operator (formerly other licensed operator or OLO) access.
User authentication	PIN access control possible on mobile but typically none for PSTN.
Terminal authentication	GSM mobile authentication vs none (typically) for PSTN.
Services accessible	POTS, International Access or Premium Rate (usually revenue share/generating).
Mobility	Mobile fraud beneficial for call selling and organised crime.
Exploitation of design weaknesses	Inadequate product testing may propagate fraud vulnerabilities, e.g. unguarded release on PBX or Auto Call Distribution systems may allow dialing through.
Interaction of features	Unexpected fraud vulnerability caused by a design feature interacting with a known fraud scenario, e.g. call diversion stacking interworking with call selling.
Customer premises equipment	Often outside the influence of the Telco operator this type of fraud is one of the crime success stories of the last century in its PBX fraud guise.
Billing frequency	May open a window of opportunity, e.g. if a revenue generate or share product pays out more frequently than billing for the originating service.

Figure 25.4   Generic and specific product attributes

It is not intended to go into the debate about the causes, social or otherwise, of the Can't Pay situation but the problem can be mitigated by

- low volume user rebate schemes;
- prepayment schemes, either deposit based or prepay calling card;
- limited access to some services, for example, international calling, until a credit history is built.

More insidious is the case of Won't Pay. This category is estimated to form about 25% of the total bad debt in a Telco. Telcos are not the only people in the business of making money from phone calls. Illegal call selling ranks alongside PBX fraud as one of the leading Telco frauds today. Most Telcos operate on the basis of periodic billing of their customers, typically monthly or three monthly. This instantly gives rise to a time based exposure. If we ignore the low volume of singleton customers who have no intention to pay but are in other respects 'normal users' we are left with the organised fraudsters, often operating as teams and seldom exposing the principal organisers. An organised call selling operation – which is essentially a subscription fraud – can cost the Telco £20,000 or more in one weekend in terms of lost revenue. This highlights the assertion that telephony is a commodity product in the criminal world as well as the legitimate Telco world as it is easily sold on. The attributes of a call selling operation are

- avoidance or defeat of credit checking systems, for example, by theft of identity;
- some early bills may be paid, untraceably;
- advance deposits which may be required by Telcos are readily paid but often untraceable;
- use of network features such as three way calling and call diversion are prevalent;
- may involve application for unusual number of access lines;
- may involve the use of a compromised PBX;
- may involve the use of a compromised calling card;
- may involve the use of a compromised credit card;
- frequently operate in teams with high mobility;
- rarely is the organiser involved in the actual operation of the fraud – 'drones' are easy to find;
- and, of course, ultimately there is no intention to pay.

Fraud detection systems do not need to be sophisticated in order to detect call selling but they need to be able to react quickly in order to minimise exposure where serious losses can amount from only a few days of operation of a scam. Simple profiling of customers and thresholding can detect all but the most sophisticated of operations and this functionality can often be integrated into existing billing systems. Countermeasures might then be early customer contact or interim bills. Experience shows that, even if no fraud is present, customers appreciate the attention that a Telco affords them. Dedicated fraud management systems may then be directed at more sophisticated or high potential cost scenarios.

Old fashioned detection methods are negated by newer network services. Antifraud practitioners will remember the days when it was possible to cruise the streets of a town or city and spot the house, or even a garage, with a queue outside. This would be the likely location of the call sell operation. Nowadays, call diversion and three way calling mean the fraudster's customer does not need to leave the comfort of his own home. The organiser's 'drone' will be the only occupant of the fraud location and he will be responsible for setting up and conferencing calls according to a timetable.

The discussion so far has been oriented towards wireline but subscription fraud affects mobile networks also. Multiple fraudulent applications for service have been seen to facilitate call selling operations. Handsets bought in a country where they are deeply discounted, such as the UK, have found ready customers in countries where the pricing regime requires full payment for the handset up front. Handsets are stolen, both in bulk and individually, with the intention of selling them – usually after reprogramming the IMEI. The mobile industry is working hard to close down their exposure.

25.6 Access network fraud

Frequently referred to as 'Clip On' or 'Teeing In' this fraud area refers to illicit connections facilitated by the wireline access network's ubiquity and discussed in Chapter 24. Illicit connection in the access network may be used as part of a call sell scam, the unfortunate renter of the affected line becoming inadvertently involved. Access network abuse has also been seen as a method to avoid legal interception of telephone traffic although the advent of Pay as You Go mobiles makes access network abuse less attractive. Aside from the physical countermeasures discussed in Chapter 24, as the mechanics of the fraud are essentially the same as for subscription fraud or call selling, the fraud management countermeasures are likewise the same. Clip on fraud is referred to in phreaker terminology as Beige Boxing. A beige box is no more than a placeholder term for any apparatus able to connect to a line pair – for example, a linesman's telephone or modified domestic telephone.

25.7 Payphones

Payphone fraud usually involves bypassing of the charging arrangements of the payphone mechanism. This is sometimes by exploitation of design errors or it may be effected by defeating the incoming charging signals from the local telephone exchange or central office. For example, if a payphone is reliant upon Subscriber Private Metering (SPM) pulses in order to charge then by blocking the pulses a free call may be made. Figure 25.5 tabulates some generic payphone fraud issues. Many of these have been fixed but they serve to teach the lesson that vigilance in fraud exposure is necessary right from the design stage, particularly as even modern payment methods using smart card technology have come under attack.

Due to their occasional isolated location, payphones have also been the target of clip on fraud, in one case involving the connection of eight auto-diallers programmed to call a range of revenue generating premium rate numbers.

Modern payphones and their operational support systems are now able to assist in the direction of Police operations against theft, fraud and vandalism. With mobile telephony eating into the profitability of the payphone business worldwide and its move into the multimedia terminal business, payphone businesses around the globe cannot afford to let their guard down.

25.8 Staff fraud

Any organisation of any size has an exposure to staff fraud. The classically quoted 80/20 figure of the FBI, 80% of crime is internally committed, seems to work. The usual countermeasure is through awareness programmes which seek to build a trust ethos within the organisation. This needs to be supported with clear evidence of the consequences of transgression which of course implies one must weigh up the pros and cons of exposing individual situations to public scrutiny and the effect that might

Generic	Specific
Defeat coin validation mechanism	Lesser value coins made to appear as higher value by covering with metal foil.
Defeat debiting mechanism	Block SPM pulses if used.
Blocking coin return chute	Stops unused credit being returned to legitimate customers. Later recovered by offenders. A variant is to force open the mechanism door and place a block inside the mechanism itself.
Software errors	May be very obscure, e.g. flash switch hooks 15 times then operate follow-on call button – decrements credit counter past zero to hexadecimal FFFF.
Inexhaustive testing	Minimum cost outgoing call made to desired number followed by <recall> to get dialling tone then clear down. Suspended call is then reconnected as an incoming call and not charged.
Security measures not implemented	Smart card cashless payphone does not have the 'extra wiring' detector implemented or full authentication, allowing the use of PIC based 'cloned' card.
Rob the cash container	Forged keys through drilling, through hydraulic jacking to thermic lance.
Good ideas go bad	Decision taken that calls to emergency services will ignore charging pulses. Offender uses DTMF keypad over the mouthpiece to dial his desired number then keys the emergency number on the payphone keypad. The local exchange connects a normal call but the payphone treats it as a free call.
Electric money	For example, in Red Boxing a call to the operator is made with a request to connect a call. When asked to insert cash, the fraudster places the red box, actually a tone generator, against the mouthpiece and simulates the coin pulses which would normally be generated by the payphone mechanism. This fraud mechanism is not universally available with modern payphones.

Figure 25.5   Some payphone fraud examples

have on brand image. In general, however, the benefits of publicising the detection of offenders outweigh the downsides. Staff fraud may affect any part of the business through processes or through technology. Figure 25.6 tabulates some of the more prevalent issues.

25.9 PBX fraud

Private Branch Exchange (PBX or PABX) fraud is one of the major fraud phenomena facing Telcos in the last century and into this one. In spite of the huge efforts made by the Telcos through partnerships between themselves and with manufacturers of PBX equipment, the exposure continues. Such is the importance of this area that it warrants detailed treatment beginning with some definitions.

PBX fraud may be defined as:

- The illegal accessing of outgoing lines via switch facilities to onward route calls at reduced or no cost to the caller.

Or more generally,

- The illegal reprogramming or abuse of a PBX and its facilities to further criminal intent.

Stakeholders

Figure 25.7 depicts the stakeholders in the PBX fraud arena and their interactions. In a liberated Telco situation, the PBX vendor may have relationships with the customer which are independent of those with the Telco providing service to that customer. This can lead to arguments over liability for fraud involving PBXs. Is the Telco liable to an apportionment of fraud experienced by a customer who has bought his own PBX? This is an untested question in some countries. Others have come down in favour of a split liability whilst some put the onus firmly on the maintainer of the PBX whether that be the customer or the Telco operating on his behalf.

Motivation

PBX fraud motivational factors are:

- to facilitate other crimes by making call tracing more difficult as depicted in Figure 25.8;
- for free telephone calls for personal use or to be sold on as part of a call selling operation;
- to facilitate computer hacking;
- to be used in conjunction with other frauds, such as Premium Rate fraud;
- as a substitute for other fraud methods as they are closed down.

Mechanics of PBX fraud

The basic through-dial mechanics of PBX fraud are depicted in Figure 25.9.

Generic	Specific
Disclosure of information	Maybe customer details, such as ex-directory numbers or business data such as product launch information. Customer credit card details might also be at risk if deterrent measures are not in place. Local Data Protection laws have a direct impact here and all employees should be aware of their responsibilities, perhaps through an entry in their job description. Telcos should adopt a proactive stance on auditing access to customer information. Automation, eliminating the human element, can also help.
Network manipulation	Misuse of customer lines. Misuse of Telco service lines. Barring access may deflect abuse to customer lines. A better solution is to allow access but to monitor for reasonable use. Premium rate fraud involvement, maybe by setting up non-charging routes or by using auto-diallers.
Manipulation of data	Alteration of customer billing
Theft	• of equipment; • of cash; • of fuel.
Call connection	Often perpetrated by Operators but has been observed from test desk positions in central office switches. Modern systems which automatically ticket calls provide for easy detection.
Social engineering	Telco staff may be subjected to sophisticated social engineering 'attack' to • extract customer information; • provide services which will not be paid for; • map out the Telco's operation for later attack or for sale to an information broker.

Figure 25.6   Staff frauds

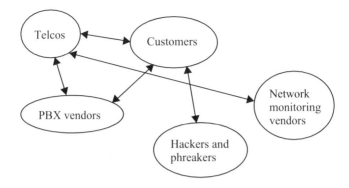

Figure 25.7   PBX fraud stakeholders

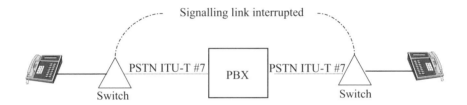

Figure 25.8   Call tracing made more difficult

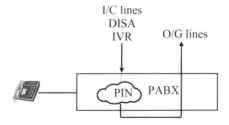

Figure 25.9   PBX fraud through-dial mechanics

To effect a connection between an incoming line and the outgoing trunk group an offender may use the feature of Direct Inward System Access (DISA), usually PIN protected or some form of Interactive Voice Response system (IVR), maybe an auto-attendant. The aim is to overcome the security measures in place to make a (usually) high cost outgoing call at the PBX owner's expense.

A development of this scenario is shown in Figure 25.10 where voicemail systems are the target. PBX sophistication has grown considerably in recent years and exotic call forwarding mechanisms are now commonplace on even small, SME sized, budget PBXs.

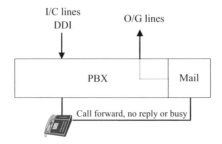

Figure 25.10   PBX voicemail abuse

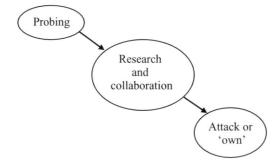

Figure 25.11   Attack sequence

In this scenario calls incoming from Direct Dialling In (DDI or DID) are directed to a compromised voicemail box which can be configured to either forward the incoming call to an outbound trunk or maybe to send a notification to a dedicated telephone number. Recently this technique has been used to launch a denial of service attack by using a number of compromised PBXs with auto-mail notification in cascade. The unfortunate target is then the recipient of huge numbers of short duration calls effectively blocking his ability to trade.

When a PBX comes under attack there is often a sequence of events under way. Figure 25.11 illustrates the basic sequence. A PBX system under attack will typically be probed both to identify its manufacturer and to seek to identify any vulnerabilities. Offenders often work in collaboration in their efforts to 'own' a customer's PBX. In extreme cases a PBX owner may find that his entire voicemail system is being used by phreakers and hackers to the extent that his legitimate users have no access at all.

Figure 25.12 tabulates some of the attack methods used by offenders.

Generic	Specific
Mis-operation of the design or a timing problem	e.g. poor release guarding of outgoing circuits allowing a trunk to be 'hijacked'. Incoming calls on night service numbers time out to dialing tone and allow through dialing.
Software patches may introduce new vulnerabilities which are lost in the importance of the issue at hand	e.g. Year 2000 patches.
Voicemail defaults not removed	Frequently these default to the extension number. They must be removed or disabled.
DISA PIN default not removed	The importance should be obvious.
Routing tables not policed	Routine checking of allowed routings should be undertaken to ensure time of day settings etc. are up to date.
Remote administration port not protected	Best protection is disconnection but can also be protected by proper dial back systems or by strong authentication mechanisms.
Social engineering	e.g. incoming caller asks operator or extension user to be connected to extension 9*xxx*, where 9 is the outgoing access number.
Probing	May manifest itself as silent incoming calls, strange requests to forward calls, or bursts of DTMF or other signalling heard by the called party.
Denial of service	Call flooding directed to a target PBX either directly or using other compromised PBXs in cascade.
Targeting PBXs which have toll free access	Many commercial concerns advertise toll free access numbers to the public and to their employees. These may be national toll free or International toll free (ITF) numbers.
Looping	Fraudsters often transit fraud calls through more than one PBX, often involving more than one carrier in more than one country. The problems of detection and evidence gathering should be apparent.
PBX network design	Network hop-on and hop-off permissions should be defined and monitored.

Figure 25.12   PBX attack methods

Offenders have a number of sources of information at their disposal:

- scanning the toll free number space – this may be illegal in some countries. In any case scanning of numbers is not 'normal' practice and is easy to detect. Potential fraudsters may attempt to hide a scan by adopting a pseudo random dialing pattern rather than a simple sequential scan. In this case other criteria may be used for detection, such as short duration, silent calls or 'wrong numbers';
- social engineering, either of operators, extension users or manufacturers;
- careless executives, who might write access details down on business cards;
- disloyal or dishonest employees;
- other PBXs with call logging, such as hotels which may record access codes and PINs;
- bulletin board systems, Usenet and other fraudsters;
- shoulder surfing; observing legitimate users of dial through facilities and noting down access codes;
- 'Dumpster diving'; trawling through company waste for access codes.

```
<snip>
> > I've only tried about 5VMB's and found that in all of them at least 50%
 > of the boxes are still using the default (box number) password!!, one
> > system in fact had 300 boxes (3 digit box number), and i only found 3
> > boxes with a p/word that wasn't default (did i have fun listening to
> > everyones messages?, yes i think i did :)
> > The 'hardest' (or most boring anyway :) part of hacking a VMB is
> > (IMHO) finding the boxes themselves, as they tend to be scattered
> > throughout the range of possible numbers, unless u find one thats set up
> > by a retarded/lazy/security unconscious admin who clumps them all
> > together (usually around 300/3000 i've found).
>
```

Figure 25.13   Example Usenet posting

Figure 25.13 is an example of a real Usenet posting which should serve as a warning to all PBX owners or maintainers.

Symptoms

PBX owners under attack are able to detect certain symptoms of an attack but they need to look for

- inflated bills;
- billed calls to unusual destinations;
- activity at unusual times of day;
- voicemail availability problems;
- customer complaints of busy signal;
- employee complaints of outgoing trunk congestion;
- nuisance calling to attendants and to extension users, also social engineering attempts.

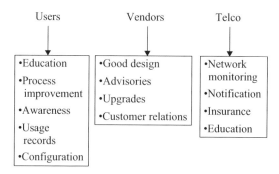

Figure 25.14   Stakeholder opportunities

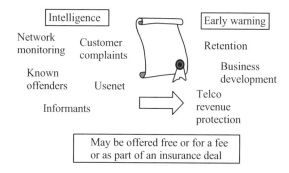

Figure 25.15   Notification and insurance

Countermeasures

The opportunities open to affected stakeholders are shown in Figure 25.14, some of which derive from Figure 25.12 and others are developed further.

The Telco serving the customer is well placed to use network based intelligence, such as Call Detail Records (CDR), in order to detect fraudulent calling patterns. This may be complicated if the customer takes his incoming and outgoing service from different providers, not an unusual situation by any means and a case for cooperation between Telcos. A further complication arises if the customer is a multinational concern with its own private network where incoming and outgoing service providers may be in different countries. These encumbrances provide potential exploitation points for fraudsters and reinforce the need for international Telco cooperation through bodies such as FIINA and Communications Fraud Control Association (CFCA).

In an effort to reduce customer churn and to increase product confidence, some Telcos have adopted a notification scheme arrangement whereby a PBX owner signs up to be notified if any anomalies in their traffic are observed by the Telco. The Telco or indeed a third party may offer insurance and audits, or health checks, as a way of mitigating or avoiding losses due to PBX abuse. Figure 25.15 summarises the situation.

In terms of a value proposition a Telco is well placed to offer its expertise – this should be second to none. Network monitoring or rapid CDR access, maybe

Figure 25.16   PBX fraud vicious circle

in conjunction with a fraud management system, should lead to rapid close down of a fraud opportunity which in turn minimises potential legal exposure due to a customer–Telco dispute.

PBX fraud has been described as a business success of the 1990s – but for the wrong stakeholders. PBX owners/maintainers have a responsibility to ensure the robustness of their systems – their business may well depend upon it. Telcos have the responsibility to deploy fraud detection mechanisms. The fight continues to reduce and eradicate PBX fraud to avoid the vicious circle depicted in Figure 25.16.

25.10   Split revenue services – premium rate, revenue share, audiotex(t)

This term is used here to encompass services in which a proportion of the revenue received by a Telco for terminating calls is paid to the renter of the destination number or group of numbers as depicted in Figure 25.17. The services provided are wide and varied, ranging from adult services through to horoscopes and horse racing tipsters. Such services are known by differing specific names such as 'premium rate', a reference to the fact that the paying customer pays at a higher than normal rate for the call. They may also be termed 'revenue share'. Internationally the services are collectively known as 'audiotex(t)', the 't' is optional.

In the International audiotex situation, the paying customer is in a distant country and the revenue from the incoming international settlement is split between the terminating Telco and the service provider. Figure 25.18 shows this situation in generic terms as the principle of payment for call termination can also be applied to intra-country interconnects. In simple terms, as a service provider the more incoming traffic you can generate, the more revenue you get. If you do not have to pay for the originating traffic then you win. In most cases, for legitimate callers and service providers, this model works fine. Fraud connected with split revenue services is predicated upon the artificial generation of traffic incoming to the services.

When intra-country services first came upon the scene a number of novel methods for inflation of traffic were devised by fraudsters. One of the major exposures related to billing frequency. It was not uncommon for service providers to be paid on a monthly basis. The problem was that the normal billing cycle for domestic customers

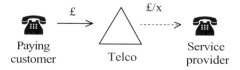

Figure 25.17 Split revenue cash flow, simple case

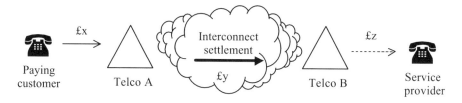

Figure 25.18 Split revenue cash flow, interconnect case

was three months. If the timing was right it was, therefore, possible to generate three months worth of calls from a single residential line to a target audiotex service, receive three months worth of incoming revenue from the service and then disappear from the residence without paying the bill. The problem of adult services and its negative effect upon the branding of Telcos has often led to the adoption of 'opt-in' to such services. That is to say that a normal customer would not have access to such services unless it was specifically requested of the Telco. This has led to the move of audiotex services offshore, where call termination settlements were the vehicle for cash flow. Not surprisingly the countries chosen to host the services were those to which high accountings rates existed. Such were the call volumes to some of these countries that the revenue received formed a major proportion of their balance of payments. A further twist came about whereby some Telcos entered into commercial arrangement with the destination country to short-stop transit calls to certain number blocks and to direct them to call answering equipment locally. The calls would therefore not occupy expensive terminating circuits with relatively long holding time calls as the caller would be more concerned with content rather than the physical terminating point. This practice attracted much negative publicity and is rarely seen today.

Some of the call stimulation or call generation methods seen are

- automatic call senders secreted in customer premises, sometimes engineered to fit inside telephone line wall sockets and with remote programming capability;
- spam e-mail campaigns;
- broadcast 'urgent' messages to pager; return number is international;
- abuse of PBXs;
- pyramid selling campaigns;
- clip on to customer lines;
- call charge information 'buried' in small print of advertisements;

- exploitation of customers' ignorance of international access prefixes, for example, in cases where there is more than one international carrier each with their own access code;
- internet content viewers;
- computer modem banks hacked.

The last two of these warrant further description. Adult content is very popular on the internet. Some web sites invite visitors to download a special viewer to see their content, usually with the promise of higher resolution pictures or faster download times. Unscrupulous web site hosts entice the visitor to download an executable program to his PC which, when run, turns out to be rather more than a simple viewer. In fact when the program is run, the visitor's computer is disconnected from his Internet Service Provider (ISP). The program then switches off the modem speaker and dials out to an international audiotex number for delivery of the content. The user is blissfully unaware that this is going on and may remain connected for a very long time to the international number. Only when the bill arrives will the unfortunate user become aware that he has been scammed. This inevitably leads to billing disputes and embarrassment.

The second example of computer involvement in split revenue fraud has been seen where a computer system was illegally accessed due to poor password security. The attacker found that the computer serviced a bank of modems and that these were accessible using the standard Hayes command set. The modem bank was then used to generate large numbers of calls to audiotex numbers.

The existence of differing accounting rates between the players in Figure 25.18 leads to the possibility of arbitrage. If the service provider can strike a wholesale deal with Telco B then Telco B will charge Telco A at its wholesale rate for call termination with a view to a profit. Telco A may not be able to recover the full cost of that outpayment from its customer if its billing system does not recognise the premium nature of the destination – that is, the normal relationship of $£x > £y > £z$ breaks down. Telcos need to be aware of this possibility which is not strictly fraud but is nonetheless unacceptable behaviour. Unscrupulous service providers may seek to exploit this arrangement by including intra-country interconnects into the picture. One solution to this problem has been for Telcos to withhold outpayments to either service providers or terminating Telcos if artificial inflation of traffic can be reasonably suspected or proven. Example criteria are

- traffic level deviation from a norm for the type of service provided;
- traffic level deviation by a threshold month on month;
- call durations suspicious – short or long holding time;
- high proportion of traffic originating from a 'point source';
- a service calls itself to excess;
- traffic arises spontaneously, without promotion;
- calls can be proven to have been made fraudulently, for example, through PBX abuse.

As a tailpiece to this section, a novel use of split revenue service has surfaced as a possibility. If one examines Figure 25.18 there is a left to right virtual cash flow.

Therefore, an individual operating as the calling customer in cahoots with a service provider in another country could use the system to transfer money. It is a lossy system as far as the caller and called parties are concerned as the Telcos take their cut of the money flow for connecting the calls. For this reason, it is not particularly attractive but it does present a relatively low risk, but is easily detectable if used to any great degree, as a means of transferring or laundering cash. The advent of relatively high cost split revenue services, in the order of $10 per minute or more develops the fraud situation further. Whilst opening up a raft of possibilities for payment for products through one's telephone bill, it is an obvious attraction for the fraudster whether by simply offering dud products or by artificial inflation of traffic – and the money launderer may well be interested.

25.11 Card fraud – calling card, prepay card and credit card

Card fraud is a perfect example of the tool-free Telco fraud. All that is required is a valid card number and PIN combination. Card numbers can be harvested by a number of means:

- hotel telephone system call logs;
- carelessly written down numbers;
- shoulder surfing, as for PBX fraud, for example in airports or railway stations even using binoculars;
- card number generators available on the internet – their use is facilitated if card validity is not checked at transaction time, not always a priority for relatively low value items like telephone calls.

Card fraud may be used as part of a call selling operation or in conjunction with other network facilities to avoid calling restrictions. Let us assume that country A has identified country B as a known fraud destination from country A involving stolen card numbers. It, therefore, blocks card access from A to B. A fraudster with a stolen card might make a call using the card to a third country C. The call is then diverted in country C to the home country direct service in country A – a free call. No charges are therefore raised in country C other than line rental. The call into country A then appears to come from country C, a non-fraud sensitive origin and the call is allowed to be completed to country B using the same stolen card. The block is, therefore, overcome until the stolen card is deactivated. Furthermore the originating country is hit by three international outpayments for the one fraud call – an obvious incentive to react.

Modern fraud detections systems are able to react very quickly to card fraud and close it down before exposure is too great.

25.12 Signalling fraud

No treatment of telecommunications fraud is complete without at least a mention of signalling fraud. It is the nature of in-band signalling systems that they are susceptible to influence if a speech path exists between a customer and the signalling elements controlling call set-up and supervision. The reader is invited to look up the exploits of Cap'n Crunch for some of the history of this activity. Nowadays, as fraud follows the money, signalling fraud focuses on international routes. There are other in-band signalling systems still in use around the world but modern day fraud activity on signalling is concentrated on abuse of ITU-T Number 5, recommendation Q.141. In summary, this signalling system uses compelled signalling for supervision and non-compelled signalling for inter-register signalling and is still found on some international circuits. It is possible for a caller to inject timed 2400/2600 Hz signals during the compelled call set up phase – usually to a toll-free international number, to cause the distant end to clear whilst the local end is in a held state. The fraudster uses a tone generator audio-coupled to his telephone line to generate the signalling codes. The device used is generically termed a Blue Box in phreaking terminology. The fraudster then seizes the distant end using his blue box then sends KP1 or KP2 followed by the digits required to complete his call terminate or transit, respectively. The fraudster is not charged for the call but the target country may face both an out-payment for the connected call and a further outpayment for terminating an incoming toll-free call. The fraud is usually detected during reconciliation of declared traffic by the two international carriers involved.

Some of the countermeasures employed are

- convert routes to Number 7 signalling which completely eradicates the problem;
- apply recommendation Q.141 para 2.1.6h which provides a partial solution;
- provide loop back routes in the originating country with digit screening so that the first digit stream sent to the destination must be of a fixed format, that is, the routing digits necessary to identify a toll-free call to the destination country;
- apply KDD's timed Proceed to Send solution;
- filter 2400/2600 Hz at the international gateway – can disrupt modem communications.

With the exception of converting to Number 7, the solutions described are not completely foolproof. The most effective solution for all eventualities is to convert to Number 7 signalling.

Blue boxing has also been seen on R1 and R2 signalling routes and although the use of these and Number 5 signalling is reducing, there are likely to be significant exposures for some years. Operators should therefore avoid sending toll-free traffic over in-band signalling routes They should be vigilant on declared traffic settlements because even if the original call uses Number 7, it may be the case that one leg of a transit call uses Number 5 and a vulnerability, therefore, exists.

25.13   International collect call fraud

The principle here is that incoming calls are made through operator positions, the caller requesting a collect or reversed charge call to a number for which there is no means of collecting the revenue for the call. The destination is often a public payphone.

Countermeasures are

- disallow incoming calls to payphones, or do not display the payphone number;
- payphones modified to play an audio 'cuckoo tone' on incoming calls, ITU-T recommendation E.180. This signals to an operator that the called party is a payphone whereupon the call should be cleared;
- payphones allocated unique numbers which are known to operating staff.

None of the above countermeasures are perfect so a residual exposure remains.

25.14   Boxes

We have so far encountered three colours of phreaker's box:

- Beige Box – aids clip on fraud;
- Red Box – payphone coin pulse simulator;
- Blue Box – in-band signalling simulator.

Boxes exist, in reality and in modern mythology in a spectrum of colours – and sometimes non-colours such as the Cheese Box which is simply a means of connecting two telephone lines together. The claims of many of these boxes range from the outrageous and simply impossible through to the trite and seemingly useless. The reader is invited to do a web search on any of the above box types for a full list.

25.15   Fraud detection systems

Fraud detection systems may be home brew, proprietary or a mix of both which is probably the healthiest approach. A web search for 'telecom fraud management' will result in a bewildering number of hits. Simple thresholding systems work well when integrated into business as usual processes and could well account for 80% or so of the fraud management load. For more complex situations, where product interaction is taking place, some form of correlation is required using neural or artificial intelligence techniques. One should also consider the near real-time nature of emerging frauds and the need to close down potential high cost fraud quickly. Globally the usual experience is that fraud management systems pay for their investment in very short times – provided they are directed at the principal sources of fraud.

Data sources for fraud detection systems are principally Call Detail Record (CDR) based, deriving usually from the host billing system. For a more real-time data source, the PSTN itself is the usual source. In the narrowband world this will mean monitoring Number 7 signalling messages at strategic points in the network. Network topology then becomes an issue. In a network with widespread use of Signal Transfer Point (STP) working, signalling is concentrated in a relatively few points and monitoring is quite straightforward. In more distributed networks, monitoring the whole network may be economically infeasible so a more focused approach is called for. The usual axiom is to follow the money. Therefore, monitoring international routes would be a priority. One could also consider monitoring the Intelligent Network platform as this usually provides a signalling concentration point and may also deal with high value number translation services such as Premium Rate.

As Voice over IP services roll out, fraud detection system manufacturers are turning their attention to the specific needs of extracting or creating CDRs from telephony servers or other network components.

25.16   What of the future for fraud?

The reader will have noted that the discussion has largely centred on Plain Old Telephony (POTS) with the possible exception of voicemail and card fraud. There is good reason for this. Most fraud derives from a desire for a product, whether it be dialling through a PBX, obtaining free telephone calls or fraudulent internet dial-up access. Therefore, the fraud itself is more or less independent of the transport mechanism – within reason and excluding signalling exploits. Fraud detection mechanisms must change as next generation telecommunications systems roll out. So the collection of IP Detail Records (IPDR) is required not only to address current fraud types but also to protect new IP dependent services which may be attractive to a fraudster. For the moment, however, voice is the 'killer application'. In the foreseeable future there is one certainty: even when esoteric technologies like quantum communications using entangled states are developed, fraudsters will be looking for an exploit.

25.17   Summary

The continuing increase in complexity of modern telecommunications networks and products has produced a very difficult situation for Telcos to keep up with fraudsters. Whilst many frauds are undeniably 'low-tech', these people appear to have unusual capacity for uncovering quite sophisticated vulnerabilities and are quite prepared to spend inordinate amounts of effort in exploiting them. They will also mix and match different frauds, for example cascading a PBX fraud with a Split Revenue fraud. A Telco whose fraud losses are running at around 0.2–1% of revenues can be considered to be doing quite well in risk management terms.

The effects on QoS are fairly clear although not part of the usual measurement set. Customers harbour (usually) unvoiced expectations of the fraud resistance of the products they buy. These range from simple wireline access through to expensive and complex customer premise equipment like PBXs. More easily visualised in QoS terms is the impact on availability of services which may be occupied with fraud traffic. An example of this was the sudden increase in circuit availability to a distant country when a large scale operator-services fraud was closed down. As a corollary it can be seen that a degradation in QoS performance might be an indicator of a fraud problem. Therefore, traditional QoS measures can provide a useful input to a fraud management system.

Modern fraud detection systems are able to detect unusual traffic patterns rapidly and create cases for a 'Mark 1 Human' fraud manager to progress. They are mostly, by design and of necessity, reactive in nature and so must be supplemented by a proactive approach to product design to minimise fraud exposure at the outset.

25.18   References

1   CFCA Communications Fraud Control Association www.cfca.org
2   QSDG Quality of Service Development Group of the ITU www.qsdg.com

Chapter 26

Management of Quality of Service

26.1   Introduction

In the previous chapters of the book the topics relevant to the Quality of Service (QoS) in telecommunications and key issues were identified and guidelines offered to address these. The parties most involved in the management of the QoS are the service provider and the standards bodies. The network provider is most concerned with network performance and to some extent non-network-related issues. By this is meant operational issues such as billing. The other parties, the regulator, the researchers, the users etc. have a non-participatory interest. These bodies contribute towards knowledge but are rarely involved in the management of service offered to the public. The manufacturer has both participatory and non-participatory interest. For successful management of QoS the service provider has to put together a portfolio of activities formulated specifically to meet its strategy. In this chapter, an attempt is made to bring together the various issues. Ultimately, the service provider has to assess all the issues and formulate its own action plan for professional management of QoS. The principal steps in the management of QoS may be grouped under the following: focusing on the issues, developing an action plan, retuning the management structure where necessary and reviewing it to meet changing needs. In this chapter, we offer a philosophical flavour that must cut through all the tasks involved in the management of QoS.

26.2   Focusing on the issues

Focusing on the issues to be addressed in the management of QoS requires a review of a number of areas. They may be grouped under the following: organisational mission, market and the service provider, and the status of the product portfolio.

26.2.1   Organisation's mission

If the organisation's mission is not specific this needs to be reviewed and the quality and QoS aspects of the mission should be clearly stated. The management of QoS requires a top-down approach for successful implementation. Sections 26.4 and 26.5 discuss further the organisational aspects for successful management of quality and QoS in telecommunications.

26.2.2   Review of service provider's status in the market

The prime measure of the service provider's status in the market is its market share for its product or service. A more refined approach would be to establish what the market share would be had it delivered the quality the market wants. The provider ought to aim to establish the elasticity between market share and quality in relation to price and ask itself at what cost it would be able to produce the level of quality for an acceptable level of profit.

The provider ought to assess the impact of the competitor's quality for the same or similar products and the relationship between it (the competitor) and its own market share – the cross elasticity. It has to assess what technological developments are in the pipeline, for example, 3G, 4G supported services and what the implications are for both customers and providers. Would offering of high quality increase the market share? How easy would it be to pitch the level of quality of a service not yet offered? Should collective research between other 3G providers be attempted? A review of the market may also involve other pertinent issues which only the provider with its intimate knowledge of its own market can identify. It would be necessary for the service provider to identify all pertinent factors in the review.

26.2.3   Product portfolio management

The service provider needs to review QoS delivered for each service in its product portfolio. It should have comparative data on performance of each product with that of the competitor. It should also be aware of what level of performance is realistically achievable by comparing what is achieved in other parts of the world should such information be obtainable. Such a review would enable the provider to assess its position in relation to its ideal position. This knowledge of the gap would be an input in the formulation of action plans for reaching the ideal position in the quality league.

26.3   Developing an action plan

Based on the review of the present position of the provider it can formulate action plans in 'modules'. These modules are chunks of management areas which can be handled in manageable units. Put together, the modules would form the total action plan for the provider to embark on. The following guidelines are offered for the development of action plans in the various management modules.

26.3.1 Customer's QoS aspirations, expectations, requirements

Ascertaining the customer wants, their aspirations, expectations and requirements are essential for the service provider to determine the plans for its business activities. This is particularly true when there is competition and the customer is sophisticated enough to evaluate the products in the market. In Chapter 4 the customer's QoS requirements were covered. Various activities may be undertaken to find out what the customer really wants. One of these is survey among targeted segments of the population to capture their concerns. Study of the complaints profile will enable the provider to be aware of the chief dissatisfiers of customers. Personal knowledge of the business and their communication needs could assist in determining the customer's true aspirations. Analysis of the 'value gap' will give insight into customer's perception of quality and their requirements in this area. Analysis of the market ought to include the disabled and those on low income. These together with the principles given in Chapter 4 would enable a comprehensive set of data on customer's concerns. An intimate knowledge of the market and the needs in the market is most useful in formulating business plans.

26.3.2 What level of quality should the service provider offer?

Business logic would demand that the level of quality a service provider should aim for would be the optimum level of quality considering the market requirements and the price the customers are willing to pay. The factors that govern the judgement on the level of quality to be offered, from the information on the market requirements, would reflect the professionalism of the service provider. The provider, therefore, needs to arrive at the offered level of quality with the consideration of the most relevant factors commensurate with its resources for implementation. The various conflicting interests, for example, cost of quality improvement versus incremental revenue arising from the improvement, would have to be addressed before the level of quality is finalised. These are then translated into planning documents, monitoring systems and personnel allocations.

26.3.3 Delivered level of quality

Estimation of delivered quality is covered in Chapter 6. One of the decisions to be made is on the type of measurements to be carried out. The pros and cons of intrusive and non-intrusive measurements are to be evaluated. Another area where caution is to be exercised is accuracy of measurements. Errors are known to have crept in unnoticed and noticed only when something very obvious has taken place in the data. It is necessary for cross correlation checks to be carried out. This will ensure the values of the same parameter measured by different systems are the same (within the limits of accuracy). If these values are significantly different investigations may be carried out before damage is done. Consideration ought to be given for the choice of measuring systems in the light of technology and emerging services. In many cases it will be far too complex to carry out individual element performances and compute statistically the end-to-end performance. It would be more practical to carry out a sample of end-to-end measurements.

Auditing of measurements would ensure accuracy for internal use and obtain customer's confidence that data published are representative. Regular comparisons with the competitor's performance and what is achieved in the rest of the world will enable a perspective of the provider's performance. Analysis of 'execution gap' will enable any action that needs to be taken to ensure that what has been planned has been achieved. Continuous review of the delivered quality is an essential part of the management of quality cycle. This review would also take into consideration how a minimum set of measurements will enable adequate quality checks to be made for all services.

26.3.4   *Managing customer's perception of quality*

Customer perception surveys can be designed to obtain a surprisingly good insight into customer's perception of the service provider's portfolio of services. The secret of obtaining such insight lies in the design and administration of the survey. The key issues on this topic are covered in Chapter 7. However, the service provider should be aware that customer perception is highly subjective and the factors influencing the subjectivity not only varies with cultures but with time. It is also highly media sensitive and sensitive to personal experiences of the service provider's relationship with the customer. It is necessary for the service provider to be aware of when customer's perception changes and to identify what factors are responsible for the changes. While it may not be possible for the service provider to always influence the factors affecting the subjectivity of the customers, it can always carry out actions to mitigate any unfounded fears. These would include education and adopting alternative strategies.

26.3.5   *CPE*

In the management of QoS of multimedia type services the performance of the customer premises equipment becomes relevant. If the service provider does not provide the terminal equipment (as is often the case) it should be aware of what the market offers and what the standards of performance are. The criteria influencing the CPE performance are quality of the monitor, modem speed, features of software, ergonomic quality of man–machine interactions etc. Where the service provider has no control over these it would be necessary to be able to isolate any degradation of quality arising from poor quality of the CPE. This would enable correct apportioning of source of poor quality.

26.3.6   *Content of multimedia services*

Multimedia services have the following core bases: voice, graphics, video and audio. The quality of recording or creation of any of these would affect the quality as received at the customer end. A poor resolution camera used in telemedicine will produce less information content in the image at the far end of the communication link compared to one with high resolution. This simple concept applies to any form of content creation. In managing content of any multimedia service the quality of the content will need scrutiny. This topic is outside the scope of this book.

26.3.7 Multimedia service provision

Service provision has the following elements: service creation, service operation, service management, portals, applications and service tools. The quality element in each of these affects the quality experienced by the user.

Service creation deals with the navigation and design of interaction for multimedia services.

Service operations comprise management of services on which online multimedia services are hosted, loading and churning the content, managing server farms, Web design and hosting etc.

Service management deals with provision of such things as billing, order provision and customer help desk.

Portals are primarily search engines but an important gateway to information and traders, able to generate a great deal of revenue from advertising banners due to the large numbers of 'eyeballs' attracted to their sites. Vortals are industry specific, vertical portals. Vertical portals (vortals) specialise in a particular industry, for example, health. E-marketplace portals provide an electronic location for trading partners to meet and transact their business over the Internet.

Applications are provided by Application Service Providers (ASP) over the network and comprise application software such as e-commerce training, payroll order provisioning etc for all industries plus vertical, industry specific applications, for example, for health etc.

Service tools include software (operating systems, middleware and groupware etc.) and hardware (servers, chips etc.).

In the management of QoS of multimedia services the service provider has to evaluate the quality contribution from each of the above sources.

26.3.8 Transport considerations

Transport has the following elements; access, main backbone, physical media, network management tools, hot spots and tail management, disaster recovery, traffic surge due to special events etc. Performance of each of these contribute towards the end quality experienced by the customer.

QoS parameters for access to multimedia type services have been studied but internationally agreed set of definitions are awaited. In the meanwhile, de facto definitions are available for inter-service provider comparisons.

The main backbone quality is the responsibility of the network provider (and perhaps also for the service provider). The performance of network is reasonably well documented. However, the inter-network performance responsibilities are perhaps not as well organised as it could be.

Physical media in which multimedia can be distributed is currently the CD-ROM.

Network management tools are for the management of the network and the access.

26.3.9 Standards

When no standards exist for a particular service or a facet of the service (as may be the case for QoS of 3G services at present) it is difficult for the service provider to pitch the level of quality when planning for service implementation. It may perhaps be useful for service providers to get together and pool their resources to produce either an internationally agreed set of standards or produce de facto standards for quick implementation of service/s.

26.3.10   Human resources

Human aspects in relation to QoS management are distributed in many of the dimensions mentioned in the book. The four viewpoints of the quality cycle, the elements of the four market model, formulation of standards, liaison with user groups, regulatory aspects all have a human element. These may be grouped under the following categories: customer relationship management, human resources for the provision and running of the business and lastly the culture and education of the staff.

Customer relationship management has been covered in Chapter 15. Considerable skill and art is required of staff in dealing with the customers over the telephone or in person. Customer's opinion is perhaps most sensitive in person-to-person communication. It is, therefore, necessary to evaluate the service provider's effectiveness in this area. It is noteworthy that in the USA one of the principal areas of concern in QoS is the people related quality aspect (see Section 20.4.2).

Allocation of resources for carrying out the network and non-network-related tasks are straightforward except for the allocation of resources for disaster recovery and resolving unanticipated problems. Personnel have to be trained for these activities and posted in normal duties, ready to be released as occasion demands.

Due to the nature of the telecommunications industry, maturity is a long way off. The rapidly changing face of it requires a calibre of staff who has the mental agility and skill to grasp the fundamental issues related to quality and to be able to arrive at a sound working solution in a short time. These solutions need to be arrived at by sound reasoning and any quick fix or fire brigade solution will have both short term and long term consequences. Staff is to be trained to discern the problems requiring quick fix from those requiring well thought out solutions.

26.3.11   Regulation

Complying with regulatory requirements on matters of QoS is often a straightforward matter for service providers. Normally the service provider is required to publish delivered quality at regular intervals.

The service provider could influence the regulator by recommending more meaningful QoS parameters to be measured and published. The provider could take the role of the leader rather than the follower thereby making use of its intimate knowledge of the market.

26.3.12   Fraud/security

No formal language exists for expressing the QoS aspects of security and fraud. From Chapters 24 and 25 it is undeniable that security and fraud failure will impact on the quality dimension of a telecommunications company. The effects will range from the hard, for example, circuit availability, through direct financial losses to soft, for example, detrimental effects on brand value. Fraud exposures of up to 6% of turnover should focus any company's efforts on identifying its propensity to fraud and security failure. By adopting a formal risk management approach to products and systems, by adhering to extant standards such as ISO17799 and by adept use of fraud management systems it is entirely possible to manage exposure down to 0.2% of turnover and below. Fraud and security management systems, either provided in-house or by outsourcing, should provide the necessary financial information to justify, in business case terms, their applicability to the products and systems they are designated to protect. Remember that fraud and security losses come off the bottom line. A company which claims to experience no fraud or security issues is not looking hard enough.

26.4   Organisational structure

There are two main schools of thought among the service providers on the best method of organising expertise within the organisation for effective management of QoS. The first is the concentration of expertise in one department. In this department network technical expertise, logistics expertise and customer relationship expertise would all be housed, together with any other support required to maintain a self contained quality department. The advantage of such a concentration of expertise is the environment for close cooperation between various disciplines required to manage the QoS. Communication would be easier and there is the greater possibility of increased synergy. However, the disadvantage is that due to basic differences in the approaches of engineers and marketers and other disciplines, people do not always interact as effectively as they were expected to.

In the second type of organisation the expertise are separated functionally. Network engineers, marketing staff and operational staff are separated in their own units. A senior manager coordinates a spokesman from each function. This type of organisational structure has its advantage of each functional area concentrating on the intricacies of that discipline and liaising with other functional areas as and when necessary. The disadvantage of this type of management is the lack of ideal liaison between departments leading to the 'the left hand does not know what the right hand is doing' syndrome. Cooperation between different disciplines is often wrought with interdepartmental warfare.

A service provider has to carry out its own evaluation and judgement to determine which management structure is most suitable for its mission.

26.5   Changing the philosophy of management

The nature of telecommunications is such that it is not a static industry but one of continuous improvement and change. It is both technology and market led and the pace of change is breathtaking. The impact on QoS management is threefold; first, QoS management should be based on proven background fabric of management practice, second, the management should rid itself of archaic working practices and embrace the culture of educated thinking and third, treat QoS as if it is not a target to be achieved, but one requiring continuous improvement.

26.5.1   Sound background for management practice

It is necessary for the service provider to implement the principles of Total Quality Management (TQM) throughout the business. The management of QoS cannot be very effective if isolated from the rest of the management functions of the company and managed professionally leaving the rest of the business to function on a cavalier basis. A company wide culture and change of attitude towards quality is essential. This can only come through education and understanding the message that 'quality is free'. Indeed quality is free and poor quality is costly to the business. Under whatever name the principles of TQM are applied the efficacy of good management has been proven and the management of QoS will thrive best if it is embedded in the backdrop of TQM within the business. The requisite for such management is education on the basic concepts of quality and its management and implementation.

26.5.2   Ensuring a culture change

There are situations where 'quick fix' or the 'fire brigade' type of management will be required. Examples of these are disaster recovery, unforeseen performance problems, such as tail management, unexpected surge in traffic giving rise to contingency plans etc. It is necessary for service providers to study the worst scenario and make plans for such emergencies. The disaster of 11 September, brownouts in the USA, are examples where quick fix and fire brigade type of actions are necessary. Indeed the industry needs 'professional fire brigade type fixers'. It will be necessary to have staff trained for this type of activity in addition to their normal functions. It would be unworkable for specialist teams to be trained for this type of activity and to be idle waiting for disasters to occur. However, quick fix or fire brigade type of management cannot substitute for reasoned study of QoS issues and its management on a day-to-day basis. It is necessary to take a step back and look objectively at all the issues relevant to the service provider and the market and to draw up plans designed to make the organisation effective in its delivery of quality of the services offered. Change of personnel, working practices, education and training are to be looked at. The EFQM model provides an effective tool to carry out an audit of any service provider and formulate an objective policy to bring about such a change.

26.5.3   Quality is not a stationary target

It is also to be understood that quality is not a target to be reached. It is to be reviewed regularly and constantly improved. The technological advances and market changes have to be closely studied and refining of management strategies constantly undertaken. A review of the products and services over the past thirty years would show how both technology and market have considerably influenced the customer behaviour. The place for quality is for it to provide its characteristics irrespective of the nature of the product or service, but offer the same high value for money concept to the consumer. This is the essence of quality management.

26.6 Summary

Management of QoS in telecommunications involves identifying the key issues to be managed and formulation of suitable strategies. Such management is not a one-off exercise, but a culture of its own. Companies that lasted the test of time have shown that a key element for their continued success is their commitment to and successful management of quality. Quality is also to be continuously improved and is not a static target to be aimed at. Successful management of quality requires education and reasoned thought. There is scope for the 'quick-fix' and 'fire brigade' type approach – these are reserved for unforeseen situations and are not recommended for normal management. Successful management of QoS is both an art and science. The science is in the educated and reasoned approach to the individual problems. The art is in combining the various disciplines in the management of QoS (telecommunications engineering, logistics, economics, statistics, customer relationship management and others) in the correct proportion and weight to develop and implement the necessary action plans. Each service provider has to draw up its own plans based on its strategy to operate in the trading environment based on its own philosophy.

Section VII

Future?

The rapid rate of change in the telecommunications industry means that QoS studies have to be reviewed at suitable intervals. In this section two concepts are addressed; one affecting the QoS today in the form of proposing an internationally agreed framework for the study and management of QoS and another, a look into the future. In the latter some educated speculation is made on the type of issues that are likely to be of interest in the management of QoS.

Chapter 27

Architectural framework for study of Quality of Service

27.1   Introduction

The need for an internationally recognised architectural framework for the study and management of Quality of Service (QoS) has been discussed in Chapter 2. A proposal for such a framework is made in this chapter. Part of this proposal has already borne fruit in the form of ITU-T Recommendation G 1000 [1] and the rest of the chapter suggests how future work may help to achieve an overall framework.

27.2   Benefits of an internationally agreed architectural framework for the study and management of QoS

A universally recognised framework, agreed on in the international forum, such as ITU-T, could result in the following benefits:

- a clearer focus on the mapping of network technical performance to the end-to-end QoS of relevance to the customer and the service provider;
- consistency in the use of terms and definitions for describing quality criteria;
- benefit of comparability of QoS on networks and services throughout the world;
- an easier grasp of the significance and relevance of each parameter in the overall map of QoS;
- identification of all relevant QoS criteria and less likelihood of wasted or duplicated effort on similar topics, resulting in a clearer focus for individual studies leading to greater depth;
- interests of various parties in the management of QoS, that is, network providers, service providers and the users could be analysed in the context of overall QoS, leading to an optimum number of performance specifications;
- clearer identification of work areas could result in fewer resources and more output than the pre-architectural era.

Potential parties concerned with the framework are the network and service providers, regulators such as Federal Communications Commission (FCC) of the USA, European Commission (EC) and Oftel of the UK, the regional standards bodies such as the ETSI and ANSI through the ITU-T, the ISO and the OECD.

27.3   Requirements of an ideal architectural framework for the study and management of QoS

An ideal architectural framework should fulfil the following criteria:

- The framework should be based on end-to-end QoS.
- The framework should be applicable to any telecommunication service (be sufficiently flexible and high level).
- The framework should be reasonably easy to understand and be usable by the main parties of the telecommunications industry, that is,
  - users and customers;
  - service providers;
  - network providers.
- The framework must assist with the identification of most, if not all, of the QoS criteria, network-related, non-network-related and the 'soft issues'.
- The framework must show the inter-relationships of QoS between users of telecommunication services and the suppliers of such services (service providers and network providers).
- The framework must show the relationship between the QoS criteria and the service provider.

27.4   Proposal for an architectural framework

An architectural framework that fulfils the requirements in the previous section is illustrated in Figure 27.1. This framework should cater for both the legacy and the emerging services.

The four viewpoints (shown in the figure) have been described in Chapter 3 and were developed originally for the legacy network. However, many aspects can be applied to IP based services. The QoS matrix is described in Chapter 4. This was also originally developed for the legacy network. However, many of its components are applicable for IP based services. The ACF model is described in Chapter 10 and complements the QoS matrix for IP based services. The four market model is described in Chapter 28. This complements the QoS matrix and the ACF model. By suitably selecting the appropriate portions of the model for a particular service it is possible to identify most, if not all, the QoS parameters and the key issues for its management. In the following sections, the key areas requiring standardisation are indicated.

The interests of the three main parties involved in the QoS management, the service providers, network providers and the users are illustrated in Figure 27.2. The

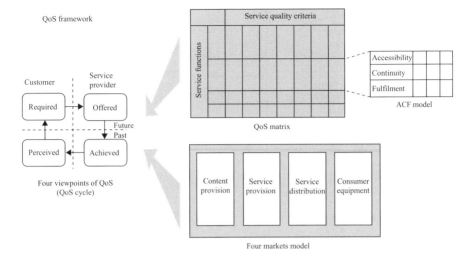

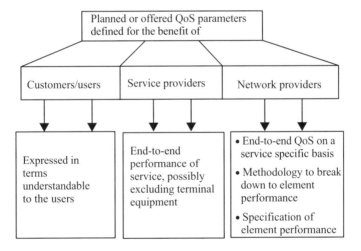

Figure 27.1   Architectural framework for the study and management of QoS in telecommunications

Figure 27.2   Relationship between performance parameters, users and providers in the study and management of QoS

areas requiring standardisation for each of the elements are given in the following sub-sections.

27.4.1   The four viewpoints

The principal feature of this framework is the logical division of the QoS studies management into four viewpoints. The issues are therefore decoupled from those other

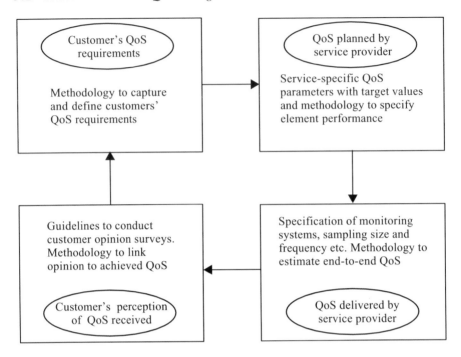

Figure 27.3   Architectural framework for the study of Quality of Service: The four viewpoints model

issues not directly relevant. The salient points of each viewpoint and the relationship with the other viewpoint/s are dealt with in the following subsections (Figure 27.3).

27.4.1.1   Customers' QoS requirements

The study and management of customers' QoS requirements should contain the following:

- a matrix to facilitate the identification of QoS parameters in a consistent and standard manner;
- cell definitions of the matrix.

A suitable matrix, the QoS matrix, one of the elements in Figure 27.1 is described in Chapter 3 and Appendix 4. It is reproduced here for convenience in Figure 27.4. Additional features to be associated with the matrix and the cell descriptions are:

- cell definitions must be generic;
- guidelines for the derivation of service-specific QoS criteria for the principal services used throughout the world should be considered;
- guidelines on the capture mechanism: these could be provided to interpret the effects of different cultures;
- principles of questionnaire design;

Service quality criteria / Service function		Speed	Accuracy	Availability	Reliability	Security	Simplicity	Flexibility
		1	2	3	4	5	6	7
Service manage-ment	Sales and precontract activities 1							
	Provision 2							
	Alteration 3							
	Service support 4							
	Repair 5							
	Cessation 6							
Connect-ion quality	Connection establishment 7							
	Information transfer 8							
	Connection release 9							
Charging & Billing 10								
Network/service management by customer 11								

Figure 27.4   QoS matrix to facilitate QoS requirements

- guidelines for the selection of samples and sizes;
- guidelines on arriving at a meaningful set of criteria from the collected requirements.

Requirements expressed in the language of the customers may need to be translated, where necessary into the language of the service provider for input to the decision making process to determine the offered QoS.

27.4.1.2   The ACF model

The components of the ACF model may be defined as accessibility, continuity and fulfilment. These have been defined in Chapter 10.

27.4.1.3   QoS offered

(i) Parameters of offered (or planned) QoS for the benefit of users, service providers and network providers: Parameters of offered QoS need to be defined to express the basic QoS of principal services. These parameters should be end-to-end performance, and for the benefit of users, service providers and network providers. Any variations of definitions of performance for these three parties would depend upon the service and the performance parameter being considered.

(ii) Methodology for the breakdown to element performance: Guidelines for the breakdown of end-to-end QoS into individual network element technical performance.

(iii) Specification of element performance: Standardisation of relevant element technical performance is desirable.

The standardisation of the above two categories would benefit equipment man-ufacturers. The planned QoS will form the basis for the specification of monitoring systems, the next viewpoint.

27.4.1.4   QoS delivered

Two principal activities make up the QoS achieved or delivered by the service providers. First, the specification of monitoring systems and second, the calcula-tion of end-to-end QoS from the performance data of elements. The first activity is schematically shown in Figure 27.5 and described in Chapter 6.

Activities associated with the monitoring systems, all of which require study in the standards bodies, resulting in specifications or guidelines are

- specification of test equipment for the different parameters;
- where the measurement is to be carried out;
- what measurements are to be carried out;
- sample size of measurements;
- when and how often (frequency) measurements are to be carried out.

Specifications of test equipment should not recommend any particular manufacturer, but state the technical principles and the characteristics of the measurements, for example, the weighting curve of the psophometer.

The second activity, computation of the end-to-end performance based on mea-surements taken, is illustrated in Figure 27.6. The principal activities, each requiring

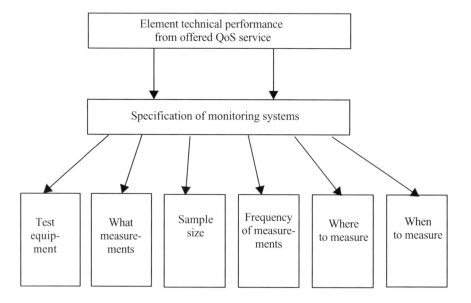

Figure 27.5   Specification of monitoring systems – constituent parts

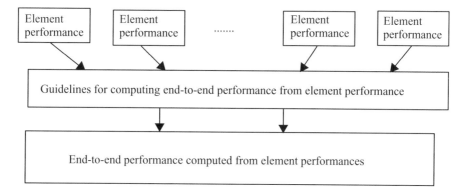

Figure 27.6   Computation of end-to-end performance from element performance or end-to-end measurement

study are:

- principles for the computation of end-to-end performance from element technical performance or end-to-end measurements;
- algorithms for the computation of end-to-end QoS for more sophisticated parameters. An example is call quality to be computed from measurements taken on elements. These measurements may be a selection of, noise, delay, slip, jitter and wander, all contributing to the transmission quality of the connection. The combined effect on a particular service is to be established and specified on a service-by-service basis;
- where relevant, guidelines on the computation of confidence limits of the end-to-end performance arrived at by the combination of element performances;
- where this is not possible, principles of end-to-end measurements to estimate the performance.

27.4.1.5   Customers' perception of QoS

The principal study areas in the customer perception of QoS are shown schematically in Figure 27.7.

(i) Rating scales

Customer opinion is rated on scales, varying from 1 to 10. The following are some of the more popular scales used: 1 to 4, 1 to 5 and 1 to 7.

Studies are necessary to choose which of these scales are most suited for both national usage and international comparisons. The standards forum could state the recommended ranges for the expression of customer perception ratings, after stating the merits of each.

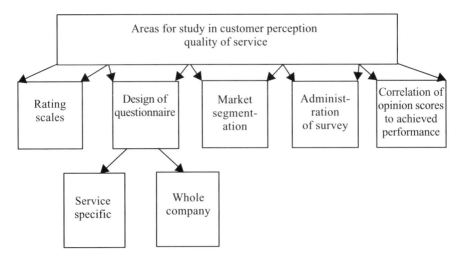

Figure 27.7   Study areas for customer perception of quality of service

(ii) Design of questionnaires

Generally, there are two types of questionnaires; one to assess customer opinion of a particular service, the other to assess the overall opinion of a service provider. Whilst it must be recognised that the questionnaire design for various business cultures could be different, the generally applicable guidelines may be established and recommended by the standards body.

(iii) Market segmentation

Market segmentation has, in general, followed industry classification. Individual experiences of service providers have also added further knowledge to the segmentation process. As competition in telecommunications is relatively new, regular review of existing knowledge could prove useful to both service providers and the regulators. It may not be wise for the standards body to recommend how market segmentation ought to take place and general guidelines may be sufficient.

(iv) Administration of survey

As discussed in Chapter 4, surveys could be administered in many ways. Studies to assess the methods of producing the most accurate customer survey are considered necessary in telecommunications. The standards body must provide guidelines for universal usage.

(v) Correlation of customer opinion rating to delivered performance

The correlation of customer opinions to delivered QoS forms the basis of the post mortem of any service provider to improve their performance. Before they carry out corrective action it is necessary to account for the discrepancy, usually present for many service providers, between the customer opinion scores and the delivered performance.

Studies in this area could be initiated both by the service providers and user groups for a better understanding of customer behaviour. The standards body could provide certain guidelines, but the service provider should be left to carry out the detailed correlation.

27.4.2   The four markets model

In the four markets model (Figure 27.8) the following areas could benefit from standardisation (see also Section 28.4).

Customer premises equipment

Standards or guidelines may be suggested on the following topics:

- browsers;
- download speed;
- encryption;
- modem;
- monitor quality of video applications classified according to service requirement – expressed in resolution (pixels) and number of colours;
- standards for search engines.

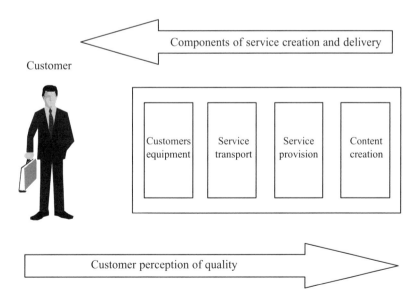

Figure 27.8   The four markets model

Service provision

The following topics may be considered for recommended guidelines:

- advertising;
- date stamp;
- encryption;
- firewalls;
- helpdesk quality;
- IP rights;
- payment security;
- piracy;
- privacy;
- service provision;
- tariff options;
- trust services;
- unsolicited mail;
- virus scan;
- vortals.

Transport

- access network;
- core network;
- broadband network;
- network management tools.

Content creation

Requirements for recording of contents, video, audio etc. depending upon the application. Telemedicine would require higher resolution than a road map. Standards such as JPEG, MPEG may be specified for different services with appropriate annotations and additional standards perhaps developed.

Usefulness of content – guidelines for despatch or otherwise of pornographic material, terrorist material etc. may be given. Isolation of unwanted material or its screening.

Across the four elements

Apart from contributions towards quality from individual elements, consideration ought to be given for quality degradations right across the four elements, for example, synchronisation could be affected from more than one element in the model. Identification of most, if not all degrading factors may be carried out and guidelines for its limits and control may be beneficial to the industry.

27.5   ITU-T Recommendation G 1000 and future work

The ITU-T has recently approved Recommendation G 1000 titled 'Communications Quality of Service: A Framework and definitions', which outlines the basis for an architectural framework. The main features of this recommendation are:

- a set of basic definitions related to QoS (including quality);
- the four-viewpoint model of QoS as described in Section 27.4;
- a matrix to determine the QoS criteria for any service as described in Chapter 3 (Section 3.3), and
- the relationship between QoS and Network Performance (NP).

By comparing the framework described in this chapter with the ITU-T Recommendation G 1000 the components requiring further standardisation may be considered.

27.6 Summary

A methodical approach is necessary to establish a sound foundation for the study and management of QoS. The rewards would be great: much greater synergy from the various studies, less resources spent to attain a resolution and the corresponding benefits being passed on to the equipment manufacturers and the customers.

The benefits of an architectural framework for the study and management of QoS should become more apparent when it is put to use. The benefits resulting from the investment of resources for the development of a framework and its subsequent adoption for further studies must be considered before further resources are expended on quality issues. Since more sophisticated service applications and issues regarding optimum economic levels of quality are likely to become more pertinent in the future, the time is considered ripe now for the development and adoption of an effective architectural framework.

27.7 Reference

1 ITU-T Recommendation G 1000: 'Communications Quality of Service: A Framework and definitions'

Exercises

1 Identify the ITU-T Recommendations that have QoS related components.
2 Map as many QoS related ITU-T Recommendations as you wish to Rec. G 1000 and identify areas where further work could be carried out which could benefit the telecommunications industry.

Chapter 28

Quality of Service, the future

28.1   Introduction

The world is entering a new information age driven by such things as technological convergence, multimedia, social changes, globalisation of business and entertainment, developments in consumer electronics, the emergence of e-commerce and the rise of the, so-called, information workers. There are already signs that the information age is beginning to overtake the traditional industrial age in terms of investment and output. The rise of the Internet, with its ubiquitous, global communication and the Web that provides a common and simple interface, has caused a paradigm shift in the conduct of trading by the introduction of e-commerce, that is, the selling/purchasing via an electronic communications medium. This can be via TV, fax, online networks and the Internet, particularly when facilitated by the ease of use made possible by the World Wide Web (WWW). The efficiency of trading is enhanced because much of the transaction processing is automated, for example, e-mail, online directories, trading support systems, customised services and goods, ordering and logistic support systems, settlement support, management information systems etc. It has also caused a paradigm shift from the traditional telecommunications function of providing a communications channel between a caller and recipient, which is evolving to connecting a buyer to a seller. Thus, perception of QoS will depend on a user's experience of the transaction in addition to that of the tradition telecommunications service. This chapter considers the evolution of the telecommunications environment and speculates on its impact on QoS.

28.2   Evolution of the environment

Examination of the history of technological progress shows that the economic environment is unstable and constantly disrupted by technological innovation or creative destruction, as shown in Figure 28.1 [1]. The fifth wave is usually known as the

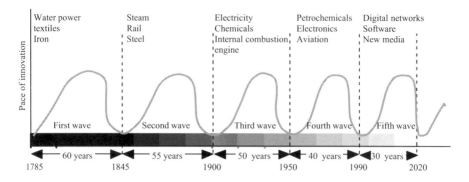

Figure 28.1   Model of production revolutions [Schumpeter's waves]

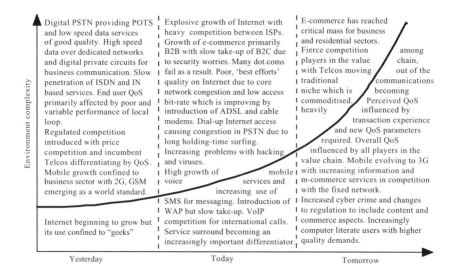

Figure 28.2   Evolution of the environment

information age, which is characterised by the emergence of cyberspace that electronically links the geographic world of physical space with the mental space of the human world, for example, a telephone call is a meeting in cyberspace. Examples of cyberspace emulations of physical space are virtual shopping arcades and chat rooms, so that people can undertake transactions or share a meeting even though they are geographically dispersed.

The convergence of telecommunications, computing, commerce and entertainment will play a key role in the emerging environment and the evolutionary phases are illustrated in Figure 28.2. In order to survive and prosper, Telcos will need to expand from their traditional communications niche, which is becoming increasingly commoditised, to other areas of the information/e-commerce value chain, in a highly complex environment. They will need to understand the QoS dimensions of the new transaction based services in order to effectively compete with other players who have much more experience in these areas.

28.3 The Internet

The Internet is an unmanaged unstructured network, which uses best efforts to deliver information, and its quality, in terms of speed of access and delivery, is highly dependant on the number of users the 'net' is serving at any particular time, that is, the World Wide Wait. Efforts to improve the situation have been made by offering mid-band access over ISDN and ADSL and caching of the more popular information, but this cannot overcome congestion bottlenecks in the PSTN access to the Internet Service Provider, Point of Presence (ISP POP); it was never designed for long holding time 'net surfing', and gridlock in the backbone network. The net evolved from a defence via academic network and, despite improvements in browser techniques, still has an unfriendly user interface and is largely populated by computer literate users, who cannot be considered as representative of the mass market.

Examination of the usage of the net would seem to indicate that it is mainly used for free information retrieval and the commercial usage is still in its infancy. Therefore, it is not a good model for the quality requirements of the future when there is likely to be a plethora of priced multimedia services, as illustrated in Figure 28.3.

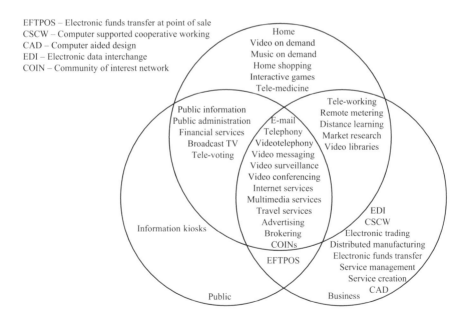

Figure 28.3   Potential information age services

Each of these services may have its own unique set of QoS parameters. The market for such services can be coarsely segmented into home, business and public. But, as illustrated in the diagram, there may be a considerable overlap between the services and markets. However, the Internet is giving an indication of some of the potential quality issues that will arise in the post-2000 information era, such as, navigation, speed of delivery, offensive (pornographic) material, security of electronic payments, guaranteed delivery, intellectual property rights etc.

The approach to quality is different from that used for telecommunication services because the end-to-end communication no longer depends upon the well managed communications link, but involves a variety of players in the information industry from information creation through delivery to the end-user's equipment. The end-user's QoS requirements will depend on the service being accessed. Therefore, the QoS parameters for each service must be determined and then the impact on each parameter from each part of the chain needs to be evaluated. For example, raw information may have the key quality attribute of timeliness but may not be useful if it lacks corroboration and context. Content packaging to sift, collate, validate, summarise, visualise etc. will give it marketable value, but only if it can be found by the search part of the chain and transmitted with minimum degradation to the end-user. The end-user may depend on PC applications to manipulate the information to maximise its value. The user and provider will rely on secure electronic payments for the information. The quality of a video service may be affected by how it is repurposed to, for example, change it from analogue to digital (MPEG) format, together with the error rate over the distribution network and the end-user's display mechanism. Since much of the key functionality is distributed across the chain it then follows that the quality contribution will be similarly distributed.

User expectation and perception will depend upon country and culture. The information service industry is global, there are no borders to the Internet, what is acceptable in one country may be totally unacceptable in another but to provide differential standards may not be possible. User expectation will also depend on benchmarking against known analogues, for example, TV for video based services, CD-ROMs for multimedia information services, home shopping catalogues for screen based shopping etc. Since multimedia information services will generally be screen based, QoS perception could depend on the familiarity with the screen interface and the suitability of the content for the screen, as illustrated in Figure 28.4.

The market for multimedia information services is largely unknown but segmentation is likely to be highly granular, individual customer QoS expectations and perceptions are likely to vary, even for the same service, far more than is experienced for the telecommunications services of today. A user interface that a technophile would find quite normal could be impossible for a technophobe. Price elasticity could be variable according to the value for money perceived by individuals and, in turn, this will depend on the value of the information as well as the quality of its delivery. Service surround should seamlessly encompass all parts of the value chain.

In terms of an individual multimedia session, quality could be affected by different aspects of its life cycle based on the Digital Audio Visual Council [3] (DAVIC) core

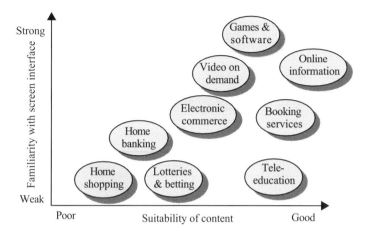

Figure 28.4   Presentation suitability and QoS

Source – Analyses [2].

functions, for example,

- initial access – user authentication and verification of credit and payment;
- navigation, programme selection and choice – to find and choose application or content;
- bit transport – at the appropriate performance and bit rate between points to be connected;
- session control – to establish or change logical connections and determine data rate and protocols to be used;
- application launch – where it is necessary to provide facilities to run an application not resident on a user's equipment;
- media-synchronisation – to link sound segments, subtitles, still and moving images, and applications to achieve a multimedia session;
- application control – for user interactive control of a session, for example, pause, rewind, content options etc.;
- presentation control – of delivery and display, for example, subtitle activation, choice of language etc.;
- billing – which could be for transport and information/service.

28.4   Information value chain

It is likely that a new information marketplace will create many new players, some of whom will challenge the traditional role of telecommunications operators as providers of the communications infrastructure and services.

The information actors have been described by ETSI [4] as shown in Figure 28.5. The relationship between these players and their involvement in end-user transactions and QoS is complex and not well understood. However, an insight of what might

Roles	Activities
Source information ownership	• Selling rights of recordable events • Using intellectual property rights
Information production	• Creation, capture and production of information • Creation and production of information services and applications of information services
Information provision	• Creation, capture and production of information • Presenting and advertising to information brokers and service providers • Selling information to service providers • Delivering information to service providers
Management of information provision	• Accounting, billing, revenue sharing • Auditing • Providing access rights • Assuring against illegal information movements
Information brokerage	• Registration of information providers and their offerings • Presenting, advertising and trading information • Provision of navigation for information service providers to information providers
Information service provision	• Searching requesting and purchasing information from information providers • Presenting and advertising to information service brokers, end-users and other information service providers • Searching, requesting and purchasing information services from other information service providers • Selling information to end-users and other information service providers • Delivering information services to end-users and other information service providers
Management of information service provision	• Accounting, billing, revenue sharing • Auditing • Providing access rights • Assuring against illegal information movements
Information service brokerage	• Registration of information service providers and their offerings • Presenting, advertising and trading information services from information service providers to end-users • Management of the access rights of end-users to information service providers • Provision of navigation for end-users to information service providers
End-user	• Consumption of services and applications • Combining information industry services with other services • Defining and specifying requirements
Regulator	• Development of regulation • Encouraging production of standards for competitive supply • Licensing of players

Figure 28.5   Information actors

Source: ETSI sixth strategic review committee.

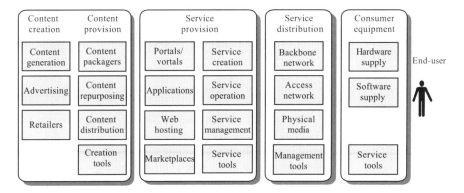

Figure 28.6   Information value chain [Four markets model]

happen can be achieved by considering the simplified value chain of Figure 28.6, which is commonly known as the 'four markets model'.

The information value chain starts with raw information that has the key attribute of timeliness but is not necessarily saleable if it lacks corroboration and context, via processing, to give it marketable contextual validity; it may then be stored pending retrieval by the broker or end-user. Information has no value unless it can be found, thus the search part of the chain is particularly important and must minimise the efforts of the end-user.

There are four segments,

28.4.1   Content creation/content provision

- Content generation – includes companies that create original content and are the rights owners, for example, film studios, graphic designers etc. (Disney and Time Warner).
- Retailers – who use multimedia as a channel to market for hard goods (clothes, books etc.) or soft goods (down loaded video games etc.) (Amazon, eBay).
- Advertising – where the information infrastructure is used, for example, for 'selling' a brand image through online multimedia pages.
- Content packagers – who add value to original content by filtering, aggregating and repackaging it (Reuters financial and news services).
- Content distribution – for the transfer of multimedia content between different companies and locations, for example, producer to facilities house to design agency (BT SohoNet).
- Content repurposing – for the change of information from one form to another, for example, changing analogue movies to digital format by MPEG coding (Electric Switch).
- Content creation tool providers – who provide the tools for the creation of multimedia content, for example, Silicon Graphics who provide special effects computer systems used in films such as Jurassic Park.

28.4.2   Service provision

- Service creation – for creation of the navigation and design of interaction for multimedia services (CompuServe).
- Service operation – for managing the servers on which online multimedia services are hosted, loading and churning the content, managing server farms, Web design and hosting etc. (USWeb, Frontier).
- Service management – for provision of such things as billing, order provision and customer help desk (America Online).
- Portals – primarily search engines but an important gateway to information and traders, able to generate a great deal of revenue from advertising banners due to the large numbers of 'eyeballs' attracted to their sites (Yahoo). Vortals are industry specific, vertical portals. Vertical portals (vortals) specialise in a particular industry, for example, health. E-marketplace portals provide an electronic location for trading partners to meet and transact their business over the Internet.
- Applications – provided by Application Service Providers (ASPs) over the network and comprise application software such as e-commerce training, payroll order provisioning etc. for all industries plus vertical, industry specific applications, for example, for health etc.
- Service tools – includes software (operating systems, middleware and groupware etc.) and hardware (servers, chips etc.) from, for example, IBM, Microsoft etc.

28.4.3   Service distribution

- Backbone network provision – of the broadband networks (Uunet, MCIWorldcom).
- Access network provision – the final network link to the consumer provided by traditional Telcos, for example, BT or their competitors such as cable TV companies.
- Physical media – on which multimedia can be distributed, including CD-ROM, which currently is the main vehicle for distribution.
- Network management tools – for the management of backbone and access networks (BT, HP and EDS).

28.4.4   Consumer equipment

- Hardware equipment supply – of computers, set-top boxes, multimedia kiosks etc.
- Software supply – browser software that runs on consumer equipment for a user-friendly interface to multimedia services (Netscape Navigator 3 and Microsoft Explorer 3). Also, application software used to manipulate and use information received from the Internet.

Value is added down the chain towards the end-user and the revenue is distributed up the chain, perhaps according to the value, to the end-user, added by each stage. This would suggest that the value to the end-user of the distribution element is a small proportion compared to the value of the information or transaction; hence, the network operator's share of the total revenue might be quite small unless more value can be implanted into the network. Other revenues generated by players in the chain complicate the value chain. For example, advertising represents a rapidly growing source of revenue for the, so-called, 'portals' such as Yahoo where small banners on the home page contain a message from the advertiser and a link to its WWW site. Banners account for a high proportion of Web advertising. Currently, revenue is relatively small, but it is a growing opportunity.

There are many companies occupying parts of the value chain, but many of the large companies in the entertainment, media, IT and telecom's sectors are beginning to position themselves throughout the chain [by mergers, alliances and acquisitions] to reap as much of the end-user revenue as possible and to improve their competitive position in the information market. Companies such as Microsoft, AOL and News Corp are particularly avaricious in this respect, but Telcos have recognised the threat and are researching means of opening their networks to provide transaction application software to third parties to increase their revenue potential, for example, the Parlay consortium [5].

The value chain for information services contains a series of customer/supplier relationships where quality requirements must be satisfied if the end-user is to receive an acceptable QoS. The basic interdependencies and feedback are illustrated in Figure 28.7.

There will be quality requirements of the 'traders' who inhabit the value chain. For example,

- Content providers (from the entertainment, media and information industries) will require their content to be packaged, repurposed (i.e. reformatted to suit recipient, e.g. MPEG for video), stored and delivered in a manner that provides quality to

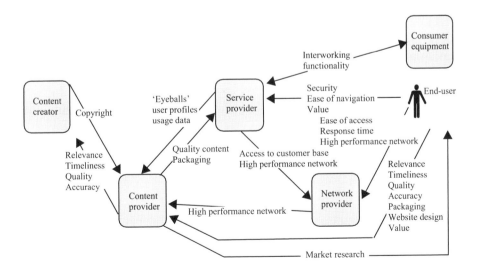

Figure 28.7   Key players' quality requirements

the end customer. Creation tools are important to obtain attractive and easy to use information. Revenue collection from users may require accurate micro-billing, which could be outsourced to Telcos, who have considerable experience in this field.

- Advertisers (e.g. Web marketing) and retailers require to attract eyeballs, so navigation is important to them. The rise of portals on the Web is an indication, and key Web sites, such as Yahoo, with high reach ratings and click through rates can capture high advertising revenues. Accurate usage data and user profiles are also important for one-to-one marketing.

- Service providers require easy access to a large customer base without PSTN congestion to their POPS. Grooming of traffic onto IP networks as near to the end-user as possible will reduce the likelihood of gridlock in the PSTN.

- Distributors [Telcos] would like to attract service providers to parent on their network. This will avoid interconnect charges to other network providers for delivery of information traffic to service providers. They would also collect revenue from interconnect charges for traffic from other networks to service providers connected to their network. Attracting service providers might be achieved by adding value in the distribution network over and above high quality bit transport. This might be achieved by open application programming interfaces (APIs) and network middleware that bridges applications and adds value between providers and users in much the same way as middleware in a PC allows many disparate applications to use common functions such as cut and paste.

- Consumer equipment relies on well-defined standards for interfaces, protocols and formats for its efficient and trouble free use.

28.5   E-commerce

There are many definitions of e-commerce, one of which is to link it to electronic trading, for example, 'e-commerce is trading by means of new communication technology. It includes all aspects of trading, including commercial market making, ordering, supply chain management and the transfer of money' [6]. It can comprise Business to Consumer [B2C], for example, Amazon.com the well know book site, or Business to Business [B2B].

28.5.1   B2C e-commerce

The process of B2C electronic trading is analogous to the manual process of purchasing an item from a shop or via mail order. Consideration of the processes involved in a simple electronic trading transaction will reveal that problems with any of the processes and actors can cause quality problems to the end-user as illustrated in Figure 28.8.

The extent to which Telcos provide transaction functionality, in addition to their traditional communications, will determine the need for their involvement in QoS matters. For example, the perceived lack of security of the Internet, hyped by the

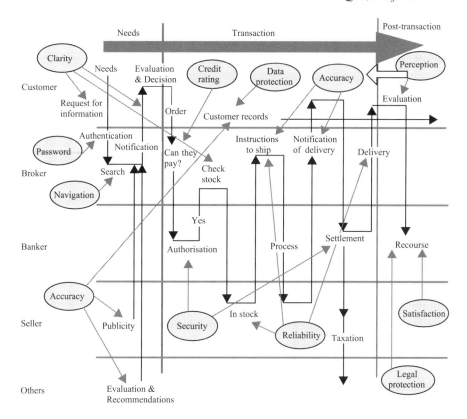

Figure 28.8   QoS problems

media, has been an inhibitor to the growth of e-commerce. Aspects that need to be addressed are given below.

- Confidentiality – data must not be visible to eavesdroppers.
- Authentication – communicating parties must be certain of each other's identity and/or credentials.
- Integrity – communicating parties must know when data has been tampered with.
- Non-repudiation – it must be possible to prove that a transaction has taken place.

Telcos, due to their known and trusted brand, are well placed to provide security services as Trusted Third Parties [7], but this carries a heavy responsibility for the quality of such things as validation, issue and distribution of encryption keys, notarisation and time stamping of transactions etc.

28.5.2   *Virtual communities or community of interest networks*

A virtual community comprises value-add of communications and exchange of information between members and partners [8]. Sometimes known as Community of

Interest Networks (COINs), which are many-to-many communities who are brought together through a community host. The QoS to community members is heavily dependent on the community host who is responsible for member acquisition, web hosting and design, content and where necessary, connection of members to appropriate vendors.

28.5.3   *Portals*

Portals are an important element in the e-commerce value chain [9]. Originally known as search engines and although navigation is still a major feature, a great deal of diversification and specialisation to increase 'eyeballs', 'stickiness' and customer loyalty, has occurred in the portal arena. The principal portal stakeholders are the following:

- Portal owners – who develop and manage the portals and whose aim is to maximise revenue by selling their own portal based services, getting commission on transactions through the portal and from advertisers.
- Portal business clients – who use portals to direct end-users to their Web site (and, hence, improve their business by improving visibility to potential customers), to gain revenue from trading with consumers and other businesses and to streamline business processes and reduce costs.
- Portal users – who are the end-users of the services offered by or through the portals.

Portals can be broadly categorised as follows:

- Navigator portals – do not provide e-commerce features but attract traffic by focussing on the location of content. Quality is determined by 'reach' that is, how many businesses a customer can connect with and how many products can be offered to those customers, and also, 'richness' which refers to the depth and detail that can be provided for customers. A vendor is concerned with the number of customers it can reach through the portal and the detail of information that can be collected about customers. Portal owners require loyal vendors and users for whom they need to provide customised interfaces and attractive product offerings.
- Marketplace portals – which connect sellers and buyers together to conduct transactions. In order to create a critical mass of sellers and buyers, most marketplaces concentrate, at least initially, on narrowly defined categories of transactions, for example, books, records etc.
- Community portals [10] – whose members build communities of a common interest or objective. They are independent of vendors but form a group of end-customers who share common purchasing profiles and have significant purchasing power. Attractive content is very important for such portals.
- Infomediary portals – which aggregate information about a larger group of users and use the resulting market power to negotiate favourable terms with vendors in exchange for customer information. It also gives access to a comprehensive range of information on product availability and price. It generates revenue from commissions on transactions.

To extract profitable revenue, portals require a critical mass of members and usage profiles to exploit the 'network effects' where the value of the service increases as the number of its end-customers and vendors grows.

28.5.4   B2B e-commerce

The concept of B2B e-commerce is not new – commercial transactions between businesses have been carried out for many years using Electronic Data Interchange (EDI). This is the exchange of structured business data (e.g. orders invoices etc.) between the computer systems of trading partners, in an agreed standard format. The messages are usually carried over proprietary value added networks (VANs) that provide store-and-collect mail services, audit trails and protocol resolution facilities. The EDI messages are often coded in a standard data format governed by X12 and EDIFACT specifications, although many industries used proprietary non-standard formats in closed user groups. The business model is based on established trading agreements between buyers and sellers, often in a 'hub-and-spoke' configuration, where the hub is the dominant buyer, who dictates the terms of relationships with a large number of small suppliers (spokes). Whilst there are clear benefits from EDI to dominant buyers, the benefits to the spokes (typically small sellers) are not so obvious due to the cost and need to subscribe to different VANs and standards when trading with a number of hubs.

Because EDI requires rigid agreements about the structure and meaning of data it is not suitable for the mass market. E-commerce, based on IP technology, removed the barriers of EDI because it is relatively cheap to employ, easy to understand and uses the ubiquitous Internet. The next phase was, therefore, initially driven by supplier Websites, initially used for marketing, but more recently to process orders. This was followed by a phase in which buyers demanded applications that helped to stream-line the selection, internal processing and ordering of suppliers goods and services. The current phase is one of B2B trading communities that are underpinned by B2B functional and vertical hubs.

Examination of the value chain processes (Figure 28.9) indicates that QoS is important for all associated activities if the transaction is to be satisfactory for both buyer and seller. Since a number of players can be involved throughout the value

Figure 28.9   Trading process and activities

chain and its activities, it follows that the overall QoS of the transaction is influenced by the contribution of these players.

Telcos are well placed to take advantage of the commercial opportunities arising from e-commerce. They have a large population of customers spanning a wide cross section of consumer types and they have a known and trusted brand. They can also build on their network infrastructure to embed value added services in it or use it to gain access to the technology used for e-commerce, thus moving from basic services to high revenue value added services.

To determine the commercial opportunities of e-commerce it is necessary to examine its value chain and the components that facilitate such trade, illustrated in Figure 28.9. Trust services are well suited to the Telco's brand as a trusted third party and can provide the foundation for other e-commerce applications. Trust services comprise a Certificate Authority to provide authenticated certificates and to manage the complete certificate life cycle, that is, issue, replacement, renewal and revocation. Other trust services include dual key support (different keys for signatures and encryption), notarisation and time stamping, cross certification (between different standards), secure messaging and smart cards.

Charging and payment services are also well suited to the Telco expertise in billing, particularly micro-payments, on behalf of other traders, which can be incorporated into the normal telephone bills. Additionally, web and application hosting together with the provision of server farms are obvious areas for Telco involvement. Moving to higher value services would lead to the provision of e-commerce (B2C or B2B) portals that provide an electronic location for buyers and sellers to meet and transact business over the Internet. These marketplace portals provide much of the functionality of the trading process and integrate the buyer's applications with those of the seller to achieve an overall integrated process as shown in Figure 28.10 [11]. In this case, the marketplace owner has a heavy responsibility for the overall quality of the transaction.

28.6   Intelligent home

Currently the majority of home peripherals are isolated from each other and connected to a variety of non-intercommunicating service networks, for example, entertainment systems and home computing/telephony networks. In the emerging information era, networking will bring many services to the residential customer the primary segments being communication and entertainment, security, energy management, home automation and home care. Examples are:

- shared Internet access for PC, TV, Web telephones and gaming consoles;
- networking of multiple devices in the home allowing, for example, appliances to be monitored to detect faults and automatically arrange for servicing, an inventory of consumables (food etc.) monitored and automatic orders to replenish when approaching exhaustion etc.;
- home heating and cooling adjusted to meet weather conditions;
- security systems using sensors and web-cams alerting police and neighbourhood watch communities;

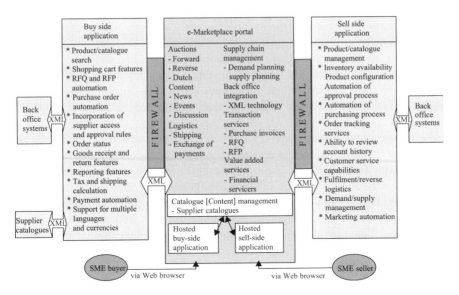

Figure 28.10   e-Marketplace

- synchronisation of family agendas;
- personal virtual private networks between relations and friends;
- home care for the elderly and infirm.

A residential server is likely to host the applications used by the family, interconnect household devices and provide the connection to the external network. It will enable equipment and appliances from different vendors with differing interfaces to be installed in an easy 'plug and play' process into a single home network. There are many appropriate standards such as (wireless) Ethernet, HomePNA, IrdA, IEEE-I394, X10 and Bluetooth for the physical connection. Higher level standards such as Jini, UpnP and Salutation enable devices to join the network and look up services that are available. But the most important standards initiative is the Open Services Gateway Initiative (OSGi) by the OSGi consortium who are defining a Java based framework and applications programming interface (API) for end-to-end service provisioning and management in the home network. The increasing complexity of home networks will therefore intensify their influence on end-to-end QoS for both basic communications and high value transaction services.

28.7   Summary

In a world of rapidly accelerating technological development, much of what is available today could not have been conceived of a decade ago. Likewise, speculating on the future in the first decade of the new millennium is bound to be wrong. Nevertheless, despite the bursting of the 'dot-com' bubble, e-commerce will continue to grow and develop, both in the commercial and residential sectors; the Internet will increasingly become the prime source of information, and is likely to become ever more convergent with leisure and entertainment. Ian Pearson, BT's futurologist, believes that knowledge of the development rates of many different technologies together with consideration of their interactions with society enables credible scenarios to be developed that foresee many potential consequences on business and social life [12,13].

The telecommunications industry can play an important role in the future, but faces major challenges as well as opportunities. The commoditisation of basic transport services together with the fragmentation of the industry by the arrival of many competing niche players will require Telcos to migrate to other areas of the information value chain, in order to survive and prosper. The QoS parameters for e-commerce, information services and entertainment are quite different from those required for traditional telecommunication services. Moreover, as people increasingly rely on the new services for their lifestyle and work, their expectations for high reliability and quality are bound to rise. QoS is therefore likely to become a major differentiator in a highly competitive market and will need to match customer's criteria for buying [14].

The customer experience will embrace the totality of the transaction, which in most cases the service provider does not control (through ownership or outsourcing), hence the need for close collaboration with other players in the value chain. Telcos have some experience in dealing with such situations since, for basic telecommunications services, multiple networks have been the norm for many years. They also have a good reputation for service management on which they can build for the future and take advantage of the many opportunities that arise, but only if they adapt to meet the changing QoS requirements.

28.8   References

1   Catch The Wave: *The Economist*, 20 February 1999
2   Analysys web site: www.analysys.com
3   DAVIC.: web site: www.davic.org
4   ETSI: 'Report of the Sixth Strategic Review Committee on European Information Infrastructure', June 1995
5   PARLAY.: web site: www.parlay.org
6   GARRETT, S. G. E. and SKEVINGTON, P. J.: 'An Introduction to e-commerce', *BT Technology Journal*, 1999, **17** (3)
7   SKEVINGTON, P. J. and HART, T. P.: 'Trusted Third Parties in Electronic Commerce', *BT Technology Journal*, 1997, **15** (2)
8   GREENOP, D.: 'Community Communications Networks', *IBTE. Telecommunications Engineering*, Structured Information Programme Issue 37, April 2001
9   WARD, H. J. and GARDNER, M.: 'Portals – Their role in the emerging networked economy', *The Journal of the Institution of British Telecommunications Engineers*, 2000, **1** (4)

10 HAGEL, J. and ARMSTRONG, A. G.: 'net. gain, expanding markets through virtual communities' (HBS Press, 1997)
11 WELLER, T. C.: 'B2B e-commerce: The Rise of eMarketplaces', Equity Research, Spring 2000, Legg Mason Wood Walker Inc., Spring 2000.
12 PEARSON, I. D.: 'Technology timeline – towards life in 2020', *BT Technology Journal*, 2000, **18** (1)
13 PEARSON, I. D. and NEILD, I.: 'Technology Timeline', *The Journal of the Institution of British Telecommunications Engineers*, 2002, **3** (1)
14 STROUSE, K. G.: 'Strategies for success in the new telecommunications marketplace' (Artech House, 2000)

Appendix 1

Quality parameters of diamond

Four parameters specify the quality of diamond. These are

- cut;
- clarity;
- colour;
- weight.

The categories for colour and clarity are reproduced here. The weight of diamond is expressed in carats, one carat being 0.2 g. Cut is too complicated to express in a book on QoS of telecommunications. Readers interested in this aspect are invited to read the manual referenced at the end of these tables.

Classification of diamond for colour in the GIA and CIBJO standards

GIA	CIBJO	General appearance
D	Exceptional white + (EW +)	Face up colourless
		Face down colourless
E	Exception white (EW)	
F	Rare white + (RW +)	
G	Rare white (RW)	
H	White (W)	Face up colourless
		Face down slightly tinted
I	Slightly tinted white (STW)	
J		
K	Tinted white (TW)	Face up slightly tinted
L		Face down obviously tinted
M	Tinted colour (TC)	Face up obviously tinted
.		Face down obviously tinted
Z		
Fancy	Fancy colour	Face up definite colour

GIA = Gemmological Institute of America; CIBJO = International Confederation of Jewellery, Silverware, Diamonds, Pearls and Stones.

Classification of clarity *of diamond (there are two standards for clarity)*

GIA		CIBJO
Flawless (FL)	Shows no inclusions or blemishes under 10× magnification	Loupe clean (LC)
Internally Flawless (IF)	Shows no inclusions and only insignificant blemishes under 10× magnification	
Very very slightly included (VVS)	Contain minute inclusions that are difficult for an experienced grader to locate at 10× magnification	Very very small inclusions (VVS)
VVS1	Inclusions are extremely difficult to see	VVS1
VVS2	Inclusions are very difficult see	VVS2
Very slightly included (VS)	Contain minor inclusions under 10× magnification. Small crystals, feathers and clouds are typical	Very Small Inclusions (VS)
VS1	Inclusions that are difficult to see	VS1
VS2	Inclusions are somewhat difficult to see	VS2
Slightly included (SI)	Contain noticeable inclusions under 10× magnification	Small inclusions (SI)
SI1	Inclusions are easy to see	SI1
SI2	Inclusions are very easy to see (may be seen with the unaided eye)	SI2
Imperfect (I)	Contain obvious inclusions under 10× magnification, which can often be seen easily face-up with the unaided eye. They may seriously affect the stone's potential durability, or are so numerous they affect the stone's transparency and brilliance	Pique (P)
I1	Beauty or durability are somewhat affected	P1
I2	Beauty or durability are seriously affected	P2
I3	Beauty and durability are very seriously affected	P3

Source: The Diamond Grading Manual – published by The Gemmological Association, 27 Greville Street, London EC1N 8SU. More information on the classification of diamond may be obtained from the above manual.

Appendix 2

Quality and TQM

Management of quality of a product, service or a process is not best done in isolation, but as part of a total quality management affecting it directly or indirectly. Such an approach is known as Total Quality Management (TQM) and a brief summary of it is given here.

TQM may be defined as philosophy and company practices that aim to harness the human and material resources in the most effective way to achieve the objectives of the organisation. TQM is basically geared for quality performance of every function in an organisation. The characteristics of TQM are embraced in

- universal participation;
- focus on customer needs;
- everything is a process which contributes to quality;
- continuous process improvement;
- better performance at lower cost;
- excellence in communication and understanding.

Among the principal contributions made to this topic are those of the following 'gurus'.

According to Crosby [1,2]:

- Quality is conformance to requirements.
- The system of quality is prevention.
- The performance standard is zero defects.
- The measure of quality is the price of non-conformance.

According to Deming [3]:

- create consistency of purpose;
- adopt the new philosophy;
- stop mass inspection;
- stop awarding contracts on price;
- find and solve problems;

- on the job training;
- re-define supervisor's role;
- drive out fear;
- breakdown departmental barriers;
- eliminate numerical goals;
- eliminate slogans and posters;
- have common terms and conditions;
- educate and re-train;
- top management co-ordination.

According to Juran [4]:

- build awareness;
- set goals for improvement;
- organise;
- provide training;
- carry out projects;
- report progress;
- give recognition;
- communicate results;
- keep score;
- continuous improvement.

The Total Quality organisation focuses on

- continuous process improvement
- everything as a process
- the use of scientific methods
- perfection as the goal,

through

- universal participation
- everyone
- everywhere
- individuals and teams,

resulting in

- customer satisfaction
- exceeding expectations

for

- internal customers and
- external customers.

More than anything else, what distinguishes a TQ organisation from an ordinary one is the way its people think and act. The value that people place on quality of performance in every activity and what they do to improve the quality of their work are key factors in a TQ organisation. It is vital that the organisation culture supports the Total Quality concepts.

For further information on TQM and its implementation readers are recommended to refer to published literature. (See also references in Chapter 1.)

A2.1   References

1   CROSBY, P. B.: 'Quality is free: the art of making quality certain' (McGraw Hill, 1979)
2   CROSBY, P. B.: 'Quality is still free: making quality certain in uncertain times' (McGraw Hill, 1996)
3   DEMING, W. E.: 'Out of the crisis; quality productivity and competitive position' (Cambridge University Press, 1986)
4   JURAN, J. M.: 'Juran on quality by design: the new steps for planning quality into goods and services' (Free Press, 1992)

Appendix 3
Template for a typical service level agreement

1   Parties in the Agreement

State the customer organisation and their names of contacts and that of Service Provider.

2   Objectives

2.1   Parties need to identify what they want to achieve through the use of a Service Level Agreement (SLA). These may include 'clarification of responsibilities', 'enhancement or effectiveness of service delivery' and 'service performance aspects'.

2.2   When the key parties have identified the main objective/objectives of the SLA, the next step is to identify needs and expectations for the specified service/s. This enables both parties to understand what service attributes are most valued by each group. A service attribute is the aspect of service quality that is most important to the client/customer.

2.3   Parties should consist of one or two representatives from each unit: service provider, clients, and other interested parties who may have an interest in the outcomes of the service.

2.4   The provider may also identify with the customer areas for improvement and state what may be done during the period of the agreement. The objectives may be mutually arrived at with discussions with the customer.

3   Services covered

Here state the services covered, for example, basic telephone service (national, intra-site, international), Frame Relay platform, provision of ISDN etc. together with the service description. Add also the service features.

4   SLAs and Terms and Conditions (one can separate the 'Terms and Conditions' in a different section if so desired)

4.1   An effective SLA acknowledges that clients and service providers have responsibilities and obligations to each other. It is important to address these points in the development phase, and acknowledge them in writing. An agreement that clearly defines the roles and responsibilities of both parties will be of particular benefit if problems arise.

4.2   The service provider needs to discuss their capacity and potential constraints on providing the service for the customer/client. This will assist both parties in gaining an understanding of each other's requirements and reduce unrealistic expectations. If there are concerns or issues regarding service levels they should be discussed at the negotiations stage, not when a SLA is close to implementation.

4.3   The critical areas of service must be identified and both client and service provider need to agree on a minimum level of service to provide client satisfaction.

4.4   Avoid setting too many service levels, as this can get complicated. It is recommended that identified service levels are measurable and are based on user satisfaction as well as service output.

4.5   Both the client and the service provider need to agree on service levels that cannot be met. A general rule for the service provider is to 'under-promise' and 'over-deliver'.

4.6   Sometimes one may choose not to specify a minimum service level, but to put in place a monitoring arrangement to establish a history and propose a timeframe for later implementation.

4.7   It is a good idea to identify and plan areas for future improvement, so parties to the SLA can foster common goals and build on the service relationship.

4.8   It is necessary at this stage, for clients to distinguish their specific needs from broader expectations regarding service, and to appropriately prioritise service requirements. It may also be appropriate for the provider to detail other services they can provide. This is in order that both parties can fully identify their needs and expectations and agree on what services will be provided.

4.9   A service level is an agreed measure and may include one or more of the following elements to describe service performance:

  • service features;
  • quantitative aspects;
  • quality aspects;
  • timeliness;
  • cost/tariff.

4.10   When both parties have established agreed service levels, the next step is to agree on measurement of service level performance. It is essential to implement a system that will provide credible results; otherwise all parties within the process may lose confidence.

4.11   It is often the service provider, who puts forward ideas regarding measurement of performance; however, this is not always the case. Performance can be difficult to measure in some cases and it is a good idea to keep it as simple as possible. Some lateral thinking by both parties incorporating a mix of output and outcome measures often generates the best results. One-off or periodical surveys may be all that is required to measure whether performance indicators have been achieved.

    As with all stages within this process, it is important that agreement is reached between client and service provider regarding service performance measures.

4.12   The SLA should include a process for flagging issues early should they arise. Issues or disputes where two parties have attempted a resolution and failed may require some form of mediation by a third party.

4.13   It is likely that reviews would be more frequent at the early stages of the SLA, and less frequent when the relationship is well established. In addition, there should be an agreed schedule for review. A twelve month review is recommended; however, parties may agree on a different schedule. Both parties should identify a representative to manage the agreement on behalf of each other. The review process should allow for adjustment of service levels where appropriate, and redefinition of targets and indicators where appropriate.

Checklist

• Consider the capability to provide the service.
• Have all service level concerns been raised?
• Have you identified what will provide client satisfaction?
• Have you agreed on the minimum level of service?
• Have you agreed on areas for future improvement?

• Are client and service provider roles and responsibilities clearly defined in writing?
• Have you established a process that will flag problems early?
• Have you established a process to resolve issues that might arise?

• Have you formulated all indicators to measure service performance?
• Have you reached agreement on the performance indicated?

• Have you identified what services are free of charge?
• Have you determined costs for additional services where appropriate?
• Have you agreed on an appropriate cost transfer mechanism where appropriate?
• Have you agreed on the frequency of billing where appropriate?

The list of performance parameters may be provided with their performance ranges, if applicable, as shown.

Parameters	Minimum	Maximum
Parameter 1		
....		
....		
Parameter 'n'		

5 Exclusions

Identify the aspects of the service provision and maintenance not covered in the agreement.

6 Review process

6.1 A review process should be put in place and each party to the agreement should discuss and monitor the agreement throughout its life.

6.2 State who will be responsible for reviewing performance.

6.3 If staff changes in either client or provider areas, the new staff responsibilities need to be communicated clearly to the new staff.

6.4 Procedure for customer and service provider feedback to be brought together.

6.5 Agree on measurement or checking measurement of performance for reporting.

6.6 If new services or service features are required the agreement should be flexible enough to allow for review and change.

6.7 Agree on a programme for adjustment of service levels if there is a deviation from agreed performance.

6.8 Penalty clauses if performance is not met and the legal formulation on how this should be implemented. (In most cases these are in the form of rebates and a promise to improve performance backed up by real improvement.)

6.9 If any unspecified issues arise, a process on how these may be resolved and suggested timescales. Names of contacts of both the service provider and the client to be provided for such matters.

Appendix 4

Cell descriptions on the matrix to facilitate capture of Quality of Service criteria

A matrix was described in Chapter 4 to facilitate identification of Quality of Service (QoS) criteria for telecommunication services. In this appendix the cell descriptions are given; these are generic QoS criteria. Service specific QoS criteria are derived from these, wherever applicable. The matrix is reproduced here for convenience.

Service function \ Service quality criteria		Speed 1	Accuracy 2	Availability 3	Reliability 4	Security 5	Simplicity 6	Flexibility 7
Service management	Sales and precontract activities 1							
	Provision 2							
	Alteration 3							
	Service support 4							
	Repair 5							
	Cessation 6							
Connection quality	Connection establishment 7							
	Information transfer 8							
	Connection release 9							
Charging and Billing 10								
Network/service management by customer 11								

A4.1 Sales

A4.1.1 Speed

Speed of sales may be indicated by the time taken from the initial contact between the customer and the service provider to the instant an effective contract is placed for a service. Not all contacts will result in sales. Where no sales result, the speed of sales will be the time elapsed from initial contact to the instant an offer is made by the service provider after all the pertinent information has been supplied to the customer.

A4.1.2   Accuracy

Accuracy is exemplified by the correctness and completeness in the description and delivery of all relevant service information, normally expected by the customer before effective contract, for example, service features, performance, charges, service support, provision time etc.

A4.1.3   Availability

This criterion deals with access to service information from the service provider. The mode of access may be writing, electronic, verbal, or in person. Also included could be the number of offices, office hours and hours the staff may be accessed for information.

A4.1.4   Reliability

Reliability of sales (information) is an indicator of how well the service provider has satisfied the customer in terms of pre-contract formalities. The degree of accuracy associated with the information together with how easily this information is available over a given period, would determine how this parameter has performed. It is also an indication of the professionalism of the members of the service provider serving the customers.

A4.1.5   Security

Confidentiality requirements of customers from the service provider on all activities related to sales form this quality criterion. This quality criterion in association with pre-contract activities is likely to be of concern during special situations, for example, dealings with government agencies and where commercial confidences are of importance.

A4.1.6   Simplicity

This concerns the ease with which all activities associated with sales (and purchase) may be carried out with the service provider. Included are ease of identification of the point of contact for sales, ease with which information supplied is understandable, the ease with which forms can be filled and ease with which orders can be placed.

A4.1.7   Flexibility

The flexibility with which the service provider will accommodate the individual customer's requirements is identified in this performance measure.

A4.2   Provision of service

A4.2.1   Speed

Speed of provision is indicated by the time taken from the instant of effective contract to the instant service is available for use by the customer.

A4.2.2   Accuracy

The correctness and completeness in the provision of any service and associated features specified or implied in the contract.

A4.2.3   Availability

This depends on access to resources by the service provider to meet provision of service or product agreed with the customer in the contract. It could also deal with the facility offered by the customer for the installation of the service.

A4.2.4   Reliability

This concerns the long-term performance of the above three performance criteria, for example, over a period of one year.

A4.2.5   Security

This concerns confidentiality requirements with regard to a service. Examples include

- ex-directory number protection facilities;
- special requirements for government installations;
- compliance to Data Protection Act;
- non-display of Calling Line Identity etc.

A4.2.6   Simplicity

The ease and convenience with which a service can normally be expected by a customer after effective contract, for example, minimum inconvenience to customer during installation, ergonomically-designed cabling and fittings, simple but effective installation practices etc.

A4.2.7   Flexibility

This concerns options normally expected by customers to accommodate special requirements on the provision of the service without departing from the terms of the contract. Examples include, timing of the provision of the service to suit the customer, provision of terminal equipment to match customer preferences (e.g. appointments) where possible etc.

A4.3   Alteration to a service

A4.3.1   Speed

This is the time taken from request to a service provider for an alteration to a service to the instant alterations are incorporated and the service is available for use.

A4.3.2   Accuracy

The correctness and completeness with which requests for alteration to service are carried out as specified or implied in the contract.

A4.3.3   Availability

This concerns access to resources by the service provider to carry out alteration to the service as requested by the customer and implied in the contract.

A4.3.4   Reliability

This is indicated by the long-term performance of the above three measures, for example, over a period of one year.

A4.3.5   Security

This concerns customer's confidentiality requirements with regard to all aspects of alterations carried out on a service. For example, a customer may require ex-directory listing which may involve a change in number. It may involve a change in status of the working of the organisation, with possibly the award of a secret contract from the government requiring secure communications.

A4.3.6   Simplicity

This concerns the ease and convenience with which alteration/s requested may reasonably be expected by the customer.

A4.3.7   Flexibility

This concerns customer requirements on the options to accommodate special require-ments relating to alteration of a service. Examples include accommodating the customer's request for metre reading at a requested time when moving to a new address and capability to accommodate a customer's request to carry his telephone number to a new address.

A4.4   Service support

A4.4.1   Speed

This criterion may be expressed as the time elapsed from the instant a customer requests service support to the instant support is provided to the satisfaction of the customer.

A4.4.2   Accuracy

This concerns the correctness and completeness in the service support as specified or implied in the contract.

A4.4.3   Availability

This concerns the presence of access facilities for the customer making service-support requests for the customer. For the service provider it is the resources to provide the service support.

A4.4.4   Reliability

This is determined by the long-term performance of the three previous criteria, for example, over a period of one year.

A4.4.5   Security

Customer's confidentiality requirements from the service provider on all matters relating to requests and provision of service support, for example, security considerations on service support for special situations, for example, government installations. There may be a need for any service support to be kept away from the public eye. The requirement would be specified by the customer.

A4.4.6   Simplicity

This concerns the ease and convenience with which the customer can normally expect to request and receive service support and any associated activities.

A4.4.7   Flexibility

This concerns customer's requirements on ease and convenience with which service support may be requested and provided. Examples are, varying levels of service support and various modes of access. Service support may be obtained verbally, in print, or by electronic mail. In other cases a flexible approach may be provided for example, credit card payments of bills over telephone.

A4.5   Repair

A4.5.1   Speed

This criterion may be expressed by the time taken from the instant repair was requested to the time it is carried out and all service features restored to normal use.

A4.5.2   Accuracy

This concerns the correctness and completeness of the repair carried out as agreed or implied in the contract.

A4.5.3   Availability

This concerns the presence of facilities for making requests for the repair service. These facilities would include hours of access as well as methods of access. For the service provider it would mean presence and access of resources to carry out the repair.

A4.5.4   Reliability

This is determined by long-term performance of the three preceding performance criteria, for example, over a period of one year.

A4.5.5   Security

This concerns a customer's requirements on confidentiality in matters relating to repair service. For example, this would include sending of approved staff for defence establishments and respecting a customer's information on installations not being divulged to a third party.

A4.5.6   Simplicity

This concerns whether the manner in which repairs are carried out is simple and effective. For example, if one person is adequate to carry out repairs it would be most unprofessional if three people turned up and caused the customer annoyance.

A4.5.7   Flexibility

This concerns customer's requirements on the options available in carrying out repairs. For example, repairs may be carried out, where possible, in the first instance, without access to customer premises or where possible remotely. Repairs may also be carried out at customer's convenience should entry to premises be required. Alternative service may be offered if service is unusable.

A4.6 Cessation

A4.6.1 Speed

This criterion may be expressed by the time taken from request for cessation of service to the instant it is carried out.

A4.6.2 Accuracy

This concerns the correctness and completeness in carrying out the cessation of a service and the associated activities of whether the cessation was initiated by the customer or the service provider.

A4.6.3 Availability

This concerns facilities offered to customers for making requests for cessation of service. For the service provider it concerns access to resources to carry out cessation of a service. This could include issuing of final bills and closing of accounts and dealing with the correspondence for the particular service.

A4.6.4 Reliability

This depends on the long-term performance of the three preceding performance criteria, for example, over a period of one year.

A4.6.5 Security

This concerns customer's confidentiality requirements regarding all activities connected with the cessation of service. For example, a business may state that cessation of services to its premises should not commence until a move to new premises is completed and the new communications services are working satisfactorily.

A4.6.6 Simplicity

This concerns customer's requirements on the ease and convenience with which activities associated with the cessation of a service is carried out. For example, the customer may wish to close an account with the service provider in one transaction.

A4.6.7 Flexibility

This concerns customer requirements to minimise inconvenience during the process of cessation of a service. For example, wrong cessation due to error of the service provider. Cessation, if requested by the customer, to be carried out at a time requested by the customer, for example, a specified hour to coincide with moving office or residence.

A4.7   Connection establishment

A4.7.1   *Speed*

This is given by the time elapsed from the input of the last address digit to the instant the signal is received from the network to indicate the status of the called party.

A4.7.2   *Accuracy*

This measure deals with

(i)   the correct and complete indication of the status of the called party when it has been reached by the address digits, and
(ii) reaching the called party identified by the address digits.

A4.7.3   *Availability*

This is the probability of a customer being able to establish a connection when requested. This parameter could be expressed additionally with (i) frequency of unavailability and (ii) maximum duration of any one unavailable period.

A4.7.4   *Reliability*

This is determined by the long-term performance of the above three parameters over a period of one year.

A4.7.5   *Security*

This concerns a customer's confidentiality requirements on connection attempts, for example, the display of calling line identify to the called party against the stated wishes of the customer.

A4.7.6   *Simplicity*

This concerns the customer's requirements on the ease and convenience with which connections may be established. Examples are

(i)   personalised numbering schemes;
(ii)   short dialing codes;
(iii) easily identifiable blocks of numbers for uniquely identifiable blocks of customers' numbers;
(iv)   providing easy-to-understand network reactions, for example, tones and announcements;
(v)   user friendly protocols during connection set-up.

A4.7.7   *Flexibility*

This concerns customer's requirements in the options available in the process of setting connections. Examples include various network options when connection to called party is not possible, for example,

  (i)   call forwarding facility offered to customers whereby the called party can trans-
        fer the call to another attended number in the event of the main number being
        unattended;
 (ii)   ring back when free and/or call-waiting indication options when the called party
        is engaged;
(iii)   voice message, giving reason for call not maturing.

A4.8   Information transfer

A4.8.1   Speed

This is measured by the rate at which information is transferred from calling to called
end (where relevant) or the time taken to transfer the information from the transmitting
to the received end.

A4.8.2   Accuracy

This is the degree of faithfulness of the received information to the information sent
over a connection. For example, in the case of speech over the telephony connections,
this is expressed by 'call clarity' or 'call quality'.

A4.8.3   Availability

This quality criterion is expressed by the probability for the established call to hold
for the intended duration of the connection. This parameter is, strictly, to be separate
from the availability at connection set-up.

A4.8.4   Reliability

This is determined by the long-term performance of the above three criteria, for
example, over a period of one year.

A4.8.5   Security

This concerns a customer's confidentiality requirements during the information transfer phase. For example a customer would be concerned about the security against other
parties 'listening in' over a mobile phone.

A4.8.6   Simplicity

This measure includes the ease and the convenience with which interactions with the
networks may be undertaken by the customer, when interactive service is provided.
Examples include

  (i)   ease with which additional codes may be inserted by the customer during the
        information transfer for optional services and facilities;

(ii)   overriding of announcements by input digits for the benefit of customers who are familiar with the use of service;

(iii) user instructions may be made simple where interactive services are offered.

A4.8.7   Flexibility

This concerns customer's requirements on the options available for the use of the service. Examples are

(i)   call waiting indication could offer visual or audible (or both) indications to the user (with facility to select a combination);

(ii) options available on the customer–network interactions.

A4.9   Connection release

A4.9.1   Speed

Customers may have specific requirements for the connection release times to make them compatible with the terminal equipment they intend to use with the network.

A4.9.2   Accuracy

Customer's requirements on faithfulness to specified logic of release protocol and accuracy of network release times of a connection.

A4.9.3   Availability

The release should take place always when requested. The customer would be concerned that charges may be incurred if calls are not cleared as requested.

A4.9.4   Reliability

The long-term performance of the above three parameters, for example, over a period of one year.

A4.9.5   Security

Customer's confidentiality requirements. These would comprise information on the call or clearing of the call made available to a third party during clearing procedure.

A4.9.6   Simplicity

Understandable and easy procedures for the release of an established connection.

A4.9.7   Flexibility

Capability of the service provider to offer options on the connection release times. For certain applications this could be important.

A4.10   Charging/Billing

A4.10.1   Speed

There are two measures, one for charging and the other for billing.

Speed of charging may be expressed by the time taken for the charging information to be made available to the customer after the charge has been incurred. Examples of instances where this might be required are

- advice of charge after a call is made;
- advice of current cumulative charges;
- advice of charges up to a certain period.

Speed of billing is expressed by the time taken for a bill to be presented to the customer from the date of request for a bill by the customer.

A4.10.2   Accuracy

The completeness and the accuracy of the billing information in reflecting actual use of the service. The tolerance or the limits of this accuracy may be specified by the maximum number of errors (expressed as a percentage) *and* the magnitude of the largest error.

A4.10.3   Availability

The probability of the billing information being available to the customer. To the service provider this quality criterion could also mean access to resources to establish charging information and produce bills.

A4.10.4   Reliability

Long-term performance of the three above criteria, for example, over a period of one year.

A4.10.5   Security

Customer's and service provider's requirements on the security of charging, for example, security of the network against fraud.

An example of fraud is the ability to make unpaid calls on payphones. Customers would wish to ensure that no other customer/s can incur charges against their or a specified account number.

A4.10.6   Simplicity

Customer's requirements on easy-to-understand billing information and formats.

A4.10.7   Flexibility

Customer's requirements for options available on the

(i)   format of the bills;
(ii)   frequency and date when billing information may be available.

A4.11   Network/service management by customer

A4.11.1   Speed

This quality criterion deals with *access* and *response* times when exercising the network/service management function by the customer.

A4.11.2   Accuracy

The correctness and completeness of execution of a network or service management request, as specified or implied in the contract, for example, accuracy of management information on call records.

A4.11.3   Availability

This quality criterion may be expressed by the probability of availability of the network/service management facility when required for use by the customer. It may also be specified by the outages as specified for the 'connection establishment – availability' cell specified earlier.

A4.11.4   Reliability

The long-term performance of the three preceding criteria, for example, over a period of one year.

A4.11.5   Security

All matters related to confidentiality requirements on the correct management of network service management facilities and functions (e.g. use only by authorised users/customers).

A4.11.6   Simplicity

This criterion deals with user-friendliness of the network or service management facilities offered to the customers.

A4.11.7   Flexibility

This criterion deals with customer's requirements in the options for customisations in the network/s service management facilities.

Appendix 5

Sample questionnaire to capture customers' Quality of Service requirements

In Chapter 4 the use of questionnaires to capture customer's QoS requirements was mentioned. In this appendix a sample of such a questionnaire is given.

Part I

1 Pre-contract activities

(i) What is the maximum time in which you would expect answers to all your enquiries on matters related to the supply of the basic telephony service and other pre-contract activities to be completed with the service provider?

_____ (days) _____ (hours) _____ (minutes)

(ii) On a 1–10 scale, please indicate the *relative priority* for this parameter in relation to the other parameters.

	Lowest									Highest
	1	2	3	4	5	6	7	8	9	10
Tick to show priority										

2 Provision

(i) What is the maximum period in which you would expect the service provider to install the basic telephony service from instant of effective contract?

_____ (days) _____ (hours) _____ (minutes)

(ii)   On a 1–10 scale, please indicate the *relative priority* for this parameter in relation to the other parameters.

	Lowest										Highest
	1	2	3	4	5	6	7	8	9	10	
Tick to show priority											

3 Time to repair

(i)   What is the maximum time (from the instant a fault is reported) in which you would expect repairs to be carried out?

_____ (days) _____ (hours) _____(minutes)

(ii)   On a 1–10 scale, please indicate the *relative priority* for this parameter in relation to the other parameters.

	Lowest										Highest
	1	2	3	4	5	6	7	8	9	10	
Tick to show priority											

4 Time for resolution of complaints

(i)   Please state the maximum time for the resolution of complaints.

_____ (days) _____(hours) _____(minutes)

(ii)   On a 1–10 scale, please indicate the *relative priority* for this parameter (in relation to the other parameters).

	Lowest										Highest
	1	2	3	4	5	6	7	8	9	10	
Tick to show priority											

5 Time for establishing a connection

(i)   While making a call what is the maximum connection set-up time you would be prepared to tolerate for a national call and for an international call?

National call _____ (seconds)
International call _____ (seconds)

(ii)   On a 1–10 scale, please indicate the *relative priority* for this parameter (in relation to the other parameters).

	Lowest										Highest
	1	2	3	4	5	6	7	8	9	10	
Tick to show priority											

6 Misrouted calls

(i) The service provider will attempt to provide 100% correct routing of calls. However, this may not always be possible. What is the maximum number of misrouted calls you will be prepared to tolerate in one year?

Maximum number of misrouted calls _____ (please quote a figure)

(ii) On a 1–10 scale, please indicate the *relative priority* for this parameter in relation to the other parameters.

	Lowest 1	2	3	4	5	6	7	8	9	Highest 10
Tick to show priority										

7 Call quality

(i) The service provider will attempt to provide 100% of connections of sufficient quality to enable you and the person at the other end of the connection to understand each other without difficulty. However, this may not always be possible. How many calls, in one year, would you be prepared to tolerate where you will have moderate difficulty in understanding the other person? Maximum number of calls with moderate difficulty in understanding the other person.

_____ (please quantify)

(ii) On a 1–10 scale, please indicate the *relative priority* for this parameter in relation to the other parameters.

	Lowest 1	2	3	4	5	6	7	8	9	Highest 10
Tick to show priority										

8 Availability of service to make a call

(i) It would be prohibitively expensive to provide 100% availability of network resources for 100% of the time. Please state the minimum outage performance you require.

Maximum number of outages _____ (please quantify)

Maximum duration of any one outage _____ (hours) _____ (minutes) _____(seconds) (please quantify)

(ii) On a 1–10 scale, please indicate the *relative priority* for this parameter in relation to the other parameters.

	Lowest 1	2	3	4	5	6	7	8	9	Highest 10
Tick to show priority										

9 Continued availability for the intended duration of the call after the connection has been established

(i)   The network resources may sometimes malfunction and an established connection may disconnect before the intended call release. Please state the maximum number of unintended connection releases you would be prepared to tolerate in one year.

Maximum number of connection releases you are prepared to tolerate (please quantify)_____

(ii)   On a 1–10 scale, please indicate the *relative priority* for this parameter in relation to the other parameters.

	Lowest									Highest
	1	2	3	4	5	6	7	8	9	10
Tick to show priority										

10 Accuracy of bills

(i)   The service provider will attempt to ensure that your bills are 100% accurate. However, there is an extremely small risk of mistakes. Please state the maximum error you would be willing to tolerate if undetected by yourself or by the service provider.

Maximum number of errors in one year_____ (please quantify)

Maximum magnitude of any one error _____ (please quantify in units of currency)

(ii)   On a 1–10 scale, please indicate the *relative priority* for this parameter in relation to the other parameters.

	Lowest									Highest
	1	2	3	4	5	6	7	8	9	10
Tick to show priority										

Part II

Other quality criteria not captured by the questionnaire

Please state here any quality criteria you consider important, but which are not covered in the questionnaire and indicate the levels of performance you would expect.

_____
_____
_____

Part III

Report on quality achieved by service provider for telephony service

1 Please state the quality parameters which you wish the service provider to include in the published report on achieved performance for the telephony service. Please state parameters from Part I of this questionnaire and/or state your own parameters. Please state as many parameters as you consider necessary.

Parameter 1 _____

Parameter 2 _____

Parameter 3 _____

Parameter 4 _____

Parameter 5 _____

Parameter 6 _____

Parameter 7 _____

Parameter 8 _____

Parameter 9 _____

Parameter 10 _____

2 Please state desired frequency of the report and the reporting period (e.g. quarterly for periods Jan.–Mar., Apr.–Jun., Jul.–Sep., Oct.–Dec.).

Frequency_____

Period _____

End of questionnaire

Appendix 6

Typical Quality of Service parameters/criteria for telecommunication services

In this appendix we give QoS parameters, considered pertinent from the user's perspective, for five services. Some or all of these, or more parameters may be considered by service providers while drawing up a list for performance reporting of their services.

A6.1 Basic telephony

1 Ease and convenience of availability of information on service provider's products, availability, tariff etc.
2 Professionalism and speed with which pre-contract information on any service may be obtained from provider, for example, tariff, service availability, service features, choice of telephone features etc.
3 Time for provision of service.
4 Number of faults per 'n' lines.
5 Speed of repair.
6 Repairs carried out right first time.
7 Availability of network resources when requiring to make a call.
8 Call set-up time.
9 Calling Line Identity suppression and/or ex-directory facility.
10 Transmission delay especially on international calls.
11 Answer time for directory enquiries, operator assisted calls and any form of operator assistance.
12 Availability of network resources to keep call for the intended duration of the conversation.
13 Number of complaints per 'm' customers.
14 Time taken for resolution of complaints.
15 Billing accuracy.

A6.2   Mobile communication services – telephony

A selection of parameters from those listed for basic telephony plus the following:

1   Radio coverage (information on the probability of coverage of mobile telephony service may be displayed by an availability map or supplied by customer help desk).
2   Accessibility of service (an indication of network resources available for making calls).
3   Availability of service (also known as call drop outs – the network being available for the intended duration of the call).
4   Transmission quality (the degrading factors on mobile communications are generally different from that of a fixed line call).

A6.3   Voice over IP (VoIP)

Customers would in due course expect the VoIP call quality to be the same as for switched network. In the meanwhile the service provider may wish to consider the implications of the following degrading factors which are particular to VoIP, in addition to that listed for basic telephony:

1   delay;
2   packet loss;
3   delay variation.

A6.4   Internet access

These parameters were taken from the report on a study carried out on behalf of the European Commission by Bannock Consulting and published in August 2000. The parameters have not been customer validated on a large number of the population. These parameters may require such validation before use by service providers. However, these give a good indication of the concerns of users and consumers on Internet access:

1   number of attempts required to achieve connection;
2   time to connect;
3   time to connect during the busiest hour of the week;
4   frequency of connection termination;
5   frequency and duration of ISP 'outages';
6   theoretical maximum speed of connection;
7   connection speed achieved;
8   latency, jitter and packet loss statistics communicating with the ISP;
9   speed of download from ISP's server(s);
10   speed of download from ISP's mail-server;

11   ratio of ISPs' bandwidth to product of number of customers able to achieve simultaneous connection and the maximum bandwidth of those connections;

12   proportion of packets travelling through the ISP's routers which are lost;

13   proportion of designated sites connected to: (a) the ISP's own backbone/backbone provider(s); (b) to the ISP through private peering arrangements; and (c) through public NAPs/IXPs;

14   proportion of time for which designated sites are unreachable;

15   latency, jitter and packet loss statistics for designated sites;

16   number of NAPs connected to and the bandwidth of the connections;

17   what are the bandwidth utilisation figures for the ISPs NAP connections and how congested are the NAPs at which the ISP peers?

18   cost of Internet access;

19   cost of website hosting;

20   annual supplemental cost for domain management;

21   cost of technical support.

A6.5   E-mail

The following list of e-mail user oriented QoS criteria were developed by a study commission under the Federation of Telecommunication Engineers of the European community (FITCE). Though perhaps dated (1997) it gives a good idea of what the customers wanted at that time. These have been validated against a representative part of the European population and thus have the validation accreditation. Any potential user of this list is advised to re-validate against the current population for the continued validity of these parameters.

1 Service Provision

1   The service provider shall specify the service features, whenever possible, using the terms and definitions in the relevant ITU-T Recommendations.

The service provider is free to specify the service features in simple and plain language to the customers. However, if customers ask for formal specifications then offerings must be stated in recognised terms in relevant ITU-T recommendations. The availability or otherwise of the following service features shall be supplied upon request by the user/customer:

- transaction time;
  - nominal;
  - what was achieved in the recent past;
- indication of whether or not mail has been delivered;
- express mail – whether a mail could be delivered by a specified time by the sender;
- receipt that the mail has been read;
- probability of misrouting;

- When the service provider analyses the e-mail for volume, performance etc. it should make public what analyses have been carried out. Customers should be made aware of the *nature of the analyses* to satisfy their concerns on traceability of their mail. The service provider need not make known the *findings of the analyses*.

2 Information sought by potential and existing customers on the e-mail service feature (e.g. tariff, conditions of use etc.) shall be provided by the service provider in any of the following forms:

- in writing;
- electronically;
- by fax;
- by telephone;
- in person.

The supplier shall provide the answers to customers' queries in any of the above forms if requested, unless this is impossible (e.g. in person when the service provider is not open to the public).

3 The service shall be provided by the service provider at a time to be specified by the customer. It should be possible for the customer to sign on electronically for the service stating the time at which the service becomes effective.

4 Customers should be offered the facility to cease an e-mail service at any time including a specified time in the future.

2 Despatch and receipt of mail

5 The time for response when customers are accessing their mail, and any other interactions, should be as low as possible. The service provider should specify typical speeds and what portion is attributable to the telecommunications company and what portion is attributable to the computers and systems of the service provider.

6 Customers shall have the facility to compose and edit the contents of the mail off-line and send it whenever they wish.

7 Where the facility is offered for a recipient of mail to acknowledge receipt, having read mail as opposed to storing without reading, these types of acknowledgements shall be at the discretion of the recipient. Where recipients have made it clear that they do not want to acknowledge a message, this fact must be made available to the senders.

8 Customers shall have the facility to store mail in categories to be specified by them. Storage shall be free for a certain period of time. The maximum size that could be stored must also be specified.

9 If unsolicited mail causes annoyance to the recipient, the recipient should be able to bring the matter to the attention of the service provider. The service provider should be able to warn the sender of the matter and, if unjustified despatch of unsuitable material continues, the service provider should have the capability to remove the offender from the system.

10   The recipient must be advised whether the mail received is complete and without error.

11   Where the recipient has requested anonymity, and if the sender chooses to send a copy of mail to this recipient, then the anonymous recipient's details shall not appear in other addressees' mail, either on the original or other copies.

12   Recipients shall be offered the facility to interrogate how many mails are waiting for them and at a second level of request, determine who these are from and, at a third level, display the full mail.

13   The customer to be provided with a set of simple statistics of mails received during a certain period, for example, the billable period.

14   Service providers should develop and provide, with the collaboration of telecommunication network providers, a paging message (bleep) when a mail has reached the recipient, thus prompting the customer to read. This facility may be restricted or categorised into normal and express mails.

15   E-mail should be made suitable for users to send legally-binding contractual documents. This means that the delivered version of the text should be an exact copy of the one sent. Additionally only the recipient must receive a copy of such a document. The role of encryption should be for secrecy between the sender and the receiver and not be necessary for correct transmission.

16   A facility which would be useful is for a pager to be activated for various classes of mail received, for example, express or ordinary mail, upon a mail being delivered. This would circumvent the user having to search for mail. When mail is delivered a pager indication may show that mail is waiting to be read. This must be optional and be under the control of the recipient.

17   The user should have the following mail options:

- receive only;
- send and receive option but the receive option, when the recipient is absent, should be delivered by the provider by telephone, fax or post to an address nominated by the recipient.

18   The recipient should be able to download mail from a public library where access may be made available from a public computer.

19   The user should have the facility to mask all the routing information from a mail.

20   E-mail should have negative or positive filtering, that is, the facility by the user to specify the addresses from which mail may be received or barred.

21   All communications (text, graphs, tables etc.) to be received without distortion.

3 Ease of use

22   The address format should be easy to

- remember;
- search e-mail directory.

23   The software used for the e-mail shall have user-friendly protocols for access, usage and release phases. The protocols shall be unambiguous and take into

consideration, where possible, the local culture. Additionally they shall also take into consideration the type of terminal equipment used by the customers.

24   There should be interoperability

- between service provider and service provider; and
- between service provider and customer.

25   The sender's name and address is to be included in the mail.

26   Reason for non-delivery or late delivery of message must be indicated.

27   Advise, when requested, that the whole file has been delivered to the distant end.

4 Service support

28   Technical and non-technical after sales support for the e-mail service shall be made available during periods of the day and week to reflect the local business and cultural requirements. Service provider could provide a choice of access for customers to seek help, for example, by telephone, in person, by fax, by post and by electronic means. Technical assistance to non-technical customers shall be provided in as simple and user-friendly a language as possible. Consideration ought also to be given to customers with special needs, for example, the partially deaf or wholly deaf, disabled etc.

29   The speed of response and resolution of queries by customers shall reflect the local custom.

30   The technical help line must be effective.

31   Instruction manuals for use of the software and all associated guidelines in the operation of the e-mail must be user-friendly.

32   When there has been a service outage attributable to the service provider (and not the telephone company) the reason for the outage and the implications, if any, to customers who may have sent mail, shall be supplied, free-of-charge, to the customers at the resumption of normal service.

5 Security

33   There shall be a facility, upon request, of confidentiality of e-mail address on similar lines to ex-directory telephone numbers. The service provider should ensure that the incoming mail is delivered only from those whose addresses have been given by the customer.

34   The software for e-mail should not limit the applications of any existing software on the computer, whether or not e-mail related.

35   To ensure delivery of mail, the service provider's computer should store information until acknowledgement from the far end has been received that the mail has been received and read.

36   The service provider shall ensure that the customer's personal details will not be divulged either intentionally or unintentionally to a third party.

37   The service provider must take steps to ensure that any financial information on the customer, for example, credit card details, shall not be made available intentionally or unintentionally to a third party.

38   The service provider must take steps to ensure that only the addressee can read the mail and that no one else can have access to its mail.

39   E-mail software should offer the facility for customers to access their mail from any other copy of the same software on any other computer anywhere in the world but through a system of passwords.

40   If, during the downloading of mail by a customer, an outage occurs, the mail should be preserved by the service provider until it has been read and acknowledged by the customer.

41   A facility for virus check should be made available. Any message delivered to the user should have been checked by the provider for virus. This may raise a legal issue on privacy.

6 Compatibility

42   The far end should advise the customer whether the message can be downloaded.

43   When a customer sends a mail the system shall not exhibit 'loop automatic reply' resulting in a number of to-and-fro messages.

44   The conditions under which customers can enter into dialogue e-mail conversation, if available, should be specified. If this facility is available the time to reach the called end should be specified.

45   Electronic mail programs should feature a standardised interface to allow other software to interwork easily with them (e.g. en/decoders, signatures etc.)

7 Tariff, charging and billing

46   All categories of charges to be clearly indicated.

47   Where there are many tariff options, customers should be given the option to change to another after a reasonable period of time. Any qualifying time is to be specified in the terms and conditions of the service.

48   One option for payment should be zero payment to the service provider for standard service. The service provider should collect the fee from the telecommunications company which provides the access to the service provider.

49   Facility to be made available for the customer to access an accrued charge from the previous chargeable period.

50   If free access time is allowed, the customer should be offered the facility to ascertain the amount of time used, or the amount of time still available free, for the current period.

51   The service provider shall offer the customers the choice of payment by non-electronic means, that is, by standing orders through the bank, cheques by post on a regular basis etc. There should also be a choice for the customer to change the method of payment, provided that requests for change are not too frequent. The choice of payments methods should be stated in the terms and conditions of service.

Appendix 7

Regulatory requirements on service performance to be reported on a regular basis in the UK

The data are published every six months. These may be viewed on the website at http://www.cpi.org.uk.

Residential – directly connected services

	Objective statistical data	Subjective data (customer satisfaction rating)
Orders completed	✓	✓
Orders completed – tail measures	✓	–
Customer reported faults	✓	–
Repeated customer reported faults	✓	–
Fault repair	✓	✓
Fault repair – tail measures	✓	–

Residential – indirectly connected services

	Objective statistical data	Subjective data (customer satisfaction rating)
Orders completed	✓	✓
Customer reported faults	✓	–
Fault repair	✓	✓
Fault repair – tail measures	✓	–

Residential – directly connected and indirectly connected services

	Objective statistical data	Subjective data (customer satisfaction rating)
Complaints handling (residential and business)	✓	✓ (Residential only)
Billing accuracy (residential and business)	✓	✓ (Residential only)

Business switched services

	Objective statistical data	Subjective data (customer satisfaction rating)
Orders completed	✓	✓
Orders completed – tail measures	✓	–
Customer reported faults	✓	–
Fault repairs	✓	✓
Fault repair – tail measures	✓	–

Directly connected business dedicated services

	Objective statistical data	Subjective data (customer satisfaction rating)
Orders completed	✓	–
Orders completed – tail measures	✓	–
Customer reported faults	✓	–
Repeated customer reported faults	✓	–
Fault repair	✓	–
Fault repair – tail measures	✓	–

Indirectly connected business services

	Objective statistical data	Subjective data (customer satisfaction rating)
Orders completed	✓	✓
Orders completed – tail measures	✓	–
Customer reported faults	✓	–
Fault repair	✓	✓
Fault repair – tail measures	✓	–

All business services

	Objective statistical data	Subjective data (customer satisfaction rating)
Complaints handling (residential and business)	✓	✓ (Business only)
Billing accuracy (residential and business)	✓	✓ (Business only)

Appendix 8

Regulatory requirements on service performance to be reported on a regular basis in Australia

A comprehensive set of parameters and the results achieved by the various service providers are on the website: http://www.aca.gov.au under the Telecommunications Performance Monitoring Bulletin. The following list gives a flavour of what is in the web site. Readers are advised that the parameters and customer satisfaction matters are regularly reviewed and the web site should be consulted for the latest information.

A8.1   Basic telephony

- Provision of connection and service
    - in various combinations from
        - with infrastructure;
        - without infrastructure;
        - divisions of urban, major rural, minor rural, remote;
        - the eight geographical regions of Australia and national.
- Restoration of service and fault reporting in the categories of
    - urban, rural and remote divisions;
    - the eight geographical regions of Australia and national.
- Call centres, in the following categories:
    - Directory Assistance calls;
    - operator assisted international and long distance calls;
    - service difficulties enquiry.
- Complaints;
- Payphone services categorised into:
    - percentage of downtime;
    - average hours to clear a fault;
    - percentage of faults cleared within one working day;

- percentage of faults cleared within two working days;
- average trouble reports per payphone per month;
- percentage available to make calls.
- Network loss in categories of: Local call network loss, National long distance call network loss and international call network loss.

Detailed statistical information may be seen at the ACA web site [1].

In addition to the statistics on the delivered quality a Customer Service Guarantee (CSG) is in place for three parameters:

- Connection of standard telephone service;
- Repair a fault or service difficulty;
- Attend appointments with customers.

Limits for the fault repair are reproduced below

Location	Definition of location	Time of repair
Urban	Areas in Australia with a population greater than 10,000 people.	Within one full working day after being notified of a fault.
Rural	Areas in Australia other than urban areas and remote areas.	Within two full working days after being notified of a fault.
Remote	Areas in Australia with a population less than 200 people.	Within three full working days after being notified of a fault.

If a service provider does not meet the above targets the customer affected is entitled to an economic benefit. The terms and conditions are laid out in the licence.

A8.2   Internet

The Australian Bureau of Statistics (ABS) compiles the performance statistics on the Internet. The following statistical information is presented:

- number of ISPs (603 in September 2001);
- number of Points of Presence (POPs);
- access lines;
- number of subscribers;
- data downloaded by subscribers.

Detailed information may be obtained from the ABS web site [2].

A8.3   Mobile telephone network services parameters

The following parameters are reported on for mobile communications:

- call dropout
- call congestion rate
- radio coverage (on population and land mass coverage)
- mobile number portability.

Consumer satisfaction surveys are carried out. A summary of the findings are also in the web site.

A8.4   References

1   Australian Communications Authority: http://www.aca.gov.au
2   Australian Bureau of Statistics: http://www.abs.gov.au

Appendix 9

Regulatory requirements on service performance to be reported on a regular basis in the USA

A9.1   National level

The Federal Communications Commission (FCC) has a complex set of rules for reporting QoS. A set of recommended parameters is given below. For details of reported values and other aspects of QoS reporting please see the web site [1].

Access services provided to carriers – switched-access

1   Percent installation commitments met
2   Average installation interval (days)
3   Average repair interval (hours).

Access services provided to carriers – special access

4   Percent installation commitments met
5   Average installation interval (days)
6   Average repair interval (hours).

Local services provided to residential and business customers

7   Percent installation commitments met
8   Percent installation – residence
9   Percent installation – business
10   Average installation interval (days)
11   Average installation – residence
12   Average installation – business
13   Initial trouble reports per thousand lines
14   Initial trouble total MSA (Metropolitan Statistical Area)
15   Initial trouble total non-MSA
16   Initial trouble total residence
17   Initial trouble total business

18   Troubles found per thousand lines
19   Repeat troubles as a percent of trouble reports
20   Repeat troubles total residence
21   Repeat troubles total business.

Customer complaints per million access lines

22   Customer complaints residential
23   Customer complaints business
24   Total access lines in thousands
25   Total trunk groups
26   Total switches.

Switches with downtime

27   Number of switches
28   As a percentage of total switches.

Average switch downtime in seconds per switch

29   For all occurrences or events
30   For unscheduled events over two minutes.

For unscheduled downtime more than two minutes

31   Number of occurrences or events
32   Events per hundred switches
33   Events per million access lines
34   Average outage duration in minutes
35   Average lines affected per event in thousands
36   Outage line-minutes per event in thousands
37   Outage line-minutes per 1000 access lines.

For scheduled downtime more than two minutes

38   Number of occurrences or events
39   Events per hundred switches
40   Events per million access lines
41   Average outage duration minutes
42   Average lines affected per event in thousands
43   Outage line-minutes per event in thousands
44   Outage line-minutes per 1000 access lines
45   Percent of trunk groups exceeding blocking objective.

Total number of outages

46   Scheduled
47   Procedural errors – telco (instal. maint)
48   Procedural errors – telco (other)

49 Procedural errors – system vendors
50 Procedural errors – other vendors
51 Software design
52 Hardware design
53 Hardware failure
54 Natural causes
55 Traffic overload
56 Environmental
57 External power failure
58 Massive line outage
59 Remote
60 Other/unknown.

Total outage line-minutes per thousand access lines

61 Scheduled
62 Procedural errors – telco (instal. maint)
63 Procedural errors – telco (other)
64 Procedural errors – system vendors
65 Procedural errors – other vendors
66 Software design
67 Hardware design
68 Hardware failure
69 Natural causes
70 Traffic overload
71 Environmental
72 External power failure
73 Massive line outage
74 Remote
75 Other/unknown.

A9.2   State level

Every state has its own regualtory requirements. Some states have few parameters to be reported on and others like Colorado have a much wider range. The parameters selected by each state are from those recommended by the National Association of Regulatory Commissioners (NARUC), that is, installation, operator handled calls, transmission and noise, network call completion, and customer trouble reports whereas the National Regulatory Research Institute (NRRI) recommended the rest. The parameters for the state of Colorado are given here. See web site [2] for performance parameters for other states.

Installation of service
1 Primary service order installation
2 Exchange carrier's service order.

Operator handled call

3   Directory assistance
4   Intercept calls
5   Local operator assistance calls
6   Toll operator assistance calls
7   Repair service calls
8   Business office calls.

Transmission and noise requirements

9   Subscriber lines (loss)
10   PBX/Multiline (noise)
11   EAS (noise)
12   Toll (noise).

Network Call completion

13   Dial tone supply
14   Dialled intra-office calls
15   Dialled interoffice calls within the local calling area
16   Dialled intra-LATA toll calls.

Customer trouble reports

17   Customer trouble reports
18   Out of service clearing time.

Major service outages

19   Scheduled interruptions
20   Emergencies
21   Reserve power.

Service disconnection

22   Grounds for suspension
23   Customer notification
24   Special protection.

Billing and collection

25   Billing period
26   Minimum billing content
27   Disputes and adjustments.

Customer satisfaction

Not currently reported in Colorado (only Delaware, Louisiana, Maine and Rhode Island report on this parameter at the time of writing this book – spring 2002).

Public pay telephones

28 Minimum number per exchange
29 Information posted
30 Services.

911 Database

31 ALI database service
32 Interconnection
33 Information provision
34 Emergency service quality
35 Call completion
36 Failures and outages.

A9.3   References

1 http://www.fcc.gov
2 http://www.nrri.ohio-state.edu

Appendix 10

Fault incidence and repair time: OECD member countries – 1999

	Fault per 100 lines per year	Fault repair within 24 hours per cent	Notes
Australia	—	83/86	Telstra fault clearance for urban/rural areas June 1999 and June 2000. Urban faults cleared within one working day. Rural faults within two working days.
Austria	—	98.0	
Belgium	4.0	90.0	
Canada	—	72.0	Figure is annual average for Bell Canada.
Czech Republic	20.0	99.0	Including CPE. Fault repair within 54 hours.
Denmark	—	—	
Finland	—	74.1	In a working day
France	—	—	
Germany	—	85.9	Within three working days.
Greece	17.0	90.5	
Hungary	17.0	93.7	
Iceland	—	—	
Ireland	—	—	
Italy	17.2	92.0	95.4% fault clearance within 48 hours
Japan	—	—	
Korea	7.0	99.0	
Luxembourg	—	—	
Mexico	2.2	73.0	
Netherlands	—	—	
New Zealand	12.1	79.5	Only 2% of faults not repaired within 96 hours
Norway	—	—	
Poland	—	—	
Portugal	11.2	88.9	% of faults repaired within 12 working hours
Slovak Republic	—	—	
Spain	15.0	95.5	
Sweden	—	—	
Switzerland	—	—	
Turkey	—	—	
United Kingdom	14.3	92.0	
United States	13.7	—	

Reproduced by the kind permission of OECD.

Sustaining supplier services

A11.1 Introduction

In the middle of the year 2002 the telecommunications industry witnessed unprecedented changes. What was once a stable and profitable industry had, over the previous two years, lurched from one crisis to the next and in one case the second largest telecoms service supplier in the world filed for bankruptcy in the USA. In another case a pan European service supplier, who until recently carried 40% of Europe's Internet traffic effectively 'closed its doors', leaving a number of business users without service. In the UK Atlantic Telecom went into administration on 5 October 2001. The impact of these changes will no doubt be discussed and analysed for a number of years, but it is evident that the financial strength of a service supplier should now be an essential component when assessing an overall Quality of Service (QoS) framework.

The principal issue to be addressed by the customers, particularly the corporates is the continuity of telecommunication services in the event of failure of a supplier.

A11.2 A case study

Atlantic Telecom went into administration on 5 October 2001 and at the time the administrators were confident that the company could be sold. Atlantic had 128,000 customers (7,000 of these were business customers) in Glasgow, Edinburgh, Dundee and Aberdeen in Scotland and Manchester in England. Most of these customers were indirect access customers who bought telephone services which Atlantic provided over BT's network.

The indirect access parts of the business were sold and, therefore, most of Atlantic's customers were transferred to other operators. However, in November the administrators concluded that they were unable to sell Atlantic's fixed wireless access network as a going concern and decided to switch it off on 25 November.

Twelve thousand residential customers and 2,000 business customers using this network needed to be switched to other networks in order to maintain their telecoms services.

Many of these fixed wireless customers were in new buildings into which alternative fixed lines had never been installed and others had old fixed lines which needed to be checked and re-serviced. It was therefore impossible for operators like BT to carry out all this work at such short notice to avoid customers being left without a telephone service when the administrators planned to switch the network off.

A rescue package was therefore put together by the Department of Trade and Industry (DTI) and the Scottish Executive to extend the operation of Atlantic's fixed wireless network to 2 January to allow these customers to switch to alternative suppliers. This ensured that customers – particularly business customers – did not face the disruption and inconvenience of a complete loss of their telecoms services with insufficient notice to make alternative arrangements. By 2 January all those customers affected had been able to switch to other operators and the Atlantic fixed wireless network was switched off.

Telecoms company failures are generally most problematic when the company has its own network. When it is using other company's networks the customers can be easily transferred to alternative suppliers. This happened with the majority of Atlantic's customers who were using indirect access services. Although the risk of other insolvencies in the sector is real, in most other cases, it would be expected that the network would be bought and would continue to be operated, which would avoid a similar impact on consumers. Nevertheless, such a repeat is not impossible. The issue of continuity of supply in the Atlantic case had two aspects:

(a)   the length of time for which Atlantic could be required to maintain a service and the question of finance in the interim; and

(b)   the speed at which a replacement service could be provided by an alternative operator.

On point (a) Public Telecommunications Operators like Atlantic are under a licence obligation to provide certain services, for example, emergency calls. However, because Atlantic was in administration the regulator would have needed the permission of the court to require the administrators to maintain the network until such time as customers could obtain services from alternative providers. This permission was unlikely to have been forthcoming as Atlantic was insolvent, and would have been unable to finance a service as continued service and, in practice, the administrators may well have applied to the court for the company to be wound up.

On point (b) BT has a Universal Service Obligation (USO) in its licence to ensure that all reasonable demands for telecoms service are met in the UK. BT cooperated fully with Oftel and put substantial extra resources into meeting requests from Atlantic customers for replacement services. There was, therefore, no reason for Oftel to invoke the USO since no more could have been done to speed up the process.

Against the background of Atlantic Telecom (and now others), the following options for change could be considered to ensure continuity of telecoms supply in any similar future case.

A11.3   Proposals for the customers

A mix and/or a hybrid of the following options may be considered.

Option 1: Make best use of financial information

Normally when assessing a service supplier for financial soundness, it is established business practice to examine audited and properly filed accounts. However, in view of recent allegations of unconventional accounting practices it is not sufficient to go merely by the published accounts of a service provider.

Not withstanding the impact upon the accountancy profession of discrepancies such as in the Worldcom accounts, what other measures can one use? In the UK, there is no ongoing duty for Oftel to monitor the financial status of service suppliers and it is unlikely that either Oftel, or future EC directives will ensure that this will be the case. In the absence of any direct guidance, for the present, the following simple 'rules' could be considered, namely:

1   Monitor the financial position of the main supplier of service in so far as it is feasibly possible.
2   Second source services from alternative supplier(s) such that business continuity can be maintained. Obviously, the second source of supply should be subjected to the same level of monitoring as the first.
3   Institute and continually test contingency plans for business continuity on the assumption that one supplier may fail.

Option 2: Creating an obligation on administrators of licensed telecoms operators to maintain a service for whatever reasonable time is necessary for alternative services to be put in place, on a basis to allow this obligation to take priority over claims of creditors

One approach to tackle this problem would be to place an obligation on administrators of insolvent telecoms operators to maintain service pending alternative arrangements being put in place. This might be modelled on existing schemes that are already in place to deal with the insolvency of vital public services such as water, social housing and rail. For example in the water industry in the UK the Water Industry Act 1991 contains provisions for a 'special administration' scheme whereby the duties of an administrator are modified so that it is subject to the statutory duty of the company to maintain a water supply, whilst it seeks to sell the company.

In such special insolvency schemes creditors' rights are generally displaced in favour of the public interest. These schemes are currently rare and generally cover the monopoly supply of essential public services. For example, if a water company failed public health would be seriously compromised if administrators did not continue to ensure that sewage was dealt with.

The disadvantages of these schemes are that they can make it more difficult for companies covered by them to obtain funding for normal trading as they represent an increased credit risk. A case would have to be made that a loss of telecoms service was serious enough to justify setting up a special insolvency scheme in the same way as for other essential public services. The question of funding also arises as the administrator would need financing to continue the service. For telecoms operators with large networks the costs of continuing to run services could be very high. No administrator would voluntarily agree to run a network to the detriment of the company's creditors, due to his potential personal liability. Therefore, it would require legislation as per the water industry.

Option 3: Amending the USO on the incumbent to specify a time limit for the provision of a telecoms service

The incumbent(s) are normally designated universal service providers and have to provide consumers with, among other things, a basic connection to the fixed network, on reasonable request. However, there is currently, no minimum timeframe specified for provision of the service.

It might be possible to amend the USO in the incumbent's licence to not only require it to provide a telephone line to all those who reasonably request one, but to do so within a specified time limit. This would be a matter for the regulator, but the intended benefit would be to seek a speedier reconnection of service and, consequently, those consumers left without a telecoms service would suffer less hardship. However, this seems unlikely to help deal with Atlantic-type problems as any time limit would need to be subject to a reasonableness test. However reasonableness was defined, this type of extra obligation would not lead to instant connections for customers as there are practical limitations as to how quickly the incumbent could make the necessary connections (e.g. numbers of available staff, possibility of needing to dig new line out to premises, effect on its ongoing urgent work for its own customers).

Option 4: Establishing an ATOL-type scheme to fund the continuing provision of telecoms services when companies go into administration or receivership for an interim period until an alternative supply is provided

ATOL (Air Travel Organisers' Licensing) exists to protect the public from losing money or being stranded abroad because of the failure of air travel firms. It is a statutory scheme, being based on a legal requirement for licensing, and is managed by the Civil Aviation Authority (CAA) in the UK.

All licensed firms who sell air travel in the UK have to lodge bonds with the CAA so that, if they go out of business, the CAA can give refunds to people who cannot travel, and arrange for people abroad to complete their holidays and fly home. The bonds are irrevocable undertakings from third parties (banks or insurance companies) that give the CAA the right to obtain the specified sum of money in the event that the licence holder is unable to meet its obligations to its customers.

This option would involve a similar arrangement for telecommunications operators, whereby – prior to providing telecoms services – each operator would have to satisfy the Secretary of State or the Director General of Telecommunications, that they had in place an adequate bond (or possibly a guarantee, or insurance policy, depending on how the scheme is structured). This bond could be called on if an operator entered administration in order to finance the maintenance of that company's service.

A central question would be how to assess the quantum of the bond sought from each operator. If the bond is based pro-rata on the size/market capitalisation of each operator, then the larger, more established operators would end up paying more. However, it is these same operators who, by virtue of their position, are less likely to fail with the same impact as in the case of Atlantic. This leaves open the accusation that the scheme – even if the bonds are based on the same percentage of turnover for all operators – is discriminatory. It is possible that such a scheme would be complex to structure, could be discriminatory, and might increase the administrative and financial burden on the industry. Introducing such powers in the telecoms sector would, in any case, not seem to be possible under European law. The new EC framework, which will come into effect from July 2003, limits the obligations which can be placed upon individual communications providers, and makes no provision for schemes that would require operators to put up bonds as a condition of being able to operate.

Option 5: Joint liability across the industry

The ATOL scheme described in more detail above envisages that, if an operator were to fail, that operator would use its own bond (or guarantee or insurance) to fund the continued provision of essential services. If that bond is insufficient, the government backed Air Travel Trust Fund (ATTF) intercedes. What it does not provide for is the whole industry taking joint responsibility for the failure of one operator, either by a pre-emptive fund created by an industry-wide levy, or by a post-event bailout whereby each operator assumes joint liability on a pro-rata basis. Under this option rather than placing the financial burden on a single entity, the burden is spread across a large number of Telcos, who are collectively better able to absorb the 'hit' of financing an ongoing service.

One of the major disadvantages of the ATOL scheme – the need to pre-quantify the level of the contribution – disappears in a post-event bailout scheme. Instead, Telcos would be expected to contribute to the cost pro-rata based on turnover or number of customers. This requires no prior assessment of risks, and the basis of charges would not fall disproportionately on new entrants or 'weaker' companies.

As with an ATOL-type scheme, it is possible that any levy, or post-event bailout, could also be used to arrange cover of the expenses involved if customers have to change their phone numbers. In order to provide the required degree of assurance, there needs to be a prior arrangement for Telcos to meet on demand the reasonable requirements of the administrator.

Other disadvantages mirror those of the ATOL scheme; for example, the increased burdens on the industry, the need to make such a scheme mandatory and the fact that the larger operators are likely to have to contribute the lion's share to any bailout, while being less likely themselves to ever need to benefit from the fund. In addition, the new EC framework makes no provision for schemes that would require operators to put up bonds as a condition of being able to operate. However, under EC law it may be possible to put in place a post-event bailout scheme whereby each operator assumes joint liability for continued financing of a service on a pro-rata basis.

In the energy sector, when a supplier has gone into administration or receivership OFGEM has the power to revoke the supplier's licence and appoint a 'Supplier of Last Resort' to take over responsibility for supplying its customers. However, this would not resolve the timing difficulties in the case of an Atlantic-type failure, nor the difficulties of users having to change to a different number. Introducing such powers in the telecoms sector would, in any case, not seem to be possible under European law. The new EC framework limits the obligations which can be placed upon individual communications providers, and makes no provision for 'supplier of last resort' schemes.

A11.4   Summary

The recent failure of a number of telecoms service suppliers has demonstrated the need to be vigilant when buying services. This suggests that a QoS framework should consider not only the supplier's financial stability, but also the steps that should be taken to ensure continuity of service. In the UK, the DTI has initiated such a discussion and the results of the consultation and the subsequent developments are awaited with interest.

Acknowledgement is made to the Department of Trade & Industry, of the UK for permission to use the information contained in their web site in this Appendix. The full report may be viewed at www.dti.gov.uk/cii/regulatory/telecomms/index.shtml

Index